HEALTH IMPACTS

OF POLYNUCLEAR AROMATIC HYDROCARBONS

HEALTH IMPACTS OF POLYNUCLEAR AROMATIC HYDROCARBONS

Edited by A.W. Pucknat

NOYES DATA CORPORATION
Park Ridge, New Jersey, U.S.A.
1981

Library of Congress Catalog Card Number: 80-28039
ISBN: 0-8155-0840-9
Printed in the United States

Published in the United States of America by
Noyes Data Corporation
Noyes Building, Park Ridge, New Jersey 07656

Library of Congress Cataloging in Publication Data

Main entry under title:

Health impacts of polynuclear aromatic hydrocarbons.

(Environmental health review ; no. 5)
Bibliography: p.
Includes index.
1. Aromatic compounds--Toxicology. 2. Polycyclic compounds--Toxicology. 3. Aromatic compounds--Environmental aspects. 4. Polycyclic compounds--Environmental aspects. 5. Air--Pollution. I. Pucknat, A. W. II. Series. [DNLM: 1. Polycyclic hydrocarbons--Adverse effects. 2. Air pollutants. QV 633 H4345]
RA577.A68H4 615.9'51 80-28039
ISBN 0-8155-0840-9

FOREWORD

This book reviews currently available information on health impacts of polynuclear aromatic (PNA) hydrocarbons. The surveillance of PNA levels is in response to government regulations requiring a determination as to whether PNA emissions in air and/or water may reasonably be anticipated to endanger the public's health.

The book will be of interest to engineers and managers involved in pollution control decision making, as well as to biomedical and environmental scientists who must assess exposure risks to public health.

Because the information in this book is taken from multiple sources, acquired from bibliographic searches, it is possible that certain portions of the book may disagree or conflict with other parts of the book. We chose to include these different points of view, however, in order to make the book more valuable to the reader.

The use of multiple sources has also resulted in several abbreviations, having essentially the same meaning, being used interchangeably throughout the book. The reader's attention is called to these abbreviations:

PNA	Polynuclear aromatic compounds
PAH	Polycyclic aromatic hydrocarbons
POM	Polycyclic organic material

Advanced composition and production methods developed by Noyes Data are employed to bring these durably bound books to you in a minimum of time. Special techniques are used to close the gap between "manuscript" and "completed book." Industrial technology is progressing so rapidly that time-honored, conventional typesetting, binding and shipping methods are no longer suitable. We have bypassed the delays in the conventional book publishing cycle and provide the user with an effective and convenient means of reviewing up-to-date information in depth.

The expanded table of contents is organized in such a way as to serve as a subject index and provides easy access to the information contained in this book. Primary sources are listed at the beginning of each chapter and a complete listing appears in the Sources Utilized section at the end of the book. The titles of additional publications pertaining to topics in any particular chapter are to be found at the end of the chapter.

CONTENTS AND SUBJECT INDEX

INTRODUCTION

Polynuclear aromatic compounds (PNA) are widely found throughout the environment. To some extent their presence can be attributed to natural processes—certain bacteria and plants synthesize polynuclear aromatics during normal development, and naturally-caused brush or forest fires contribute additional amounts. These levels have always been present. However PNA are also formed as a result of the incomplete combustion of organic compounds. Inefficient combustion of fuels used, for example, for residential or industrial heating and for highway vehicles, as well as burning of coal refuse, forest and agricultural wastes, and municipal wastes, are some of the major sources of PNA.

Since some of these multiring compounds have been found to be carcinogenic, there is increasing concern about their presence and their occupational and environmental health impacts. While extensive experimental animal studies are being pursued by many research groups, the carcinogenicity of some PNA, particularly benzopyrene, has been clearly demonstrated.

After a preliminary chapter on the physical and chemical properties of polynuclear aromatics, the next three chapters focus on the sources of PNA and their fate in the environment.

An extensive chapter on health effects covers the various types of human exposure to PNA as well as their carcinogenic, toxic, teratogenic and mutagenic effects. This is followed by a chapter on the steps being taken to minimize the dangers posed by PNA such as decreasing and controlling the levels of these compounds in the environment.

Since detection, identification and quantification are essential in order to control the level of PNA exposure, the final chapter is a compilation of information on the various analytical techniques which can be used to accomplish this.

It should be noted that throughout this book, which is based on many different sources, the terms polynuclear aromatic compounds (PNA), polycyclic aromatic hydrocarbons (PAH) and polycyclic organic material (POM) are used essentially interchangeably.

PHYSICAL AND CHEMICAL PROPERTIES OF PNA

The information in this chapter contains physical and chemical property data from *Health Assessment Document for Polycyclic Organic Matter*, External Review, Draft Number 2, June 1980 (hereinafter referred to as "EPA Health Assessment Document"), prepared by The Syracuse Research Corporation for the U.S. Environmental Protection Agency.

Notice: This document is a preliminary draft. It has not been formally released by EPA and should not be construed to represent Agency policy. Not to be quoted or cited.

DEFINITION AND FORMATION

Polycyclic organic matter (POM) can include many chemical groups as indicated by the following list developed by an EPA Task Force in 1975 (1):

Polycyclic aromatic hydrocarbons (PAH) (See Table 1.1)
Aza arenes (arenes containing a ring nitrogen) (See Table 1.2)
Imino arenes (ring nitrogen with a hydrogen) (See Table 1.2)
Carbonyl arenes (See Table 1.3)
Dicarbonyl arenes (quinones) (See Table 1.3)
Hydroxy carbonyl arenes
Oxa arenes and thia arenes
Polychloro compounds
Pesticides (e.g., aldrin, chlordane, DDT)

Chemically, any organic compound that contains two or more rings could be considered a POM. However, of major concern are the carcinogenic polycyclic aromatic hydrocarbons (PAH), such as benzo[a] pyrene, and their nitrogen analogs, aza and imino arenes, which are formed during organic combustion processes (2)(5)(6). Because of their common sources, their existence in urban air, and the considerable experimental data on their carcinogenic effects, the PAH and the nitrogen analogs have received the most attention.

However, some of the other arenes (for example, 7H-benz[de]anthracen-7-one) have been detected in urban atmospheres (3).

POM may be formed in most combustion or elevated temperature processes which involve compounds containing carbon and hydrogen. The amounts and types of individual chemicals that are formed are dependent upon the starting hydrocarbons and the conditions of combustion or pyrolysis. Although the exact intermediates that are postulated vary considerably, free-radical paths are commonly suggested at the high temperatures attained in the flame front (500° to 800°C) or under pyrolysis conditions.

Badger (4) first postulated the synthetic route to POM formation outlined in Figure 1.1. His pyrolysis studies were conducted by passing the hydrocarbon vapor in nitrogen through a silica tube at 700°C.

Figure 1.1: Mechanism of Benzo[a]pyrene Formation

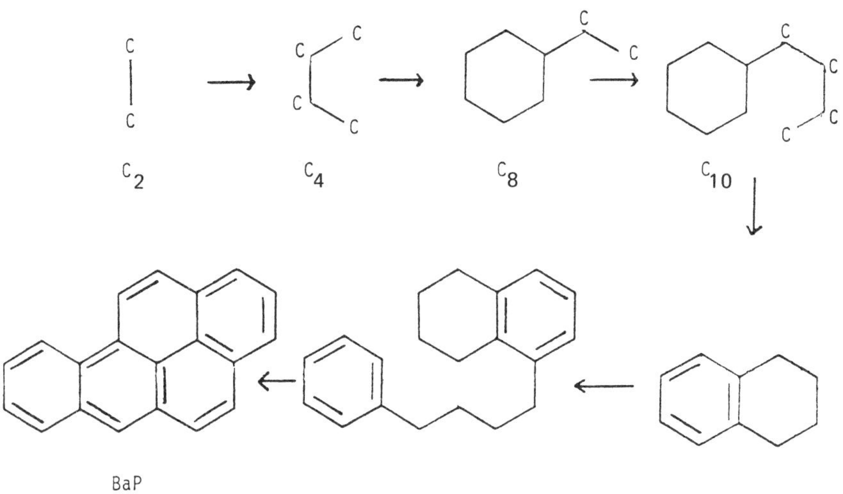

Source: Reference (4) as cited in EPA Health Assessment Document

Although the use of nitrogen atmospheres has been criticized as lacking relevance to actual combustion (5), the reducing conditions are similar to those of the oxygen-deficient environments of a flame, and the data are in good qualitative agreement with the POM combustion products formed (2). For example, Boubel and Ripperton (7) found that benzo[a]pyrene (BaP) is produced during combustion even at high percentages of excess air, but that the amount of BaP is greater at lower percentages of excess air.

Lending support to the postulated route to POM, Badger and Spotswood (8) pyrolyzed toluene, ethylbenzene, propylbenzene, and butylbenzene and obtained the highest yields of benzo[a]pyrene with butylbenzene. Badger and Spotswood (9) also found that when 1,3-butadiene is pyrolyzed with pyrene at 700°C, no increase in the yields of benzpyrenes is observed, which suggests that Diels-Alder type reactions are probably not important.

More recent studies tend to confirm most of the mechanism proposed by Badger (4). Crittenden and Long (10) determined the chemical species at various flame heights of oxy-acetylene and oxy-ethylene flames. Compounds identified suggest that the C_2 species react to form C_4, C_8, and C_{10} species, and that reactions involving styrene and phenylacetylene are probably important in the formation of polycyclic aromatic hydrocarbons. Also, a $C_{10}H_{10}$ species was detected in the gases of both flames and, although its identity is uncertain, it may correspond to the C_6-C_4 species (butylbenzene, phenylbutadiene, or related radicals) postulated by Badger (4).

With the Badger step-wise sequence mechanism, various free radical intermediates and aliphatic and aromatic fuels, hundreds of POM could be formed.

By far, the most significant source of atmospheric POM is pyrolytic or combustion processes, although some investigators have suggested that some polycyclic aromatic hydrocarbons in the environment may be synthesized by plants and microorganisms (11) and that part of the PAH could be from natural combustion sources (12). However, recent studies of aqueous sediment cores (13) suggest that the concentrations detected correspond relatively well with energy production from various fuels, which indicates that anthropogenic combustion may be the major source of PAH even in aqueous media.

STRUCTURE AND PROPERTIES

Chemical Structure

The nomenclature of POM compounds has suffered from considerable ambiguities in the past due to different peripheral numbering systems. For example, carcinogenic benzo[a]pyrene (BaP) was named "3,4-benzpyrene" by American scientists and "1,2-benzpyrene" by European workers, while the noncarcinogenic isomer, benzo[e]pyrene (BeP) was named "1,2-benzpyrene" by American scientists and "4,5-benzpyrene" by European scientists. Thus, 1,2-benzpyrene could be the carcinogenic or noncarcinogenic isomer, depending upon where a scientist worked (14).

The currently accepted nomenclature is that adopted by the International Union of Pure and Applied Chemistry (IUPAC) and by Chemical Abstracts Service (2). Of major importance are the following rules that determine the orientation from which the numbering is assigned (2)(14)-(18):

(1) The maximum number of rings lie in a horizontal row;

(2) As many rings as possible are above and to the right of the horizontal row; and

(3) If more than one orientation meets these requirements, the one with the minimum number of rings at the lower left is chosen.

The carbons are then numbered in a clockwise fashion, starting with the first most counterclockwise carbon which is not part of another ring and is not engaged in a ring fashion. Letters are assigned in alphabetical order to faces of rings, beginning with "a" for the side between carbon atoms 1 and 2 and continuing clockwise around the molecule; ring faces common to two rings in the parent compound are not lettered.

Thus, benzo[a] pyrene would have a benzene ring fused to the "a" bond of the parent pyrene structure (Figure 1.2).

Figure 1.2: Accepted Nomenclature for Benzo[a] pyrene

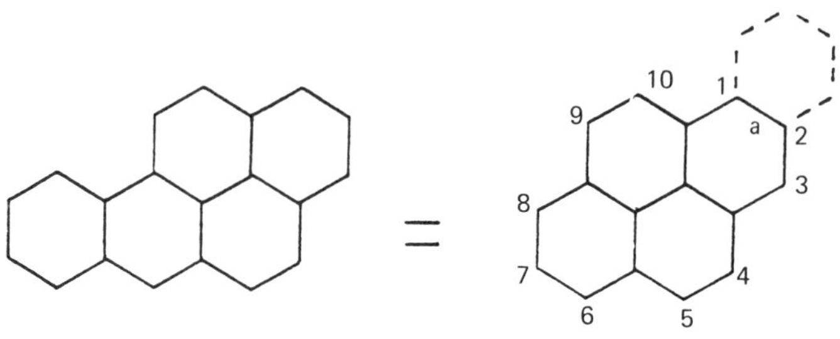

Benzo[a] pyrene   Pyrene

Source: EPA Health Assessment Document

The following tables provide lists of the POM that, in most instances, have been detected in air. However, some compounds (e.g., 7,12-dimethylbenz[a] anthracene and 3-methylcholanthrene) have been included in the table even though they are synthesized compounds which are not formed during combustion and are not present in ambient air. These synthetic compounds were included because they are used as model compounds in experimental carcinogenesis studies and have provided fundamental information concerning mechanisms of tumor formation. The chemical names, synonyms, structures, numbering system and acronyms, along with their relative carcinogenicity, are presented.

Tables 1.1 through 1.3 are divided into polycyclic aromatic hydrocarbons (PAH), aza and imino arenes, and miscellaneous other POM that have been detected in air. The structures are presented in the same way as those in the NAS report, *Particulate Polycyclic Organic Matter* (2), in that aromatic rings are shown as plain hexagons and a methylene group as CH_2.

The indications of carcinogenicity, taken from the NAS report, were obtained from several Public Health Service (PHS) surveys of compounds tested for carcinogenicity (19)-(23). In some instances, indications of carcinogenicity were obtained from individual references; in those cases, the references are provided. The indications of carcinogenicity which are a summary of all routes correspond to the following simple code:

Not carcinogenic	–
Uncertain or weakly carcinogenic	±
Carcinogenic	+
Strongly carcinogenic	++, +++, ++++

For the most part, this system of indicating carcinogenicity is very approximate. Additional compounds besides those listed in the NAS report were added when they have been detected in air. Older nomenclature is included in parentheses, and names with an asterisk indicate disagreement with standard numbering.

Table 1.1: Polycyclic Aromatic Hydrocarbons

Compound	Anthracene	Benz[a]anthracene	7,12-Dimethylbenz[a]anthracene (9,10-dimethyl-1,2-benzanthracene)
Structure	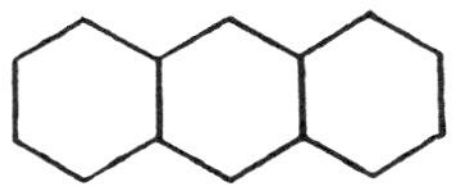	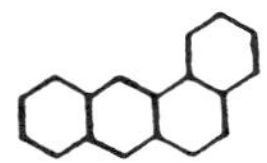	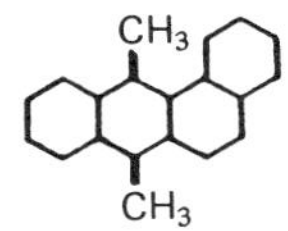 (not an air pollutant)
Formula	$C_{14}H_{10}$	$C_{18}H_{12}$	$C_{20}H_{16}$
Carcinogenicity	-	+	++++
Acronym	A	BaA	DMBA
Physical properties			
Melting point, °C	216	159.5-160.5	122-123
Boiling point, °C	340	400	
Vapor pressure, torr	1.95×10^{-4}	1.10×10^{-7}	
λ_{max},nm	308 323 338 355 375	314 327 341 359 376 386	296 345 364 384
Log ϵ	3.15 3.47 3.75 3.86 3.87	3.67 3.81 3.87 3.72 2.73 3.86	4.90 3.83 3.94 3.83
Solvent	EtOH-MeOH	EtOH	EtOH

(continued)

Table 1.1: (continued)

Compound	Dibenz[a,j]anthracene (1,2-6,7-dibenzanthracene)	Dibenz[a,h]anthracene (1,2-5,6-dibenzanthracene)	Dibenz[a,c]anthracene (1,2-3,4-dibenzanthracene)
Structure	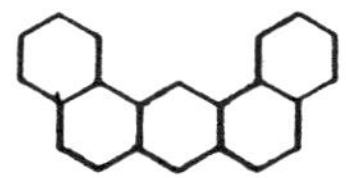	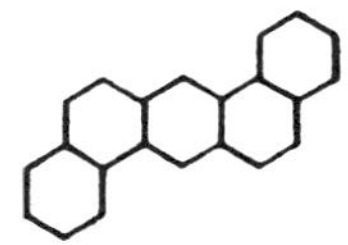	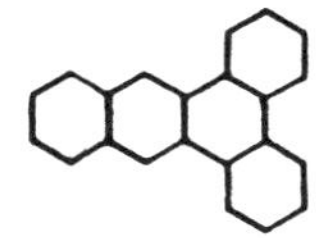
Formula	$C_{22}H_{14}$	$C_{22}H_{14}$	$C_{22}H_{14}$
Carcinogenicity	+	+++	+
Acronym		DBahA	
Physical properties Melting point, °C Boiling point, °C		 266-266.5	
Vapor pressure, torr			
λ_{max}, nm Log ϵ Solvent		299 322 335 350 374 384 5.20 4.28 4.23 4.16 3.00 2.61 dioxane	

(continued)

Table 1.1: (continued)

Compound	Phenanthrene	Benzo[c]phenanthrene (3,4-benzphenanthrene)	Fluorene
Structure	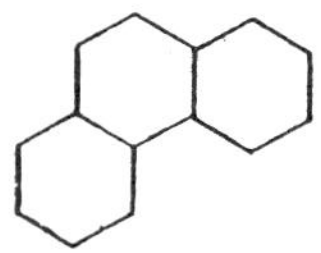	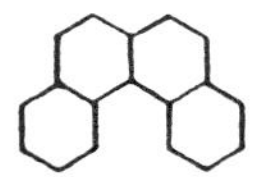	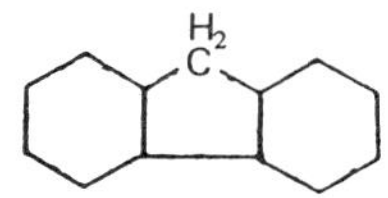
Formula	$C_{14}H_{10}$	$C_{18}H_{12}$	$C_{13}H_{10}$
Carcinogenicity	–	+++	–
Acronym	PA	BcP	F
Physical properties			
Melting point, °C	101	68	116
Boiling point, °C	340		293
Vapor pressure, torr	6.8×10^{-4}		
λ_{max}, nm	309 314 323 330 337 345	296 303 315 325 353 372	
Log ϵ	2.40 2.48 2.54 2.52 3.40 3.46	4.13 4.07 4.06 3.67 2.63 2.38	
Solvent	EtOH-MeOH	EtOH	

(continued)

Table 1.1: (continued)

Compound	Benzo[a]fluorene (1,2-benzfluorene)	Benzo[b]fluorene (2,3-benzfluorene)	Dibenzo[a,h]fluorene (1,2-6,7-dibenzfluorene)
Structure	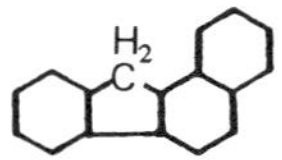	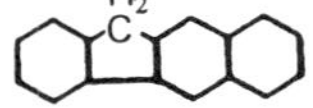	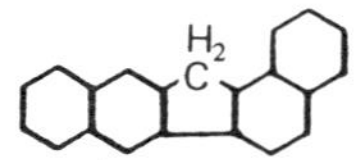
Formula	$C_{17}H_{12}$	$C_{17}H_{12}$	$C_{21}H_{14}$
Carcinogenicity	-	-	±
Acronym			
Physical properties			
Melting point, °C	189–190	208–209	
Boiling point, °C			
Vapor pressure, torr			
λ_{max}, nm			
Log ϵ			
Solvent			

(continued)

Table 1.1: (continued)

Compound	Dibenzo[a,g]fluorene (1,2-5,6-dibenzofluorene)	Benzo[c]fluorene (3,4-benzfluorene)	Dibenzo[a,c]fluorene (1,2-3,4-dibenzfluorene)
Structure	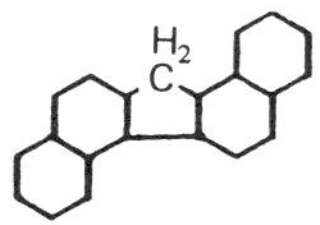	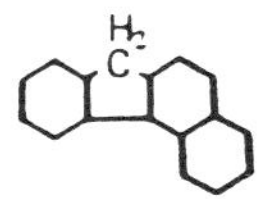	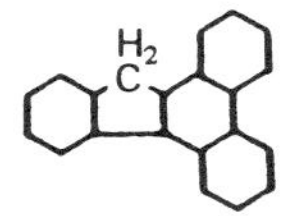
Formula	$C_{21}H_{14}$	$C_{17}H_{12}$	$C_{21}H_{14}$
Carcinogenicity	+	–	±
Acronym			
Physical properties Melting point, °C Boiling point, °C			
Vapor pressure, torr			
λ_{max}, nm Log ϵ Solvent			

(continued)

Table 1.1: (continued)

Compound	Fluoranthene	Benzo[b]fluoranthene (2,3-benzfluoranthene)	Benzo[j]fluoranthene (7,8-benzfluoranthene)
Structure	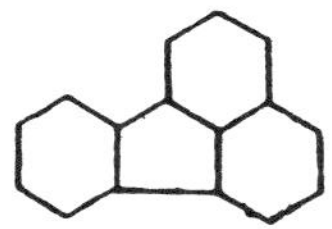	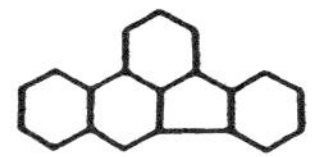	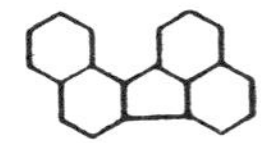
Formula	$C_{16}H_{10}$	$C_{20}H_{12}$	$C_{20}H_{12}$
Carcinogenicity	-	++	++
Acronym	Fluor	BbF	BjF
Physical properties			
Melting point, °C	110	167	165-167
Boiling point, °C	393		
Vapor pressure, torr			
λ_{max}, nm			
Log ϵ			
Solvent			

(continued)

Table 1.1: (continued)

Compound	Benzo[k]fluoranthene (8,9-benzfluoranthene)	Benzo[ghi]fluoranthene	Aceanthrylene
Structure	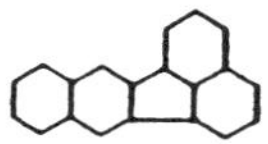	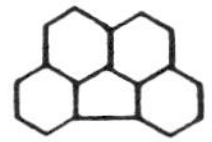	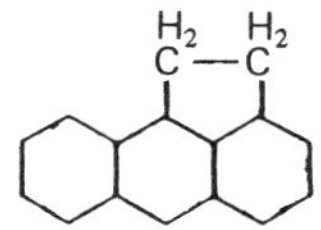
Formula	$C_{20}H_{12}$	$C_{18}H_{10}$	$C_{16}H_{12}$
Carcinogenicity	-	-	-
Acronym	BkF		
Physical properties Melting point, °C Boiling point, °C	215-216		113
Vapor pressure, torr	9.59×10^{-11}		
λ_{max}, nm Log ϵ Solvent			

(continued)

Table 1.1: (continued)

Compound	Benzo[j]aceanthrylene (cholanthrene)	3-Methylcholanthrene	Naphthacene (benzo[b]anthracene)
Structure	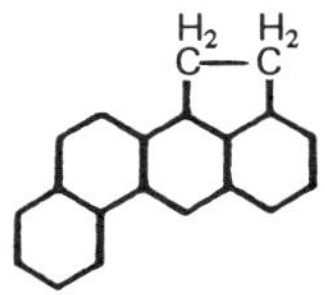	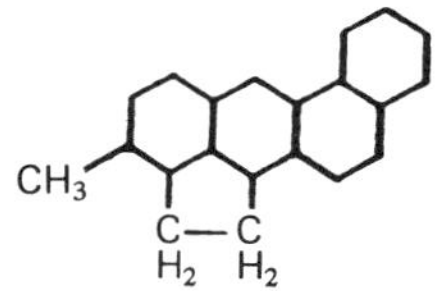	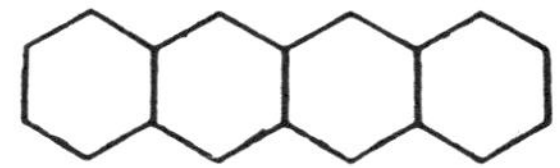
Formula	$C_{20}H_{14}$	$C_{21}H_{16}$	$C_{18}H_{12}$
Carcinogenicity	++	++++	-
Acronym		MCA	
Physical properties			
Melting point, °C		179	341
Boiling point, °C			sublimes
Vapor pressure, torr			
λ_{max}, nm		no absorption above 300 nm	
Log ϵ			
Solvent			

(continued)

Table 1.1: (continued)

Compound	Naphtho[2,1,8-qra]naphthacene (naphtho[2,3-a]pyrene) (2,3-naphtho-1,2-pyrene)	Pyrene	1-Methylpyrene
Structure	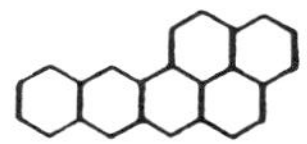	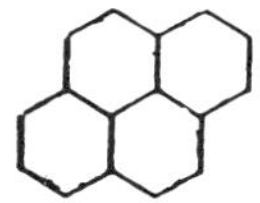	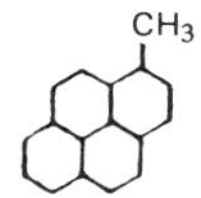
Formula	$C_{24}H_{14}$	$C_{16}H_{10}$	$C_{17}H_{12}$
Carcinogenicity	-	-	
Acronym		P	
Physical properties			
Melting point, °C		15	
Boiling point, °C		>360	
Vapor pressure, torr		6.85×10^{-7}	
λ_{max}, nm			
Log ϵ			
Solvent			

(continued)

Table 1.1: (continued)

Compound	Benzo[a]pyrene (1,2-benzpyrene) (3,4-benzpyrene*)	Benzo[e]pyrene (4,5-benzpyrene) (1,2-benzpyrene*)	Dibenzo [a,l] pyrene (2,3-4,5-dibenzpyrene) (Dibenzo[def,p] chrysene) (1,2-3,4-dibenzpyrene*)
Structure	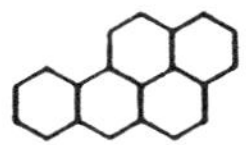	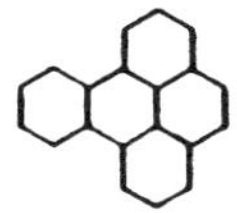	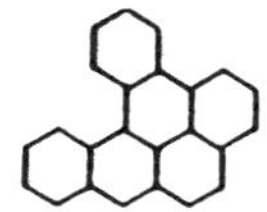
Formula	$C_{20}H_{12}$	$C_{20}H_{12}$	$C_{24}H_{14}$
Carcinogenicity	+++	-	±
Acronym	BaP	BeP	
Physical properties Melting point, °C Boiling point, °C	178.8–179.3	178.9	
Vapor pressure, torr	5.49×10^{-9}	5.54×10^{-9}	
λ_{max}, nm Log ϵ Solvent	296 330 347 364 384 403 4.76 3.76 4.12 4.38 4.48 3.60 EtOH		

(continued)

Table 1.1: (continued)

Compound	Dibenzo[a,h]pyrene (1,2-6,7dibenzpyrene) (Dibenzo[b,def]chrysene) (3,4-8,9-dibenzpyrene*)	Dibenzo[a,i]pyrene (2,3-6,7-dibenzpyrene) (4,5-8,9-dibenzpyrene*)	Dibenzo[a,e]pyrene
Structure	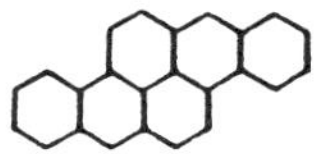	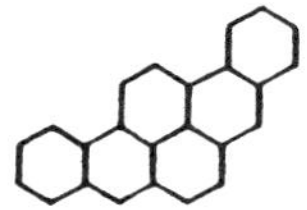	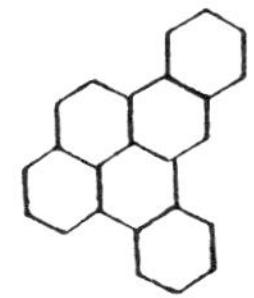
Formula	$C_{22}H_{14}$	$C_{22}H_{14}$	$C_{22}H_{14}$
Carcinogenicity	+++	+++	++ (24)
Acronym	DBahP	DBaiP	
Physical properties Melting point, °C Boiling point, °C	 320–320.5	 281–282.5	
Vapor pressure, torr			
λ_{max}, nm Log ϵ Solvent	290 301 313 379 401 424 451 4.54 4.94 5.23 3.60 4.03 4.40 4.59 benzene		

(continued)

Table 1.1: (continued)

Compound	Dibenzo[cd,jk] pyrene (anthanthrene) (dibenzo[def,mno] chrysene)	Indeno[1,2,3-cd] pyrene (o-phenylenepyrene)	Chrysene
Structure		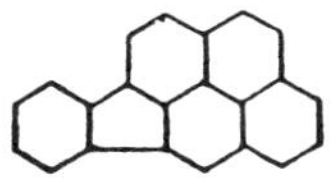	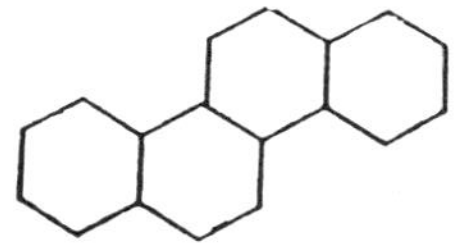
Formula	$C_{22}H_{12}$	$C_{22}H_{12}$	$C_{18}H_{12}$
Carcinogenicity	-	+	+ (25)
Acronym	ANT	IP	CH
Physical properties			
Melting point, °C	257		254
Boiling point, °C			448
Vapor pressure, torr			
λ_{max}, nm			344 351 360
L Log ϵ			2.88 2.62 3.00
Solvent			EtOH

(continued)

Table 1.1: (continued)

Compound	Perylene	Benzo[ghi]perylene	Coronene
Structure	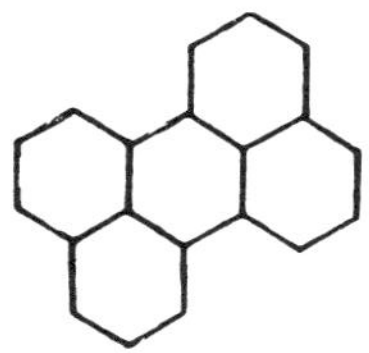		
Formula	$C_{20}H_{12}$	$C_{22}H_{12}$	$C_{24}H_{12}$
Carcinogenicity	-	-	-
Acronym	Per	BghiP	Cor
Physical properties			
Melting point, °C	273	273	438
Boiling point, °C	~500		525
Vapor pressure, torr		1.01×10^{-10}	1.47×10^{-12}
λ_{max}, nm			
Log ϵ.			
Solvent			

Source: EPA Health Assessment Document

Table 1.2: Aza and Imino Arenes

Compound	Acridine	Benz[a]acridine (1,2-benzacridine)	Benz[c]acridine (3,4-benzacridine)
Structure	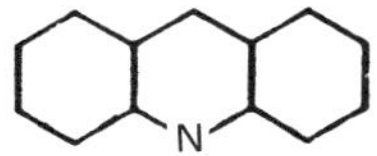	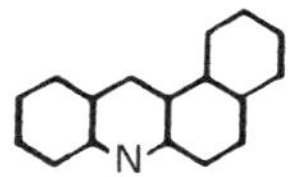	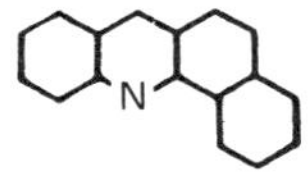
Formula	$C_{13}H_9N$	$C_{17}H_{11}N$	$C_{17}H_{11}N$
Carcinogenicity	–		
Acronym			
Physical properties			
Melting point, °C	111		
Boiling point, °C	>360		
Vapor pressure, torr			
λ_{max}, nm			
Log ϵ			
Solvent			

(continued)

Table 1.2: (continued)

Compound	Dibenz[a,j]acridine (1,2-7,8-dibenzacridine)	Dibenz[a,h]acridine (1,2-5,6-dibenzacridine)	Dibenz[c,h]acridine (3,4-5,6-dibenzacridine)
Structure	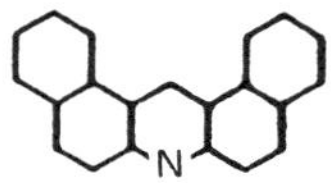	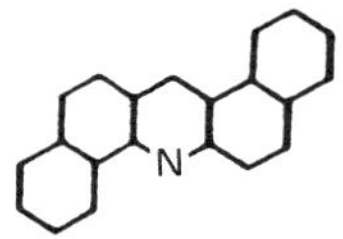	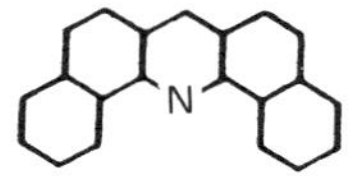
Formula	$C_{21}H_{13}N$	$C_{21}H_{13}N$	$C_{21}H_{13}N$
Carcinogenicity	++	++	±
Acronym			
Physical properties			
Melting point, °C	220	228	
Boiling point, °C			
Vapor pressure, torr			
λ_{max}, nm			
Log ϵ			
Solvent			

(continued)

Table 1.2: (continued)

Compound	Carbazole	Benzo[a]carbazole (1,2-benzcarbazole)	Dibenzo[a,g]carbazole (1,2-5,6-dibenzcarbazole)
Structure	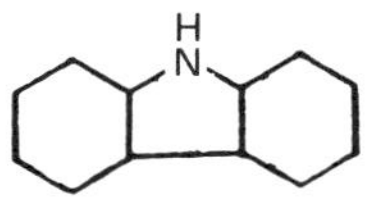	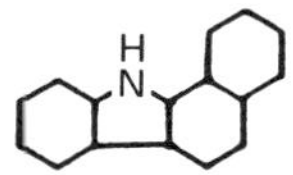	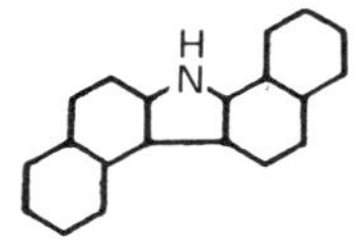
Formula	$C_{12}H_9N$	$C_{16}H_{11}N$	$C_{20}H_{13}N$
Carcinogenicity	–	±	±
Acronym			
Physical properties			
Melting point, °C	246		
Boiling point, °C	355		
Vapor pressure, torr			
λ_{max}, nm			
Log ϵ			
Solvent			

(continued)

Table 1.2: (continued)

Compound	Dibenzo[c,g]carbazole (3,4-5,6-dibenzcarbazole)	Dibenzo[a,i]carbazole (1,2-7,8-dibenzcarbazole)	Quinoline
Structure	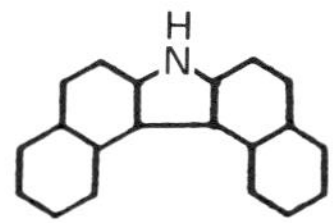	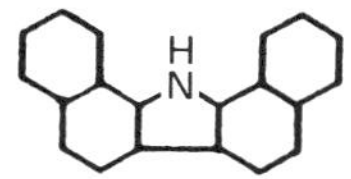	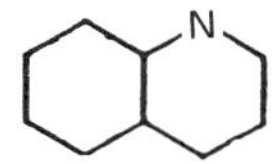
Formula	$C_{20}H_{13}N$	$C_{20}H_{13}N$	
Carcinogenicity	+++	+	+ (53)
Acronym	DBcgC		
Physical properties			
Melting point, °C			
Boiling point, °C			
Vapor pressure, torr			
λ_{max}, nm			
Log ϵ			
Solvent			

(continued)

Table 1.2: (continued)

Compound	Benzo[f]quinoline	Benzo[h]quinoline	11 H-Indeno(1,2-b)quinoline
Structure	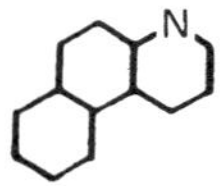	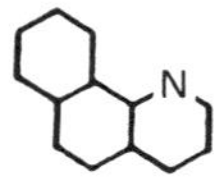	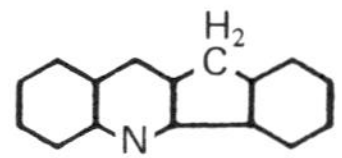
Formula			
Carcinogenicity	– (54)		
Acronym			
Physical properties Melting point, °C Boiling point, °C			
Vapor pressure, torr			
λ_{max}, nm Log ϵ Solvent			

(continued)

Table 1.2: (continued)

Compound	Isoquinoline	Benzo[f]isoquinoline	Phenanthridine
Structure	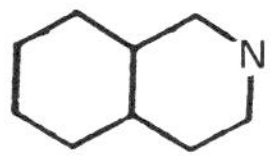	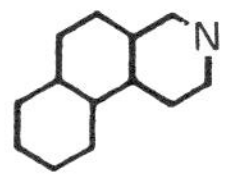	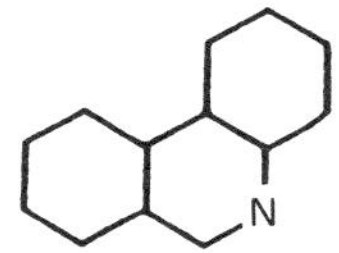
Formula			
Carcinogenicity			
Acronym			
Physical properties			
Melting point, °C			
Boiling point, °C			
Vapor pressure, torr			
λ_{max}, nm			
Log ϵ			
Solvent			

(continued)

Table 1.2: (continued)

Compound	4-Azafluorene (5-H-indeno[1,2b]pyridine)	4-Azapyrene (Benzo[lmn]phenanthridine)	1-Azafluoranthene (Indeno[1,2,3-ij]isoquinoline)
Structure	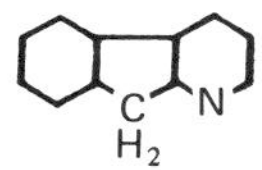	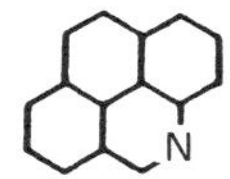	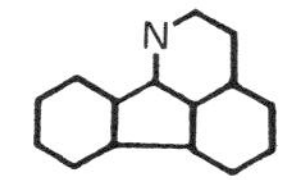
Formula			
Carcinogenicity			
Acronym			
Physical properties			
Melting point, °C			
Boiling point, °C			
Vapor pressure, torr			
λ_{max}, nm			
Log ϵ			
Solvent			

Source: EPA Health Assessment Document

Table 1.3: Miscellaneous Polycyclic Organic Matter

Compound	Phenalene-1-one	Anthanthrone	Xanthene-9-one	7-H-Benz(de)anthracene-7-one (benzanthrone)
Structure	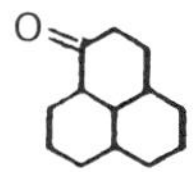	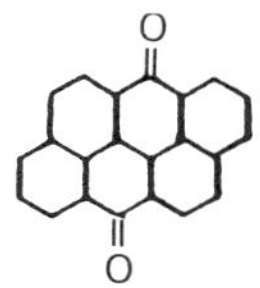	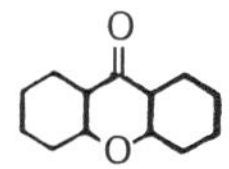	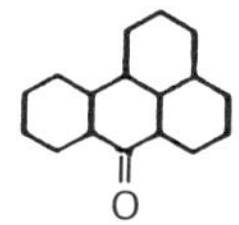
Formula				
Carcinogenicity				
Acronym				
Physical properties				
Melting point, °C				
Boiling point, °C				
Vapor pressure, torr				
λ_{max}, nm				
Log ϵ				
Solvent				

Source: EPA Health Assessment Document

Physical Properties

Most of the POM compounds are high melting/high boiling point solids that are extremely insoluble in water (55). With the exception of the basic nitrogen heterocyclics (e.g., quinolines, acridines), the major POM compounds do not dissociate and, therefore, are found in the neutral fractions of extracts of particulate material (5)(6).

Only limited physical data are available on the numerous POM compounds. The information that is available is tabulated in Tables 1.1 through 1.3. Melting points are available for most compounds and most values are considerably over 100°C. [Phenanthrene (101°C) and benzo[c]phenanthrene (68°C) are unusual exceptions.]

Vapor pressures of the pure chemicals are available for nine PAH. Depending upon the ring size and molecular weight of the compounds, the vapor pressure at 25°C of the pure compound varies from 6.8×10^{-4} torr for phenanthrene (3 rings and 14 carbons) to 1.5×10^{-12} torr for coronene (7 rings and 24 carbons) (24)-(26). This property has considerable impact on the amount of PAH that remains adsorbed on particulate matter in the atmosphere and is retained on particulate matter during collection of air samples on glass fiber filters.

In one study, the following compounds experience the indicated losses at the various times when filtered air is drawn over particulate matter at 1.2 m^3/min (1)(27): (1) tetracyclic or larger PAH, 2 hours, no loss; (2) tetracyclic PAH, 24 hours, some loss; and (3) pentacyclic or larger PAH, 3 weeks, no loss. The actual retention of POM on particulate during collection is likely to depend upon temperature, face velocity of the air with respect to the filter, and the adsorption characteristics of the individual POM.

The ultraviolet absorption spectra are also available for many compounds (Table 1.1). Most of the PAH absorbed light strongly at wavelengths longer than 300 nm (28). The absorption spectra for benzo[a]pyrene (Figure 1.3) are typical. Because these compounds absorb light at wavelengths found in sunlight (>300 nm), it is possible that they may photochemically react by direct excitation. However, the available spectra were taken for PAH in organic solvents, while PAH in the environment are usually found associated with particulate matter and adsorbed PAH may have considerably different absorption properties.

Atmospheric Forms of POM

Relatively little information is available on the exact form of the various POM in the atmosphere. Because of the high melting points and low vapor pressures of most POM, the compounds are generally considered to be associated with particulate matter, either as pure material or adsorbed to other particulate matter (2).

Some information is available on the relationship of POM to particle size, and this has particular significance with respect to respiratory tract entry, retention and deposition. The aerodynamic size of particles will determine how much and where the particles are deposited. Relatively large particles (>10 to 15 μm diameter) are deposited in the nasal and oral cavities. Smaller particles will pass into the tracheo-bronchial regions. The actual deposition efficiencies within the respiratory tract will vary considerably. They are dependent upon the dimensions and configurations of the individual air path, the pattern and depth of respiration, and the characteristics of the particles (29).

Figure 1.3: Absorption Spectrum of Benzo[a] pyrene in Ethanol

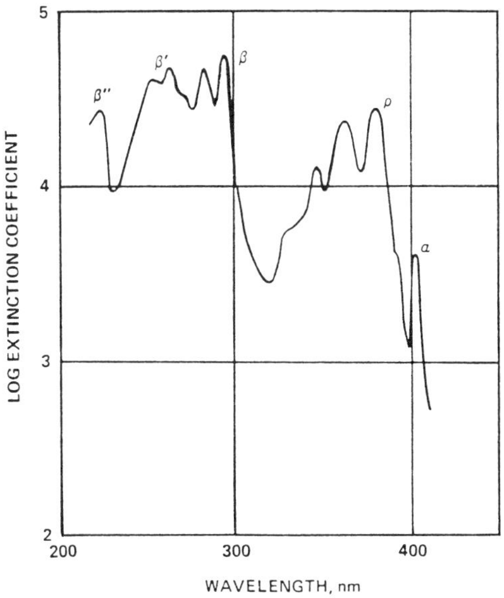

α, 4030 (3.60); ρ, 3845 (4.44), 3635 (4.36), 3470 (4.10), 3300 (3.76); β, 2965 (4.76), 2843 (4.66), 2740 (4.50); β', 2655 (4.66), 2540 (4.60); β'', 2250 (4.44).

Source: Reference (28) as cited in EPA Health Assessment Document

The particle size also affects the residence time of particulate matter. Urban aerosols appear to have a residence time without precipitation of 4 to 40 days for particles less than 1 μm in diameter and 0.4 to 4 days for particles 1 to 10 μm in diameter (30).

Tebbens et al (31) have separated particulates into different size fractions before extraction. They found a constant weight of BaP per unit weight of carbonaceous particles for all particle sizes.

DeMaio and Corn (32) collected particles with a two-stage elutriator. The first stage collected 50% of particles with a diameter of 4.6 μm and 100% of 6.5 μm particles. The second stage, which consisted of a glass fiber filter paper, collected 100% of all particles greater than 0.3 μm. The amount collected in the two stages varied considerably for the six compounds analyzed and depended upon the sampling period. Only 5% of the BaP was detected in stage one, while more than half the pyrene was found in stage one.

Kertész-Sáringer and coworkers (33) used a four-stage Casella cascade impactor backed up with a glass fiber filter and determined the BaP on various size fractions at various times (September 11 through September 24, October 15 through November 14, and November 21 through December 16) of the year in Budapest.

Their results (Figure 1.4) agree with those of DeMaio and Corn (32) in indicating that the most BaP particles are in the respirable size range. In winter, when the concentration of BaP is the highest, the pollutant is associated more with smaller particles. Also, about half the weight of atmospheric BaP is associated with particles of radii less than 0.15 μm.

Figure 1.4: Size Distribution of Particles Containing Benzo[a]pyrene

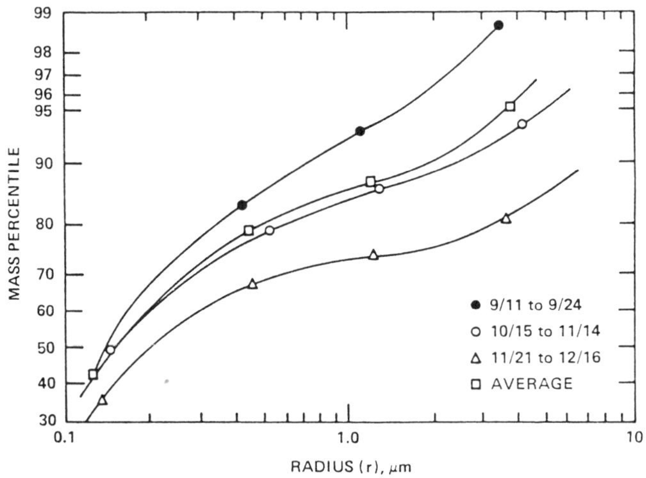

Source: Reference (33) as cited in EPA Health Assessment Document

Katz and Pierce (34)(35) have conducted a detailed study of the relationship of PAH to relative size of particulate matter in Toronto, Ontario. Particulates were collected at five locations by a five-stage Anderson Hi-Vol cascade impactor. The size ranges (μm) for the five stages were: 1 $\geq$ 7.0; 2 = 3.3 to 7.0; 3 = 2.0 to 3.3; 4 = 1.1 to 2.0; and 5 $\leq$ 1.1. Twenty PAH compounds were monitored, but since the data for the other compounds were similar to data for BaP (Figure 1.5), only a few data on other isomers were presented.

A seasonal relationship to submicron BaP particulates was found, with the highest mass fractions of submicron BaP particles (~60% total aerosol mass) found during winter. At two locations, 40 to 60% of the PAH was associated with submicron particles; at another location only 6 to 14% PAH was thus associated. Differences were also found between the fractional mass of submicron BaP particulates samples taken at 50 and 300 ft above the ground at different locations. Similar size distribution curves were also found for two oxygen heterocyclic PAH. These results demonstrate that BaP and PAH are associated primarily with small, respirable particles.

Figure 1.5: Percent Mass of Benzo[a]pyrene in Relation to Stage Number for Five Sampling Sites in Toronto, Canada

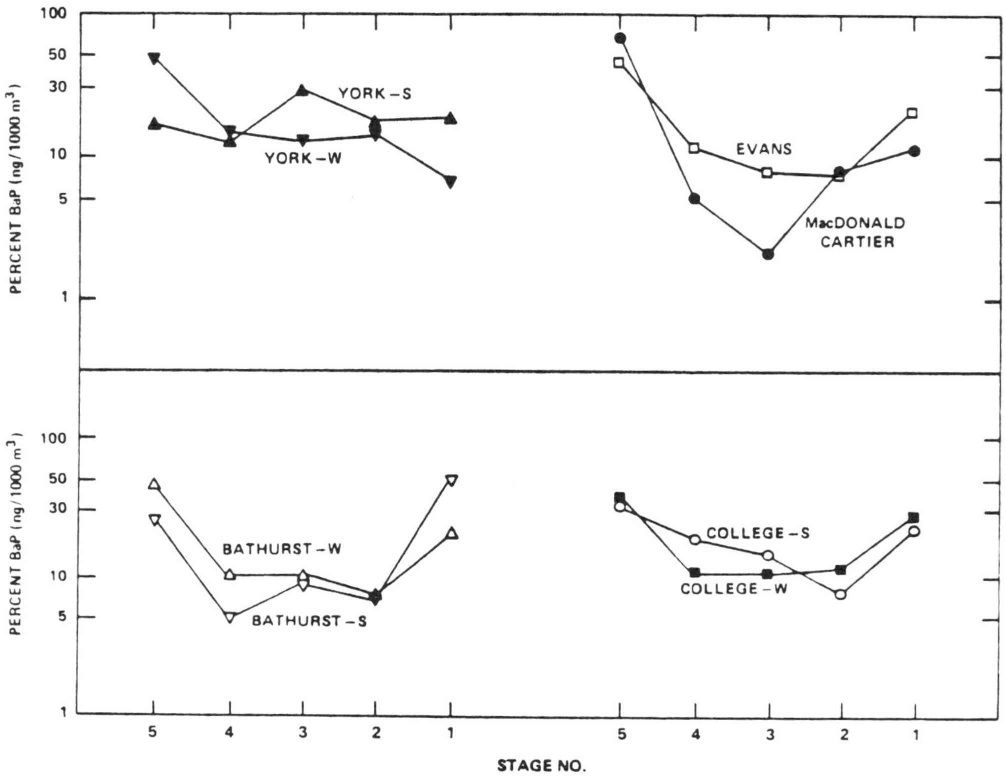

Source: Reference (35) as cited in EPA Health Assessment Document

In elevated-temperature combustion systems, substantial amounts of POM probably exist in the vapor state. However, they appear to condense or adsorb on other particles rapidly once they are away from the high-temperature areas.

Thomas et al (36) determined the BaP in deposited soot in the upper portion of the glass chimney of a combustion system and compared it to the BaP in soot collected on a filter. No difference was detected, suggesting that BaP rapidly adsorbs onto particulate matter as soon as it leaves the combustion area.

Thomas et al (36) also determined the amount of BaP from the combustion system collected on a filter (maintained at 80° to 90°C) or collected in a water and cold trap. Even at the elevated temperature of the filter, 96.3% of the total BaP was found on the filter. Sublimation of BaP off the particulate trap was ruled out.

Krstulovic et al (37) have placed polyurethane plugs behind a standard glass fiber filter (1.0 μm retention with 98% efficiency) to collect particles that pass through. These particles are referred to as being in the gas phase because they act as gas or vapor in that they follow fluid streamlines and are subject to Brownian movement.

Concentrations of several individual compounds in air samples taken in Rhode Island were greater on the foam plug than on the filter. For example, the concentration of dibenz[a,c]anthracene in Providence was 3,709 ng/1,000 m^3 on the foam plug, but 806 ng/1,000 m^3 on the filter. These results are consistent with those of Pierce and Katz (34)(35) which indicated that sizable amounts of PAH will be associated with particles less than 1 μm in diameter.

Natusch and Tomkins (38) have developed a mathematical description of the process of adsorption of PAH vapors on fly ash particles. Their calculations indicated that PAH would almost certainly exist as vapors at temperatures (>150°C) encountered in the stack systems of fossil-fueled power plants. However, at ambient temperatures (-10° to 30°C), "essentially quantitative adsorption (to particulates) would be expected" (38).

Their theoretical study also suggests that since "the extent of adsorption of PAH is proportional to the frequency of collision with the available surface area, then adsorption will result in small respirable particles being preferentially enriched (per unit mass) in potentially carcinogenic PAH" (38).

ATMOSPHERIC CHEMISTRY

The POM compounds emitted to the atmosphere have been shown to have considerably different stability. Unknown, however, is whether the degradation products are more or less toxic. Recent evidence by Pitts and coworkers (39) suggests that in some instances (e.g., nitrogen dioxide reaction with benz[a]pyrene to form 6-nitrobenzo[a]pyrene and a mixture of 1-nitro and 3-nitrobenzo[a]-pyrene) the products are more mutagenic.

The atmospheric stability of different POM compounds is dependent upon such factors as molecular structure, amount of available light, and presence of oxidizing pollutants (1). Also, because POM in the atmosphere are mostly found in the atmosphere adsorbed to particulate matter, such factors as particle size (affecting rate of deposition and surface area exposed), porosity of the particle, and adsorption factors are also likely to be important to environmental stability. The rates that have been measured vary considerably depending upon the experimental conditions.

Photooxidation

Most investigators have used light sources that simulate sunlight (>300 nm). However, the form of the POM being photolyzed has been much more varied. Solutions of compounds are one of the most convenient forms for photolysis studies, but the relevance of solution photochemistry to photolysis of POM adsorbed on particulate matter is probably remote.

Kuratsune and Hirohata (40) photolyzed thirteen PAH in cyclohexane and dichloromethane with sunlight and with fluorescent lamps filtered through glass to provide wavelengths longer than 300 nm. They found that naphthacene was the most unstable of the PAH tested and that benzo[a] pyrene and alkyl derivatives of benz[a] anthracene were labile to light.

The following compounds were stable under the experimental conditions: phenanthrene, fluorene, pyrene, fluoranthene, chrysene, benz[a] anthracene and benzo[k] fluoranthene.

In another study by Kuratsune (41), photolysis of BaP in benzene (light >280 nm and oxygen was present) allowed the isolation of 6,12-, 1,6- and 3,6-benzo[a]-pyrenequinones.

Because POM in the environment are usually found adsorbed to particulate matter, many investigators have studied the photolysis of POM naturally occurring on particulates from combustion, POM coated on some solid particles and POM coated in thin films on a smooth surface. Typical of photolysis studies of pure POM adsorbed on solids is the work of Inscoe (42) who examined photodecomposition of PAH during thin layer chromatography.

Fifteen PAH adsorbed on four different adsorbents (silica gel G, aluminum oxide G, cellulose powder and acetylated cellulose) were exposed to ultraviolet light and ordinary room light. No changes were noted for phenanthrene, chrysene, triphenylene and picene, but pronounced changes were detected for the other eleven compounds when silica gel and aluminum oxide were used; less dramatic changes were found with the other two adsorbents. Exposure to ordinary room light gave slower, but similar results.

The eleven compounds that were photosensitive were anthracene, naphthacene, benz[a] anthracene, dibenz[a,c] anthracene, dibenz[a,h] anthracene pyrene, benzo-[a] pyrene, benzo[e] pyrene, perylene, benzo[ghi] perylene and coronene. Inscoe (42) also identified 1,6-pyrenedione and 1,8-pyrenedione as photoproducts of pyrene.

Some of these compounds appear to be fairly stable when adsorbed on natural particulate matter and, therefore, the above results may be only indicative of what occurs during thin layer chromatography.

Andelman and Suess (43) attempted to simulate photodecomposition of PAH in water by adsorbing BaP on calcium carbonate and irradiating aqueous suspensions with white fluorescent light. BaP was found to be unstable under laboratory illumination with 80 to 90% lost in 17 hours in 15 to 50% dioxane-water; but it is apparently somewhat dependent upon the liquid system since BaP is stable for 17 hours in 100% dioxane or 50% methanol-water.

A couple of investigators have examined the effect of some common air constituents on the photochemistry of adsorbed POM. Jager and Rakovic (44) adsorbed pyrene and BaP to fly ash and alumina from an acetone solution and irradiated the adsorbed compounds in a quartz tube filled with 10% SO_2 in air at room temperature and at 60° to 70°C.

From irradiation of pyrene at room temperature, many compounds could be isolated containing sulfur, including 1-pyrenesulfonic acid (15%); from BaP, benzo[a]pyrene-4-sulfonic acid was detected. Whether this type of reaction occurs in ambient air samples is unknown since the experimental concentration of SO_2 is much higher than is normally present in ambient air.

Geacintov (45) studied the photolysis of 20 PAH adsorbed on solid polystyrene fluffs in the presence of oxygen and nitric oxide. Negligible photochemical degradation was found compared to quenching by the paramagnetic gases, O_2 and NO. However, the PAH were efficient photocatalysts for the formation of singlet oxygen, a reactive oxidizing agent which may react with the PAH.

Perhaps the earliest study of POM photochemistry was conducted by Falk and coworkers (46). It was noted that the amount of BaP in ambient air compared with other PAH was greater than would be expected on the basis of amounts emitted and it was thought that this could be explained by a difference in atmospheric stability.

The stability of two forms of PAH was studied: "pure" (dissolved in solvent and applied to a filter) and "adsorbed" (combustion particles collects on filter and sample divided into control and exposed sample). The samples were exposed to air and to a synthetic smog (1 hour of exposure equivalent to approximately 100 hours of exposure to natural smog) with and without irradiation. Their results are presented in Table 1.4.

For most compounds, destruction of the "pure" samples is essentially similar in dark or light. However, BaP stability is significantly different with and without light. Also, the lower rate of reaction of the adsorbed form of BaP suggests that adsorption may provide some protection from photooxidative processes.

The results for pure unadsorbed chrysene with light are difficult to explain. Falk et al (47) concluded that the three-ring compounds disappear rapidly and that differences in atmospheric stability can explain why the ratio of pyrene to BaP is reversed following atmospheric residence. Also, adsorption on particulate matter seems to have a stabilizing effect compared to coating on a filter.

Tebbens et al (31) also used the filter paper procedure of Falk et al (46) to study photolysis of BaP by sunlight or simulated sunlight. With BaP deposited on various filters from an ether solution, different filter material resulted in different losses (60 to 92% loss).

Table 1.4: Percent Destruction of Polycyclic Aromatic Hydrocarbons Under Atmospheric Conditions

	. . . . Pure Unadsorbed Compounds.					. Adsorbed on Soot .	
	. . . Air in Dark . .		. . . Air in Light . .		Smog in Light	Air in Light	Smog
	24 Hours	48 Hours	24 Hours	48 Hours	1 Hour	48 Hours	1 Hour
Anthanthrene	44	49	42	44	–	5	55
Phenanthrene	39	61	34	60	–	–	–
Pyrene	24	43	20	42	83	1	58
Fluoranthene	17	20	16	24	59	4	59
Benzo[a]pyrene	0	0	21	22	50	10	18
Benzo[e]pyrene	–	–	–	–	–	7	51
Benzo[ghi]perylene	0	0	0	0	27	0	67
Coronene	0	0	0	0	5	–	–
Chrysene	0	0	11	0	15	–	–

Source: Reference (46) as cited in EPA Health Assessment Document

Tebbens et al (31) also studied the chemical modification of BaP and perylene in smoke passed through a 22 ft dynamic flow chamber (Pyrex pipe surrounded with fluorescent lamps) by comparing the difference between irradiated and non-irradiated samples.

It was found that irradiation caused disappearance or modification of 35 to 65% of the original content and that oxygen, although its presence increases the rate of photodegradation, is not necessary for photoalterations.

Thomas et al (36) used a dynamic flow-through chamber that was coupled to a furnace and controlled lighting having one-quarter the intensity of an average July noonday sun. The BaP/soot was measured at the entrance and exit of the chamber. A 58% decrease in the ratio was noted with the light turned on compared to when the light was off (retention time of 40 minutes).

Thomas et al (36) also examined the effect of particle size by coating glass beads (1 g of either 220 or 28 μm diameters) by evaporating the solvent. The beads were then exposed to sunlight in flasks and agitated hourly to expose new surfaces. BaP on small beads decreased 1% in 12 hours; on large beads, 1% in 7 days.

It was concluded that the 60% decrease in the 40 minutes retention time of the BaP/soot in the dynamic chamber may be related to the much greater surface area of the soot exposed to light.

Lane and Katz (48) have recently concluded that the previous experiments were not representative of actual atmospheric conditions, since the oxidant levels used were often too high and fluorescent lamps are poor simulators of sunlight. A dynamic flow reaction chamber (20 ℓ glass jar) with O_3 levels of 0 to 2.28 ppm and irradiation from a Quartzline lamp were used. The disappearance of 500 ng of BaP, benzo[b]fluoranthene (BpF), and benzo[k]fluoranthene (BkF), distributed in a thin dispersion in a petri dish was studied.

The results presented in Table 1.5 show that BaP reacts rapidly with ozone compared to BbF and BkF, even in the absence of light. This is particularly significant because this type of nonphotochemical oxidation could occur during the normal 24-hour, high-volume procedure used for sampling ambient air.

Table 1.5: The Half-Lives of BaP, BbF and BkF Under Various Reaction Conditions

.....Reaction Conditions		.... Half Lives*		
Irradiation	Ozone Concentration (ppm)	BaP	BbF	BkF
		.........(hr).........		
None	0.19	0.62	52.7	34.9
	0.70	0.4	10.8	13.8
	2.28	0.3	2.9	3.3
Quartzline	0.0	5.3	8.7	14.1
Q500T/CL	0.19	0.58	4.2	3.9
lamp	0.70	0.2	3.6	3.1
	2.28	0.08	1.9	0.9

*The half-life was determined from the linear regression analysis of the initial linear portion of each decomposition curve.

Source: Reference (48) as cited in EPA Health Assessment Document

It was also found that the rate of decomposition slowed in time, suggesting that the surface-exposed layer reacted quickly, but hindered the decomposition of subsurface material. However, it should be kept in mind that these results are based upon PAH coated in a thin film on glass. Whether these are the same results that would be obtained with PAH adsorbed on soot is unknown. Falk et al (46) have shown that soot adsorption can have a stabilizing effect on BaP.

Relatively few studies have attempted to identify the photooxidative products. As indicated earlier, Masuda and Kuratsune (41) and Inscoe (42) have identified quinones of BaP and pyrene. This is consistent with the monitoring results of Pierce and Katz (49), who tentatively identified five polycyclic quinones in air samples in Toronto. The seasonal increase of polycyclic quinones during the summer is compatible with a photochemical mechanism.

Oxidation and Nitration

As indicated in some studies, light is not necessarily required for the atmospheric decomposition of POM. Falk et al (46) found that for many PAH (excluding BaP) the amount of loss of an individual compound coated on a filter was the same in air in dark as in light. Also, the amount of loss of PAH adsorbed on soot was very high in a synthetic smog that had a high oxidant concentration.

In addition, Lane and Katz (48) found that BaP reacted rapidly in reasonably low ozone concentrations (0.19 to 0.70 ppm) without irradiation and that the addition of irradiation had little effect on the rate. However, the other two PAH (BbF and BkF) were not nearly as susceptible to ozone oxidation. Thus, it appears that POM will be susceptible to air oxidation without light irradiation and that the reactivity is likely to vary for different structures.

Pitts et al (39) have demonstrated that BaP deposited on glass fiber filters will react with O_3 and PAN to form oxygenated products and with NO_2 to form nitro derivatives when these gases are drawn through the filters. Exposure of BaP to 1 ppm NO_2 containing traces of nitric acid results in formation of compounds that were assigned the structure 6-nitro-, 1-nitro- and 3-nitro-BaP. (The latter two isomers were unresolved by TLC.) Both the 6-nitro-BaP and the mixture of 1-nitro-acid 3-nitro-BaP were direct-acting, frameshift mutagens in the Ames assay. With BaP exposed to O_2, PAN or smog, oxidation products, including hydroxy-, dihydroxy- and dihydrodiol derivatives, were detected.

Perylene under similar experimental conditions (1 ppm NO_2) also formed a nitro derivative, 3-nitroperylene, which was a directly active mutagen in the Ames assay, even though the parent compound, perylene, shows no mutagenic activity with or without activation.

All of these reactions were carried out on PAH deposited on glass-fiber filters. Whether these compounds would react similarly in the form found in the atmosphere (adsorbed to particulate matter) is unknown.

Commins (27) also provided suggestive evidence of nonphotochemical decomposition of PAH by collecting PAH particulate matter from air and comparing the results of immediate analysis to results from analysis after 1 year of storage in a sealed envelope (Table 1.6).

Table 1.6: Comparison of PAH Content of Air Solids Before and After Storage

PAH	Vapor Pressure*	. . . PAH Concentration, μg/g Smoke . . . Analyzed Immediately After Collection	Analyzed 1 Year After Collection	Percent Loss
Fluoranthene	–	225	18	92
Pyrene	6.83×10^{-7}	328	38	88
Benzo[a] pyrene	5.49×10^{-9}	111	76	32
Benzo[e] pyrene	5.54×10^{-9}	71	55	23
Anthanthrene	–	70	55	21
Benzo[ghi] perylene	1.01×10^{-10}	252	226	10
Coronene	1.47×10^{-12}	142	140	1

*Of pure compound (25) (26).

Source: Reference (27) as cited in EPA Health Assessment Document

Significant loss of lower molecular weight PAH was noted, but little loss with the higher molecular weight compounds was found. This loss could be explained by evaporation of the lower molecular weight compounds rather than by oxidation.

Atmospheric Stability

The above results would suggest that many POM oxidize or photodegrade at a significant rate under atmospheric conditions. However, no satisfactory quantitative rate data are available.

In contrast to the results reported above, Lunde and Bjørseth (50) published suggestive evidence that many POM are stable enough in the atmosphere to travel long distances. Air samples, which had different trajectories, were monitored in Norway. Twenty different PAH were identified and it was determined that samples with trajectories from western Europe contained about 20 times more PAH than samples with trajectories from northern Norway or stationary air from southern Norway.

The PAH found in high concentrations under appropriate wind trajectories were phenanthrene, anthracene, methylphenanthrene, methylanthracene, fluoranthene, dihydrobenzo[a&b] fluorenes, pyrene, benzo[a] fluorene, benzo[b] fluorene, 1-methylpyrene, benzo[c] phenanthrene, benz[a] anthracene, chrysene/triphenylene, benzo[b&k] fluoranthene, benzo[e] pyrene, benzo[a] pyrene, perylene, indeno-[1,2,3-cd] pyrene, benzo[ghi] perylene, anthanthrene, and coronene. These results suggest that the above compounds are relatively stable in the atmosphere in the form in which they are naturally present.

Summary

The POM chemical groups most commonly found in ambient air are polycyclic aromatic hydrocarbons (PAH), such as benzo[a] pyrene, and their nitrogen analogs, aza- and imino arenes. Other related compounds such as carbonyl arenes and dicarbonyl arenes (quinones) are less commonly detected.

The major environmental sources of POM appear to be from combustion or pyrolysis processes which use materials containing carbon and hydrogen. Current theory suggests a free-radical, step-wise, sequence mechanism which, with various fuels, could result in the formation of hundreds of POM.

Most of the POM are high melting/high boiling point solids that are very insoluble in water. At 25°C, the available vapor pressures of the pure compounds vary for the individual compounds from 10^{-4} torr (3 rings) to 10^{-12} torr (7 rings). The PAH are strong absorbers of ultraviolet light of wavelengths from 200 to 400 nm.

It is generally agreed that most POM compounds are associated with suspended particle matter, with a large portion found with particles smaller than 1 μm. However, two- or three-ring compounds may be found partially in the vapor phase. Whether POM are condensed into discrete particles after cooling or are adsorbed on surfaces of existing particles is still unknown.

Urban aerosols appear to have a residence time without precipitation of 4 to 40 days for particles less than 1 μm in diameter and 0.4 to 4 days for particles 1 to 10 μm in diameter. Thus, the atmospheric lifetime of a POM particle will be closely related to its particle size.

Experimental chemical and photochemical reactivities of POM on particulate matter vary considerably from half-lives of less than a day to several days. Particularly interesting are the different reactivities noted for the various compounds studied and the indication that reaction rates of compounds on smaller particle sizes are likely to be faster than those of larger particles. Light irradiation considerably accelerates the decomposition of some compounds, but not others. Also, various experimental conditions have given contradictory results.

One report has indicated that oxidation by ozone is the most important decomposition process for BaP in air, but many other reports have documented the photochemical susceptibility of BaP. However, monitoring data suggest that at least 20 PAH are stable enough in the form found in the atmosphere that they can travel long distances with particulate matter. Thus, although an exact residue time of POM is difficult to estimate from the available data, many of the POM compounds are stable enough to remain at significant concentrations during transport from the source to where they are inhaled. Although the smaller particles will have a longer atmospheric residence time, they will probably also react faster because of the greater surface area exposed to oxidants and light.

All the available information does support previous conclusions that at least the outer layer of some POM adsorbed on particles is relatively reactive. However, this may mean that there is an inner layer of POM that would be relatively stable and perhaps available for elution by biological fluids following inhalation.

The relatively long stability of several PAH based upon monitoring data is consistent with this possibility. Only limited information is available on the products of decomposition or their toxicologic properties. Quinones appear to be common atmospheric degradation products of PAH.

REFERENCES

(1) U.S. Environmental Protection Agency, *Scientific and Technical Assessment Report on Particulate Polycyclic Organic Matter (PPOM),* EPA-600/6-75-001, Washington, D.C. (1975).

(2) National Academy of Sciences, Committee on Biological Effects of Atmospheric Pollutants, *Particulate Polycyclic Organic Matter,* Washington, D.C. (1972).

(3) Sawicki, E., "Airborne Carcinogens and Allied Compounds," *Arch. Environ. Health,* 14(1):46-53 (1967).

(4) Badger, G.M., *Mode of Formation of Carcinogens in Human Environment,* Natl. Cancer Inst., 9:1-16 (1962).

(5) Hoffman, D. and Wynder, E.L., "Chemical Analysis and Carcinogenic Bioassays of Organic Particulate Pollutants," *Analysis, Monitoring, and Surveying: Air Pollution,* Vol. II, 2nd ed., pp. 186-247, A.C. Stern (editor), Academic Press, Inc., New York and London (1968).

(6) Hoffman, D. and Wynder, E.L., "Organic Particulate Pollutants–Chemical Analysis and Bioassays for Carcinogenicity," *The Effects of Air Pollution: Air Pollution,* Vol. II, 3rd ed., pp. 361-455, A.C. Stern (editor), Academic Press, New York and London (1977).

(7) Boubel, R.W. and Ripperton, L.A., "Benzo[a]pyrene Production During Controlled Combustion," *J. Air Pollut. Contr. Assoc.,* 13:553-557 (1963).

(8) Badger, G.M. and Spotswood, T.M., "The Formation of Aromatic Hydrocarbons at High Temperatures. Part IX. The Pyrolysis of Toluene, Ethylbenzene, Propylbenzene and Butylbenzene," *J. Chem. Soc.,* pp. 4420-4427 (1960).

(9) Badger, G.M. and Spotswood, T.M., "The Formation of Aromatic Hydrocarbons at High Temperatures. Part XI. The Pyrolysis of Buta-1,3-diene and of Buta-1,3-diene with Pyrene," *J. Chem. Soc.,* pp. 4431-4437 (1960).

(10) Crittenden, B.D. and Long, R., "The Mechanisms of Formation of Polynuclear Aromatic Compounds in Combustion Systems," *Carcinogenesis, Vol. I, Polynuclear Aromatic Hydrocarbons: Chemistry, Metabolism, and Carcinogenesis,* R.I. Freudenthal and P.W. Jones (editors), Raven Press, New York (1976).

(11) Suess, M.J., "The Environmental Load and Cycle of Polycyclic Aromatic Hydrocarbons," *Sci. Total Environ.,* 6:239-250 (1976).

(12) Blumer, M.J. and Youngblood, W.W., "Polycyclic Aromatic Hydrocarbons in Soils and Recent Sediments," *Science,* 188:53-55 (1975).

(13) Hites, R.A., LaFlamme, R.E. and Farrington, J.W., "Sedimentary Polycyclic Aromatic Hydrocarbons: The Historical Record," *Science,* 198:829-831 (1977).

(14) Dipple, A., "Polynuclear Aromatic Carcinogens," *Chemical Carcinogens,* ACS Monograph 173, pp. 245-314, C.E. Searle (editor), American Chemical Society, Washington, D.C. (1976).

(15) Patterson, A.M., Capell, L.T and Walker, D.F., *The Ring Index. A List of Ring Systems Used in Organic Chemistry,* 2nd ed., American Chemical Society, Washington, D.C. (1960).

(16) Capell, L.T. and Walker, D.F., *The Ring Index,* Supplement I (1957-1959), American Chemical Society, Washington, D.C. (1963).

(17) Capell, L.T. and Walker, D.F., *The Ring Index,* Supplement II (1960-1961), American Chemical Society, Washington, D.C. (1964).

(18) Capell, L.T. and Walker, D.F., *The Ring Index,* Supplement III (1962-1963), American Chemical Society, Washington, D.C. (1965).

(19) Hartwell, J.L. and Shubik, P., *Survey of Compounds Which Have Been Treated for Carcinogenic Activity,* Public Health Service Publication 149, 2nd ed., U.S. Government Printing Office, Washington, D.C. (1951).

(20) Shubik, P. and Hartwell, J.L., *Survey of Compounds Which Have Been Tested for Carcinogenic Activity,* Supplement 1, Public Health Service Publication 149-1, U.S. Government Printing Office, Washington, D.C. (1957).

(21) Shubik, P. and Hartwell, J.L., *Survey of Compounds Which Have Been Tested for Carcinogenic Activity,* Supplement 2, Public Health Service Publication 149-2, U.S. Government Printing Office, Washington, D.C. (1969).

(22) Tracor Jitco, Inc., *Survey of Compounds Which Have Been Tested for Carcinogenic Activity* (1970-1971), Public Health Service Publication 149, U.S. Government Printing Office, Washington, D.C. (1974).

(23) Tracor Jitco, Inc., *Survey of Compounds Which Have Been Tested for Carcinogenic Activity* (1972-1973), Public Health Service Publication 149, U.S. Government Printing Office, Washington, D.C. (1976).

(24) Radding, S.B, Mill, T., Gould, C.W., Liu, D.H., Johnson, H.L., Bomberger, D.C. and Fojo, C.V., *The Environmental Fate of Selected Polynuclear Aromatic Hydrocarbons,* Stanford Research Institute, Menlo Park, Calif., Environmental Protection Agency, EPA-68-01-2681, PB-250 948, Washington, D.C. (1976).

(25) Murray, J.J., Pottie, R.F. and Pupp, C., "The Vapor Pressures and Enthalpies of Sublimation of Five Polycyclic Aromatic Hydrocarbons," *Can. J. Chem.,* 52(4):557-563 (1974).

(26) Pupp, C., Lao, R.C., Murray, J.J. and Pottie, R.F., "Equilibrium Vapor Concentrations of Some Polycyclic Aromatic Hydrocarbons, Arsenic Trioxide and Selenium Dioxide, and the Collection Efficiencies of These Air Pollutants," *Atmos. Environ.,* 8(9):915-925 (1974).

(27) Commins, B.T., *Interim Report on the Study of Techniques for Determination of Polycyclic Aromatic Hydrocarbons in Air*, Natl. Cancer Inst. Monograph No. 9, pp. 225-233 (1962).

(28) Clar, E., *Polycyclic Hydrocarbons,* Vols. 1 and 2, Academic Press, New York and London; Springer-Verlay, Berlin-Göttingen-Heidelberg (1964).

(29) National Academy of Sciences, Subcommittee on Airborne Particles, *Airborne Particles,* Washington, D.C. (1977).

(30) Eswea, N.A. and Corn, M., "Residence Time of Particles in Urban Air," *Atmos. Environ.,* 5:571-578 (1971).

(31) Tebbens, B.D., Thomas, J.F. and Mukai, M., "Fate of Arenes Incorporated with Airborne Soot," *J. Amer. Ind. Hyg. Assoc.,* 27:415-422 (1966).

(32) DeMaio, L. and Corn, M., "Polynuclear Aromatic Hydrocarbons Associated with Particles in Pittsburgh Air," *J. Air. Pollut. Contr. Assoc.,* 16:67-71 (1966).

(33) Kertész-Sáringer, M., Mészáros, E. and Várkonyi, T., "Technical Note on the Size Distribution of Benzo[a]pyrene-Containing Particles in Urban Air," *Atmos. Environ.,* 5:429-431 (1971).

(34) Katz, M. and Pierce, R.C., "Quantitative Distribution of Polynuclear Aromatic Hydrocarbons in Relation to Particle Size of Urban Particulates," *Carcinogenesis, Vol. 1, Polynuclear Aromatic Hydrocarbons: Chemistry, Metabolism and Carcinogenesis,* R.I. Freudenthal and P.W. Jones (editors), Raven Press, New York (1976).

(35) Pierce, R.C. and Katz, M., "Dependency of Polynuclear Aromatic Hydrocarbon Content on Size Distribution of Atmospheric Aerosols," *Environ. Sci. Technol.,* 9(4):347-353 (1975).

(36) Thomas, J.F., Mukai, M. and Tebbens, B.D., "Fate of Airborne Benzo[a]pyrene," *Environ. Sci. Technol.,* 2(1):33-39 (1968).

(37) Krstulovic, A.M., Rosie, D.M. and Brown, P.R., "Distribution of Some Atmospheric Polynuclear Aromatic Hydrocarbons," *Amer. Lab.,* pp. 11-18 (1977).

(38) Natusch, D.F.S. and Tomkins, B.A., "Theoretical Consideration of the Adsorption of Polynuclear Aromatic Hydrocarbon Vapor onto Fly Ash in a Coal-Fired Power Plant," *Carcinogenesis, Vol. 3, Polynuclear Aromatic Hydrocarbons,* pp. 145-153, P.W. Jones and R.I. Freudenthal (editors), Raven Press, New York (1978).

(39) Pitts, J.N., Jr., Van Cauwenberghe, K.A., Grosjean, D., Schmid, J.P., Fitz, D.R., Belser, W.L., Knudson, G.B. and Hynds, P.M., "Atmospheric Reaction of Polycyclic Aromatic Hydrocarbons: Facile Formation of Mutagenic Nitroderivatives," *Science,* in press (1978).

(40) Kuratsune, M., and Hirohata, T., *Decomposition of Polycyclic Aromatic Hydrocarbons Under Laboratory Illuminations,* Natl. Cancer Inst. Monograph No. 9, pp. 117-125 (1962).

(41) Masuda, Y. and Kuratsune, M., "Photochemical Oxidation of Benzo[a]pyrene", *Air Wat. Pollut. Int. J.,* 10:805-811 (1966).

(42) Inscoe, M.N., "Photochemical Changes in Thin-Layer Chromatograms of Polycyclic, Aromatic Hydrocarbons," *Anal. Chem.,* 36:2505-2606 (1964).

(43) Andelman, J.B. and Suess, M.J., "The Photodecomposition of 3,4-Benzpyrene Sorbed on Calcium Carbonate," *Organic Compounds in Aquatic Environments* by S.D. Faust and J.V. Hunter, pp. 439-468 (1971).

(44) Jager, J. and Rakovic, M. "Sulfur-Dioxide-Induced Qualitative Changes in Polycyclic Aromatic Hydrocarbons Adsorbed on Solid Carriers," *J. Hyg. Epidemiology, Microbiology, and Immunology,* 18(2):137-143 (1974).

(45) Geacintov, N.E., *Reactivity of Polynuclear Aromatic Hydrocarbons With O_2 and NO in the Presence of Light,* Prepared under Grant No. R801393, U.S. Environmental Protection Agency, EPA-650/1-74-010, Washington, D.C. (1973).

(46) Falk, H.L., Markul, I. and Kotin, P., "Aromatic Hydrocarbons. IV. Their Fate Following Emission into Atmosphere and Experimental Exposure to Washed Air and Synthetic Smog," *A.M.A. Arch. Ind. Hlth.,* 13:113-17 (1956).

(47) Falk, H.L, Kotin, P. and Miller, A., "Aromatic Polycyclic Hydrocarbons in Polluted Air as Indicators of Carcinogenic Hazards," *Intl. J. Air Poll.,* 2:201-209 (1960).

(48) Lane, D.A. and Katz, M., "The Photomodification of Benzo[a]pyrene, Benzo[b]-fluoranthene, and Benzo[k]fluoranthene Under Simulated Atmospheric Conditions," *Fate of Pollutants in the Air and Water Environments,* Part 2, I.A. Suffet (editor), Wiley-Interscience, New York (1977).

(49) Pierce, R.C. and Katz, M., "Chromatographic Isolation and Spectral Analysis of Polycyclic Quinones Application to Air Pollution Analysis," *Environ. Sci. Technol.,* 10(1):45-51 (1976).

(50) Lunde, G. and Bjørseth, A., "Polycyclic Aromatic Hydrocarbons in Long-Range Transported Aerosols," *Nature,* 268:518-519 (1977).

(51) Hoffman, D. and Wynder, E.L., "On the Carcinogenic Activity of Dibenzopyrenes," *Z. Krebsforsch.,* 68:137-149 (1966).

(52) Hecht, S.S., Bondinell, W.E. and Hoffman, D., "Chrysene and Methylchrysenes: Presence in Tobacco Smoke and Carcinogenicity," *J. Nat. Cancer Inst.,* 53:1121-1133 (1974).

(53) Hirao, K., Shinohara, Y., Tsuda, H. , Fukushima, S., Takahashi, M. and Ito, N., "Carcinogenic Activity of Quinoline on Rat Liver," *Cancer Res.,* 36:329-355 (1976).

(54) Shear, M.J. and Leiter, J., "Studies in Carcinogenesis. XVI. Production of Subcutaneous Tumors in Mice by Miscellaneous Polycyclic Compounds," *J. Nat. Cancer Inst.,* 2:241-258 (1941).

(55) May, W.E., Wasik, S.P. and Freeman, D.H., "Determination of the Solubility Behavior of Some Polycyclic Aromatic Hydrocarbons in Water," *Anal. Chem.,* 50:997-1000 (1978).

ENERGY SOURCES OF PNA

The information in this chapter is based on:

Energy Sources of Polycyclic Aromatic Hydrocarbons, NTIS CONF-770130-2, 1977, prepared by M.R. Guerin of Oak Ridge National Laboratory for U.S. Environmental Protection Agency and the U.S. Energy Research and Development Administration under contract with Union Carbide.

Polycyclic Organic Materials and the Electric Power Industry, NTIS EPRI EA-787-SY, December 1978, prepared by J.M. Allen, A. Levy, P.W. Jones and R.I. Freudenthal of Battelle Columbus Laboratories for Electric Power Research Institute.

INTRODUCTION

The United States consumed an amount of energy (73.7 Quads) equivalent to 13,266 million barrels of petroleum in 1972 (1). 1 Quad = 1 x 10^{15} Btu ≡ 180 million barrels of petroleum ≡ 42 million tons of bituminous coal ≡ 0.98 trillion cubic feet of natural gas ≡ 293 billion kilowatt hours of electricity. Of this, 45% was consumed as crude oil, 19% as coal, and 31% as natural gas. Nuclear power and hydropower constituted most of the remaining 5% of energy resources consumed. Current energy production utilizes essentially the same distribution of resources. Energy consumption in 1985 is estimated (1) to range from 96.97 to 107.30 Quads. In the process, approximately 200 x 10^8 lb of particulates and 130 x 10^8 lb of hydrocarbons are estimated to be emitted into the air and approximately 20,000 tons of nondegradeable organics will be emitted as water pollutants.

The importance of fossil fuels to the nation's energy base is shown by their constituting 95% of current consumption. The emerging preeminence of coal is due to three factors (a) its abundance, (b) dwindling supplies of petroleum and natural gas, and (c) the dependence of society on liquid fuels for transportation. As is illustrated in Table 2.1, almost three times as much "coal-energy" is available as of energy from natural gas, petroleum, shale oil, and uranium combined. Domestic crude oil and natural gas liquids production is projected to decrease from approximately 11 x 10^6 bbl/day in 1974 to 5 x 10^6 bbl/day in 2020. Natural gas production is estimated to decrease from approximately 23 x 10^{12} cubic feet per year to 6 x 10^{12} cubic feet per year in 2020. The equivalent of 1.7 billion barrels of petroleum were imported in 1972 with the current import level approximating one-half of the national annual consumption of oil. While supplies of liquid fuels decline, levels of automobile, bus, truck and rail transportation are expected to more than double by the year 2000 as compared to 1972 and air transportation is projected to increase by a factor of six.

Table 2.1: Selected Recoverable Domestic Energy Resources

Resources	Amount	Equivalent Quads (10^{15} Btu)
Coal	600 billions of tons Indeterminately greater with in situ utilization	12,000
Natural gas	750 trillions of cubic feet 1,030 Quads with enhanced recovery	775
Petroleum	142 billions of barrels 1,100 Quads with enhanced recovery	800
Shale oil	200 billions of barrels 5,800 Quads using low yield shale	1,200
Uranium	3.6 millions of tons U_3O_8 130,000 Quads using breeder reactors	1,800

Source: Reference (1) as cited in NTIS CONF-770130-2

Combustion is the predominant end-process by which fossil fuels are converted to energy. Combustion, particularly when inefficient, is also the primary technological source (2)-(4) of polycyclic aromatic hydrocarbons (PAH) released into the environment. The need for liquid fuels to supply the transportation industry and for nonpolluting fuels for heat and power generation provide the incentive to commercialize processes to convert coal to substitute natural gas and oil. These processes represent a potentially massive new source of environmental PAH.

There are several factors which affect the quantities of polycyclic organic material (POM) released from combustion, and the form in which they occur. POM may occur as persistent fuel fragments from incomplete combustion of a fuel containing polycyclic constituents. Coal contains many of these ring-type fragments, and coal-derived liquid fuels and solvent refined coal may have even more in a form conducive to the ultimate formation of POM in the combustion products stream. Low temperature combustion, especially with poor mixing, has been observed to increase the yield of POM. Low intensity burning with the products at intermediate temperatures for prolonged time periods has also been observed to increase the likelihood of POM synthesis from single and 2-carbon radicals. Neither of these combustion situations are prevalent, however, in normal operation of large power generation plants.

The POM species which might be formed in boiler furnaces generally condense into particulates in the temperature range encountered before leaving the power plant boilers. Most POM formed in a power plant would thus be expected to condense before leaving the stack, and be released to the atmosphere, either as a very fine particulate emission, or as small POM particles or molecules attached to other larger particles such as fly ash. The condensation onto other suspended particles, being a surface phenomenon, would be expected to result in a higher POM concentration per weight of particulate in the very fine sizes than in the coarse sizes. This correlation of POM emissions with size range of total particulate emissions has not yet been conclusively demonstrated. The POM being formed in connection with soot formation, as described later, would also indicate they might be concentrated on or within any soot particles persisting from any combustion changer operating in a soot-forming manner.

Hangebrauck, von Lehmden, and Meeker (2) conducted a study of sources of atmospheric polynuclear hydrocarbons several years ago, which still serves as one of the best comparative assessments of several diverse sources. Benzo[a]pyrene (BaP) and several other specific hydrocarbons were measured and BaP results interpreted as representative of POM emissions. From that study, BaP emissions from heat generation equipment burning fossil fuels (Figure 2.1) were interpreted as follows:

(1) Coal-burning equipment produces much more BaP per Btu fired than oil- and gas-fired equipment.

(2) The specific emission rates vary inversely with equipment size to a marked degree, and

(3) There is a very wide range of specific emissions from coal-fired equipment at any size.

All three of these trends are presumably associated with the improved and more efficient burning in larger coal-fired units, and in units burning gas and oil. Any incomplete combustion and nonuniform oxygen distribution within the burning region in the smaller coal-fired units presumably permits more of the polycyclic fuel fragments to pass through the unit without being completely broken down and oxidized. On the basis of these observations, the large modern electric power generating stations, with high turbulence levels and high temperature in the combustion region, and good air distribution, are expected to emit relatively low amounts of POM.

The product of the specific emission rates for each process and the quantity of fuel consumed nationally in that manner represents the nationally averaged emissions according to source characterization. These data, as reported by a National Academy of Sciences Committee (5) and an EPA report (6) are summarized in Table 2.2, again using BaP as an indicator of all POM emissions.

The relationship between cancer, polycyclic aromatic hydrocarbons, and fossil energy sources is best documented by epidemiological studies of coke oven workers (7)-(9), gas works operators (10)(11), and an experimental coal liquefaction plant (12). The association is also indicated by observations (4) of higher incidences of cancer in urban and industrial areas than rural areas and by biological studies (13)(14) of energy-related materials known to contain high concentrations of PAH. Results of systematic studies (2)(15)(16) of airborne PAH and extensive reviews (3)(4) of the subject are available. A large number of papers (17)-(20) continue to appear on the PAH composition of airborne particulates. Contamination of the aquatic environment by PAH has received less attention but has also been excellently reviewed (21)-(23).

Figure 2.1: Range of Benzo[a]pyrene Emissions from Coal, Oil, and Natural Gas Heat-Generation Processes

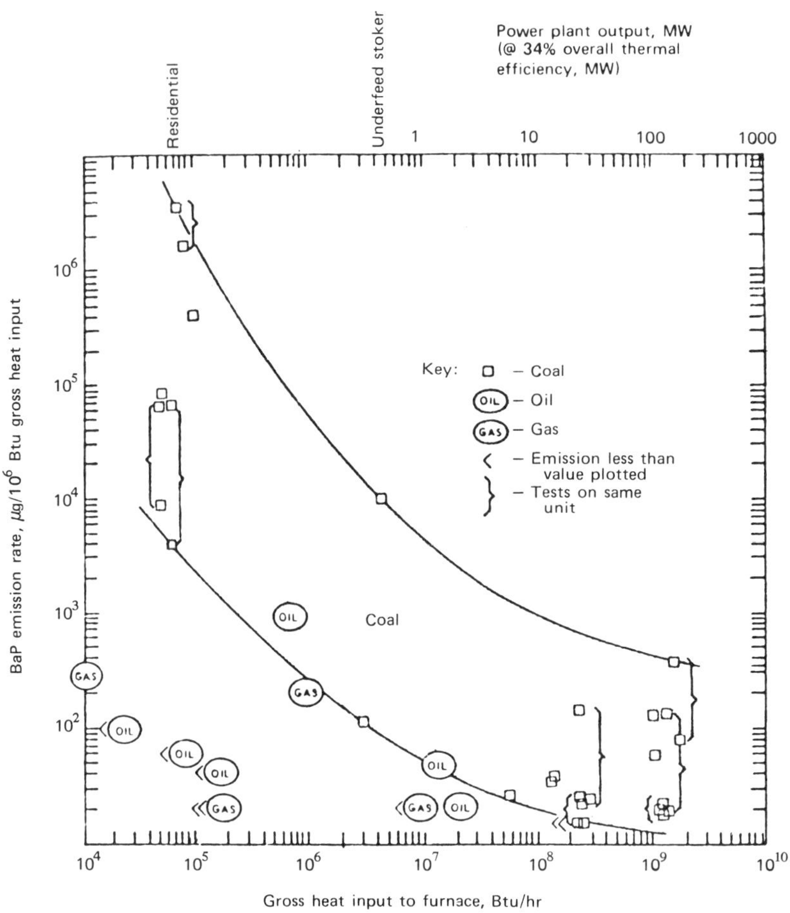

Source: NTIS EPRI EA-787-SY

Table 2.2: Estimated BaP Emissions in the U.S.

Type of Unit	BaP Emissions (tons/yr)*
Electric generation, steam-electric (~95% from coal firing)**	0.8
Commercial and institutional heating (~50% from coal firing)**	1.7

(continued)

Table 2.2: (continued)

Type of Unit	BaP Emissions (tons/yr)*
Industrial heating (~80% from coal firing)**	5.0
Petroleum refineries, catalytic cracking***	6
Highway vehicles (1968 data, being reduced significantly by controls)***	22
Enclosed refuse burning***	34
Open refuse burning, municipal, residential, vehicle disposal***	75
Forests and agricultural waste burning***	140
Coal refuse fires (uncontrolled low intensity burning)***	340
Residential heating (~97% from coal firing)**	390
Coke production***	200

*National total from all sources.
**Reference (6).
***Reference (5).

Source: NTIS EPRI EA-787-SY

PETROLEUM

Approximately 73 Quads of energy were consumed (24) in the United States in 1974 with petroleum accounting for 45.8% of the total. Approximately 6 billion barrels of petroleum are required to produce this quantity of energy. Assuming crude oil constituted the entire petroleum supply (an overestimate since natural gas liquids are frequently catalogued as petroleum), that one barrel contains 140 kg of crude oil, and that the crude oil contains 1.0 wt % polycyclic aromatic hydrocarbons and 1.0 ppm BaP, the petroleum industry is currently handling 84 million metric tons of PAH and 840 metric tons of BaP annually. An understanding of the fate of PAH through the petroleum energy cycle may provide the basis upon which to judge the relative carcinogenic hazard of coal- and shale-conversion processes being developed to supplant dwindling petroleum reserves.

Polycyclic Aromatic Hydrocarbon Content of Crude Oils

There is no conclusive evidence that the extraction and transportation of petroleum crude oils are accompanied by a major polycyclic aromatic hydrocarbon (PAH)-related cancer risk. The PAH content of crude oils is of interest in occupational carcinogenesis primarily through its possible influence on subsequent refining operations and the composition of process streams. PAH content of crudes is of interest in environmental carcinogenesis as it may affect the composition of refinery waste streams and of emissions from the end-use (e.g., combustion) of final products. Estimates (25) that more than 4 x 10^6 metric tons of crude oils and petroleum products are introduced into the oceans annually further suggest a need to characterize the PAH component of crude petroleum.

More than 7,000 crude oils have been analyzed (26) over the years using the Bureau of Mines Routine Method for Crude Petroleum (27) which determines physical properties and the weight distribution of distillate fractions. A joint effort of the Bureau of Mines and American Petroleum Institute carried out principally at the Bartlesville and Laramie Energy Research Centers has resulted in a standardized procedure (28) for characterizing "heavy ends" (high boiling distillates) of crude oils. These distillates are enriched in polyaromatic constituents including polycyclic aromatic hydrocarbons.

The procedure, involving anion and cation exchange chromatography, ferric chloride treatment and finally combined silica gel and alumina adsorption chromatography provides a weight percentage distribution for acids, bases, neutral nitrogen compounds, saturates, monoaromatics, diaromatics, and polyaromatic/polar compounds. The latter four fractions are further separated by gel permeation chromatography and the subfractions are analyzed directly using nuclear magnetic resonance and mass spectrometry. A weight percentage distribution by compound type (e.g., alkyldinaphthenopyrenes, alkyldiphenyldisulfides, etc.) is obtained for approximately 100 "probable type" classifications on analysis of the polyaromatic/polar concentrate.

Data on compound types is difficult to interpret in terms of carcinogenic potency because biological activity is dependent on the specific isomer considered. The weight percentage of the distillate and of the crude oil accounted for by each major fraction provides some insight into the polyaromatic content of the materials and of the matrix in which the polycyclic aromatic hydrocarbons reside in the distillate. These data are summarized in Table 2.3 for the five crude oils studied (29)(30) in detail using the USBM-API procedure (28).

Table 2.3: Composition of the 370° to 535°C Distillates of Crude Oil

	Wt % of Crude Oil (of Distillate)				
	Gach Saran Iran	**Swan Hills Canada**	**Wilmington California**	**Recluse Wyoming**	**Prudhoe Bay Alaska**
Acids	0.34 (1.7)	0.33 (1.8)	1.38 (5.6)	0.25 (1.3)	0.61 (2.8)
Bases	0.42 (2.1)	0.40 (2.2)	1.67 (6.8)	0.21 (1.1)	0.53 (2.4)
Neutral nitrogen compounds	0.23 (1.2)	0.08 (0.4)	1.04 (4.2)	0.17 (0.9)	0.19 (0.9)
Saturates	9.54 (48.5)	12.10 (65.9)	9.12 (36.9)	10.16 (54.7)	10.45 (48.3)
Monoaromatics	3.31 (16.8)	2.30 (12.5)	4.15 (16.8)	2.09 (11.2)	3.67 (17.0)
Diaromatics	2.30 (11.7)	1.15 (6.3)	3.05 (12.4)	0.94 (5.0)	2.58 (11.9)
Polyaromatics-polar	3.54 (18.0)	1.99 (10.8)	4.28 (17.3)	1.14 (6.1)	3.59 (16.6)
Whole distillate	19.67 (100)	18.35 (100)	24.69 (100)	18.58 (100)	21.63 (100)

Source: References (29)(30) as cited in NTIS CONF-770130-2

The polyaromatic/polar contents of the crudes range from approximately 1 to 3.5 wt % (2 to 6 wt % if diaromatics are included). Correcting for the reported (30) presence of sulfur and nitrogen compounds in the fractions, an estimate of 1 to 2 wt % polycyclic aromatic hydrocarbons (2 to 4 wt % if diaromatics are included) seems reasonable. This is in general agreement with our experience of a 3.2 wt % yield of polycyclic aromatics isolated by solvent extraction and adsorption chromatography from a mixed crude oil for mouse dermal bioassay (31). The PAH isolate included diaromatics.

Norris and Hill analyzed (32) three base-stock oils (partially refined crudes) for the presence of ten specific PAH by isolating the constituents using extraction, column chromatography and thin layer chromatography followed by ultraviolet spectrometric measurement. None of the compounds could be detected in one of the oils. Results for the other oils are summarized in Table 2.4.

Table 2.4: Individual Polycyclic Aromatic Hydrocarbons in Base Stock Oils

	Oil			Oil	
	"B"	"C"		"B"	"C"
Compound	.... ppb		**Compound**	.... ppb	
Benz[a]anthracene	–	–	Pyrene	19	–
Benzo[a]pyrene	29	106	Triphenylene	5	–
Benzo[e]pyrene	370	831	Chrysene	1	–
Benzo[ghi]perylene	48	–	Fluoranthene	3	–
Coronene	28	61	Perylene	–	666

Source: Reference (32) as cited in NTIS CONF-770130-2

The finding (32) of 8 to 10 times as much benzo[e]pyrene (BeP) as BaP illustrates the need for additional characterization at the isomeric level and the ease with which carcinogenic potency can be overestimated by results of analyses capable only of "class" resolution.

The results summarized in Table 2.4 and the fact that a third oil examined by the investigators produced negative results for all the individual constituents sought also illustrates the high degree of variability in PAH content between crude oils. This is not suggested by the USBM-API 60 procedure which indicates a variability in total polyaromatic and diaromatic content no greater than a factor of three across the five crudes studied to date. The total weight of PAH found in sample C accounts for only 1.7 mg/kg of the 10 g/kg of PAH which would be expected if the oil contained 1% polycyclic aromatic hydrocarbons. The great majority of PAH constituting these oils remain unidentified.

Values of 0.04, 1.3, and 1.6 ppm BaP have been reported by Graf and Winter (33) for crude oils from the Persian Gulf, Libya, and Venezuela, respectively. Analyses of crude oils at Oak Ridge National Laboratory using multiple extraction, multiple chromatographic purification, gas chromatographic and fluorescence measurement with carbon-14 BaP to determine recoveries have produced results ranging from 1 to 4 ppm BaP. Multicomponent gas chromatographic analyses of PAH-enriched fractions reveal the presence of a wide variety of other PAH at concentrations 1,000 times that of BaP.

Petroleum Processing and Products

Crude oil must be processed into useable products. Each oil refinery is practically unique in terms of its crude oil capacity and variety of products produced, but all rely on distillation for the primary separation. Depending on the sophistication of the refinery, the distillates are further upgraded by purification, catalytic cracking, hydrotreating, and other processes to produce a variety of products. The products may be processed further to produce, for example, chemicals from the olefins or petroleum pitch from the refinery stock by other industries. PAH are generally found to concentrate in the higher boiling point distillates and solid residues as would be expected.

Oil refineries constitute a major energy-related source of both aquatic (34)-(36) and airborne (3)(36) PAH. Reid (35) reports that 1,000 gallons of water are required to process each barrel of crude oil suggesting that some water contamination is inevitable. Lebowitz (34) has identified PAH as likely contaminants of several aqueous streams of the refining operation. Refining and petrochemical plant waste oils have been estimated (25) to contribute 195,402 metric tons to the annual source of crude oils and petroleum products introduced into the ocean. Fugitive emissions are suggested (34) as the primary source of hazardous constituents from petroleum refining.

The single most important source of airborne PAH from petroleum refining is catalytic cracking (2)-(4). Approximately 6 tons of BaP are released annually as a result of combusting the coke on the catalyst surface to regenerate the catalyst.

The nature of airborne emissions are further defined in a study of a Russian petroleum refinery. Kireeva (37) reports air concentrations of 258 μg/100 m^3 for BaP and 61.5 μg/100 m^3 for 1,12-benzperylene in the vicinity of process stages dealing with high-boiling products (fuel oil, tar, cracking residues). Air concentrations of these PAH were lowest in the first stages of the refining and at stages involving lower boiling distillates. Especially high concentrations, 3,659 μg/100 m^3 and 996 μg/m^3 for BaP and 1,12-benzperylene respectively, were found in the vicinity of the coking equipment. Evidence is reported for at least 15 other PAH in the coking environment and anthracene, 1,2-benzanthracene, 1,2,5,6-dibenzanthracene, and 1,2,3,4-dibenzanthracene are identified.

The concentrations of several PAH as determined by Norris and Hill (32) in common petroleum products are summarized in Table 2.5. Results for the BaP content of gasolines are in good agreement with concentrations of 0.48 and 0.21 μg/g reported (36) for nationwide composite premium and regular gasolines, respectively. Concentrations of BaP in gasolines ranged to 3.1 μg/g in the study. Graf and Winter (33) report finding 0.03 ppm BaP in a diesel fuel, fuel oil, and unused motor oil.

Table 2.5: Some Polycyclic Aromatic Hydrocarbons in Petroleum Products

	Gasolines . . . μg/ml (ppm) . . .				Kerosene . . μg/g (ppm) . .	Diesel Fuel	No. 2 Heating Oils . . . . μg/g (ppm)				Motor Oils μg/ml (ppm)	
	A	B	C	D			A	B	C	D	New	Used
Anthracene	–	–	–	–	0.04	2.9	4.2	1.0	6.7	2.4	–	–
Phenanthrene	–	–	–	–	ND	ND	ND	ND	ND	ND	–	–
Methylanthracene	–	–	–	–	<0.01	9.3	6.6	20.7	15.2	20.7	–	–
Benz[a]anthracene	1.0	11.5	ND	0.54	<0.01	0.13	0.03	0.03	0.06	0.02	ND	2.2
Pyrene	2.0	10.6	0.15	5.1	0.16	0.37	ND	3.0	0.58	0.45	1.3	11.6
Fluoranthene	1.8	20.8	0.06	3.2	0.09	0.57	0.47	3.2	ND	3.6	–	–
Chrysene	0.57	4.9	ND	0.54	ND	0.45	0.50	0.51	0.37	0.81	–	–
Triphenylene	–	–	–	–	ND	3.3	ND	0.76	0.23	1.2	–	–
Benzo[a]pyrene	0.46	6.2	0.03	0.55	<0.01	0.07	0.04	0.01	0.05	0.01	ND	2.4
Benzo[e]pyrene	0.37	2.0	0.03	0.85	<0.01	0.18	0.02	0.02	0.02	<0.01	0.67	2.7
Benzo[ghi]-pyrene	0.83	6.4	0.04	1.40	<0.01	0.03	0.07	0.02	0.01	0.01	0.28	1.8
Coronene	0.22	0.53	0.01	0.34	–	–	–	–	–	–	ND	0.57

Note: ND means not detected.

Source: Reference (32) as cited in NTIS CONF-770130-2

The high concentration of BaP in gasolines compared to higher boiling distillate products may be due to upgrading steps in processing. Gasoline B (Table 2.5), containing catalytically reformed product, is found to contain ten times more BaP and most other PAH than do the other gasolines. While initial distillation undoubtedly produces a gasoline fraction depleted in BaP and PAH relative to the starting material, upgrading steps result in gasolines containing approximately the same concentrations of these constituents as is found in petroleum crudes. The range of concentrations of PAH in gasolines is sufficiently large to overlap concentrations found in kerosene, diesel fuels, and No. 2 heating oil.

Data (Table 2.5) on motor oils (32) represent an interesting case of the end use of a petroleum product enriching its PAH content. Graf and Winter (33) report a 200-fold increase in BaP content of a motor oil, from 0.03 μg/g to almost 6 μg/g after only 1,400 miles of use. It has been estimated (35) that 350,000,000 gallons of used motor oils are discarded (much into municipal sewer systems) annually. A worst-case calculation (assuming a concentration approximately 6 ppm BaP) suggests that up to 8 metric tons of BaP is released into the environment annually as waste motor oil. The total PAH released by this route would be expected to be many times higher. The complexity of the PAH content of motor oils is well illustrated by the capillary column gas chromatographic profiles reported by Lee et al (38).

Petroleum pitch is an example of a petroleum product heavily enriched in PAH content. Methylphenanthrenes, pyrene, methylpyrenes, chrysene, benz[a]anthracene, methylchrysenes, dimethylchrysenes, benzo[a]pyrene, and benzo[e]pyrene constitute (39) almost 2.5 wt % of one pitch studied. The petroleum pitch volatiles from the same materials were composed of almost 5.5 wt % of these PAH. The pitch contained 0.2 wt % BaP and the volatiles 0.4 wt % BaP.

Additional information about the PAH content of petroleum products, process streams, and process effluents is particularly important to the development of coal conversion technologies. Concerns about the carcinogenic impact of these processes, largely due to documented occupational exposure hazards and high concentrations of PAH inherent to the processes, are destined to slow the commercialization of these needed technologies. Industrial hygiene (40) and environmental control methodologies which have proven effective in minimizing the hazards of exposure to petroleum related PAH can only be translated to coal conversion processing if the relative quantities of PAH in the materials and the carcinogenicity of these materials are known.

Combustion of Petroleum Products

Approximately one-half of the petroleum presently consumed in the United States is consumed in the form of gasoline. An additional quantity of petroleum is also consumed for transportation as diesel fuel. Annual consumptions of gasoline and diesel fuel in 1968 were estimated (3) as 80.6 x 10^9 and 5 x 10^9 gal/yr, respectively. Higher-boiling distillates of petroleum crudes are consumed to produce process- or space-heat and for electric power generation. Combustion is clearly the predominant end-use of petroleum as an energy source.

Combustion processes are among the most intensively studied energy-related sources of environmental PAH. Concern about air pollution in the late 1950's prompted a study of the polycyclic aromatic hydrocarbon content of 28 rural and 103 urban atmospheres by Sawicki et al (15)(16) which found BaP present in all of the urban areas sampled. Concentrations ranged from 0.1 to 61 ng/m^3 with a geometric mean of 6.6 for the urban areas and from 0.01 to 1.9 ng/m^3 with a mean of 0.4 for the nonurban areas.

Continuing studies have produced extensive tabulations (41)(42) of the BaP content of air and of natural and man-made materials. Systematic investigations of the polycyclic aromatic composition of gasoline and resulting vehicle exhausts have been reported by Gross (43)-(45). Hangebrauck (2) includes motor vehicle emissions and fuel oil combustion in his study of sources of polycyclic aromatics in the atmosphere, basic investigations of conditions leading to the formation of polycyclic aromatics have been reported (46)-(49) and reports continue to appear on the composition of airborne particulates (17)(20)(50)-(52), diesel (53)(54), and automobile exhausts (55)-(58) and all facets of combustion-generated polycyclic aromatic hydrocarbons.

The reviews prepared by the National Academy of Sciences on Particulate Polycyclic Organic Matter (3) and Vapor Phase Organic Matter (59) point to energy sources as a primary contributor of polycyclic aromatics to the environment.

The PAH content of vehicle exhausts and of stack effluents is related to the PAH content of the fuel. Begeman (49) found higher emissions of BaP when a test engine was operated using fuels higher in PAH content than commercial gasolines. The NAS review (3) reports Gross (43)-(45) as finding that increasing the aromaticity of the fuel from 12 to 46% resulted in an increase of from 8 to 74% in the PAH emissions depending on the vehicle and emission control conditions.

The data reported by Hangebrauck et al (2) indicates that 2 to 5 times more BaP, perylene, and fluoranthene is produced on combustion of a No. 6 fuel oil than No. 2 fuel oil to produce process heat under similar operating conditions. No. 6 fuel oil would be expected to contain higher concentrations of PAH and higher molecular weight organics which could serve as precursors for PAH production. The importance of PAH precursors in the fuel is underscored by emission factors for BaP of 11,000 μg/gal for two-cycle engines (e.g., outboard motors, motorcycles, lawnmowers) which operate on oil:fuel mixes (typically 1:30) as compared to 170 μg/gal for automobiles (3).

While aromaticity of the fuel clearly influences PAH emissions, combustion characteristics are far more important. A major factor is the air:fuel ratio used for combustion. The NAS review (3) reports Begeman as finding a 30 times greater production of BaP at an air:fuel ratio of 10:1 than at 14:1 for automobiles. It is also pointed out that BaP emissions increase with increasing engine use such that emissions are 5 times higher after 50,000 miles of use than at 5,000 miles and 10 times higher when oil consumption reaches 200 mi/qt rather than 1,600 mi/qt. The deteriorating engine alters its fuel composition and resulting combustion characteristics by drawing lubricating oil into the cylinders.

The importance of combustion and combustor characteristics is readily apparent in the data provided by Hangebrauck et al (2) who reports emission factors of from 40 to 470 μg BaP/gal for eight automobiles tested, from 40 to 1,450 μg BaP/gal for four trucks, and from approximately 1 to 8 μg BaP/lb of fuel oil for six conditions of process or space heating. Emission factors are more consistent than is indicated by these extremes but, with the exception of inadvertent combustion of lubricating oil, the influence of fuel aromaticity is relatively minor in comparison. The efficient, controlled combustion of coal as is sought in electric power generation can produce less BaP/lb of fuel than does the less efficient combustion of oil or even natural gas (2)(4) as might occur in space heating. Oil and natural gas produce significantly lower quantities of PAH than does coal when conditions of comparable combustion efficiency are considered.

BaP emission factors (Table 2.6) are useful for identifying major sources of environmental contamination by PAH. Expressed per quantity of material combusted, the factors may also suggest sources of high localized concentrations of BaP and thus regions and periods of significant carcinogenic potential. Measurement of BaP alone obviously provides an underestimate of total PAH emissions and is likely to be inadequate for estimating the relative carcinogenicity of the emissions.

Table 2.6: BaP Emission Factors for the Combustion of Petroleum Products in Motor Vehicles

	. . μg BaP/gal fuel . .		. . . . tons BaP/year		
	Hangebrauck	NAS	Hangebrauck	NAS 1968	EPA 1972
Gasoline-Automobiles	170	170	8.6	10	10 (automobiles and trucks)
Gasoline-Trucks	>460	~500	>10	~12	
Diesel	690	62	2.0	0.4	<1
			>20.6	~22.4	~11

Source: References (2)-(4) as cited in NTIS CONF-770130-2

Table 2.7 shows that pyrene, a documented (60) cocarcinogen, is present in motor vehicle emissions and fuel oil-generated process heat emissions at concentrations much exceeding that of BaP. BaP constitutes 0.5 to 4.2% of the ten PAH reported and presumably a much smaller percentage of the total PAH present. While BaP quantities (Table 2.7) emitted by two home heating units are similar, quantities of pyrene, phenanthrene, and fluoranthene differ by two orders of magnitude.

Table 2.7: Distribution of PAH in Petroleum Combustion Effluents

Source	Benzo[a] pyrene	Benzo[a] pyrene	Pyrene	Benzo[e] pyrene	Perylene	Benzo[ghi] perylene	Anthanthrene	Coronene	Anthracene	Phenanthrene	Fluoranthene
	μg/gal	. . . . micrograms per vehicle mile.									
Automobiles (1956-1964)	170	12	111	~16	1	49	1	20	5	32	74
Trucks (1956-1964)	>455	40	745	>48	6	199	–	105	78	485	475
Diesel Powered	690	–	–	–	–	–	–	–	–	–	–
	μg/lb	. . . . micrograms per million Btu heat input									
Process Heat											
No. 2 Fuel Oil	<0.3	<20	49	–	–	–	–	–	–	–	56
No. 6 Fuel Oil	0.9	47	300	–	–	–	–	–	–	1,800	270
Home Heating											
Sectional Boiler	<0.9	<40	1,800	–	–	–	–	–	–	8,900	5,000
Hot Air Furnace	<1	<60	15	–	–	–	–	–	–	–	76
Institutional Heating											
Low Pressure Air-atomized Firing	18	900	6,100	–	–	300	–	2,100	3,900	3,500	1,900

Source: Reference (2) as cited in NTIS CONF-770130-2

A more recent study (56) of automobile exhausts reports the presence of chrysene and 2,3-benzofluorene at levels of approximately 120 to 240 μg/gal and 160 to 1,000 μg/gal for the same fuels. Other recent examples of the chemical complexity of emissions are found in high resolution chromatographic profiles of automobile exhausts (55)(57) and in a high resolution mass spectrometric analytical study (54) of diesel exhaust particulates. The latter study identified constituents with the molecular formulas corresponding to 48 known hazardous compounds including naphthalene, anthracene/phenanthrene, pyrene, and chrysene/benzo[c] pyrene/benzo[b] fluoranthene/benzo[j] fluoranthene in diesel exhausts. One sample produced a molecular formula corresponding to 7,12-dimethylbenz-[a] anthracene.

An additional factor important to judging the carcinogenic potential of emissions is the morphology of the carcinogenic constituents—particle size distribution and the nature of the particles. Several studies (61)-(63) have shown that PAH tend to be concentrated on the smaller, more respirable, airborne particles indicating that the hazard may be greater than predicted by total BaP mass particulate loading. The work of Stanton (64) has clearly demonstrated the importance of a continuous localized dose of PAH to lung epithelium as a factor in the etiology of experimental lung cancer. The vehicle (dissolved in a liquid particle as with tobacco smoke, adsorbed on a solid particle as with fly ash, etc.) with which the carcinogenic polycyclic aromatic is dosed may be as important as is the total dose of PAH.

Boyer and Laitenen (55) have demonstrated that respirable and nonrespirable particulates generated in automobile exhausts differ widely in elemental composition. Particle characteristics, including surface properties, must be known before the specific carcinogenicity of an emission can be estimated.

With the exception of the low efficiency combustion of coal for space heating, the combustion of petroleum products contributes as much BaP to the respiratory environment as does any other energy source.

COAL

Approximately 18% of the energy consumed (24) in the United States in 1974 was supplied by coal while coal constitutes 63% of the national energy reserve. The increasing cost of oil and increasing dependence on foreign sources of oil are providing strong incentives to increase coal utilization. Plans are being made to double the current level of coal production to one billion tons per year as soon as practical. Coal will continue to supply heat and electric power and is likely to become a major source of liquid fuels for transportation and of crude oils for the chemical industry.

The conversion of coal to gaseous or liquid fuels is further attractive in that it provides an alternative to expensive environmental control devices now required at power plants. A 1977 study (65) concludes that the production of electricity from pipeline quality natural gas produced from coal is less expensive than electricity produced directly from coal in certain areas. The predominant end-use of coal will continue to be through combustion. As is evident from the Wiser (66) conceptualization of coal, combustion and conversion are accompanied by the possible production of polycyclic aromatic hydrocarbons.

Coal Extraction and Conventional Processing

Freudenthal et al (67) have reviewed the carcinogenic potential of types of coal processing and conclude that there is little to no carcinogenic risk associated with the mining operation. This is in spite of the fact that coal dusts are inhaled (as is evidenced by the incidence of pneumoconiosis) and known to contain PAH. The presence of PAH including BaP is reported to have been determined (67) in mine dusts collected after blasting operations. Mentser and Sharkey (54) have recently surveyed mine dusts by high resolution mass-spectrometry and report the presence of constituents with the formula $C_{18}H_{12}$ and $C_{20}H_{12}$ (possibly chrysene and BaP, respectively). Evidence of the presence of other PAH and of aliphatic hydrocarbons was also found, particularly from the section of the mine which employed diesel rather than electrical haulage.

Lebowitz et al (34) has surveyed other stages (e.g., crushing and grinding) of conventional coal processing. Coal dusts are itemized among the potentially hazardous emissions but PAH were not.

Clearly the most important source of PAH [a major contributor (3)(4) to the national airborne BaP burden, and a documented (9)(68) occupational hazard] from conventional coal processing is the coking operation. Coke is produced by heating coals to temperatures as high as 1400°C in the absence of air. Ammonia, coke-oven gas, a light oil, tar, and coke result from the process. As a measure of the magnitude of the industry, approximately 87 x 10^6 tons of bituminous coal were processed to produce 60 x 10^6 tons of coke in 1972.

Coke oven batteries are difficult to sample because emissions are diffuse rather than from a point source. Most studies examine samples from personal monitors or from the general vicinity of the coke ovens. Results are generally expressed as coal tar pitch volatiles (CTPV), the cyclohexane or benzene soluble constituents of the particulates, or BaP is determined. Concentrations of BaP above the gas works retorts of a coke oven have been reported (69) as high as 216,000 ng/m^3. Jackson et al (70) report concentrations up to 16,000 ng/m^3 on the coke oven battery.

Of the references surveyed for this review, only Hangebrauck (2) reports quantitative data on other PAH. Air samples taken from the general vicinity of a combined steel and coking industry ranged from 440 to 8,500 ng/m^3 in concentrations of BaP, benzo[e]pyrene, pyrene, perylene, benzo[ghi]perylene, anthanthrene, coronene, anthracene, and fluoranthene. BaP was present at the greatest concentration of the constituents determined.

Coal Conversion

Coal conversion (71)-(73) is a generic term encompassing more than 100 (74) specific processes designed to produce liquid, gaseous, or refined solid fuels as energy sources or substitute petrochemical feedstocks from coal. Undesirable ash and sulfur are removed and the coal is converted into a form suitable for direct use as low Btu or high Btu gas or for further refining to produce gasoline, fuel oil, and other products typical of petroleum processing. Processes designed to produce liquid fuels are optimized to maximize production of saturated hydrocarbons. Processes designed to produce petrochemical feedstocks are optimized to generate low molecular weight aromatics. Most processes generate a mixture of gaseous, liquid, and solid products, some of which become an integral part of the process, being consumed at the plant to enhance the yield of the specific product of interest.

Cochran (71) points out that the common chemical objective of the processes is to increase the atomic hydrogen-to-carbon ratio of the coal from its typical value of 0.8:1 to a ratio of 1.75:1 as is typical of oil. Higher ratios, ideally 4:1 as for methane, are sought in the production of gaseous fuels. It is necessary to depolymerize the coal and hydrogenate the resulting fragments to adjust the H/C ratio.

Cochran (71) divides the many conversion technologies into four basic categories, (a) carbonization, (b) direct hydrogenation, (c) extraction, and (d) synthesis. A coal is carbonized by heating (typically at 1000°F) in the absence of air causing the release of gases and tars. Hydrogen may be added at relatively low pressure, e.g., 600 psi, to result in "hydrocarbonization" to increase the yield of volatiles and tars. The remaining solid char is used to generate synthesis gas for the production of process hydrogen.

Direct hydrogenation involves heating a coal slurry to approximately 850°F in the presence of hydrogen at pressures of 3,500 psi and, usually, a catalyst. The extraction process requires that the coal be dissolved in a petroleum solvent. Dissolution occurs at approximately 500°F and 300 psi pressure and the resulting liquid is hydrogenated to produce a crude oil and additional solvent for extraction. The Fischer-Tropsch "synthesis" involves burning the coal in a gasifier in the presence of oxygen and steam to produce primarily carbon monoxide and hydrogen. The gas is purified and passed over a catalyst yielding liquid products ranging from methanol through waxes and oils. Substitute natural gas, liquid petroleum gas and liquid fuels are produced.

State of the art engineering technology is required to make these conceptually straight-forward operations commercially practical. Process flow diagrams are reminiscent of those for large scale petroleum refineries. Adequate technology is at hand for commercially viable small scale conversion as is evidenced (71)(74) by the successful operation of the Sasol synthetic fuel plant for 20 years in South Africa. It is the size of the projected conversion plants and conversion industry in the United States that challenges engineers and concerns environmentalists.

The Sasol plant processes approximately 3,500 tons of coal per day while U.S. planners visualize at least 20 plants, each processing approximately 25,000 tons of coal per day, preferably 40 plants processing a total of 10^6 tons/day and possibly as many as 100 such plants. One plant processing 25,000 tons of coal each day would produce (71) 205 x 10^6 scf of gas, 5,000 barrels of liquid petroleum gas, 40,000 barrels of motor fuels, and 5,000 barrels of fuel oil daily for a capital investment of $1.5 billion. Twenty such plants would produce one million barrels of liquid products and 4.1 x 10^9 ft^3 of gas daily to provide 6% of the nation's present energy demand.

An industry of this size might be expected to have some impact on the environment and, considering the similarity to coking and coal tar works, on the health of the worker. An increasing number of documents are appearing which outline and evaluate specific processes for environmental impact. General studies have been reported by Glazer et al (75) and Katz (76). The Energy Research and Development Administration (ERDA) coal gasification program is undergoing an environmental assessment at Carnegie-Mellon University (77). Gasification processes which have been reviewed include the CO_2 acceptor (78)(79), Synthane (80), Bi-Gas (81), Lurgi (82), Koppers-Totzek (83), and Hygas (78) processes. The solvent refined coal (SRC) process (81) and COED pyrolysis process (80) are among the specific liquifaction technologies which have been evaluated. A comprehensive review of the environmental and health aspects of coal conversion has been prepared (84) by the Information Center Complex of Oak Ridge National Laboratory. The results of these and other studies are forming the basis of Environmental Development Plans (85), documents being prepared by the USERDA to guide energy-related environmental and health research.

The threat of an increase in occupational and/or environmental carcinogenesis is the major health impact receiving attention (86)(67). A firm epidemiological basis for this concern exists for occupational cancer. As is summarized in Table 2.8 technologies closely related to those comprising the projected coal conversion industry are known to be accompanied by an elevated risk of occupational carcinogenesis. Laboratory, clinical and epidemiology studies (12)(87)-(90) in support of the Union Carbide Corporation coal conversion plant at Institute, West Virginia confirm [Coomes (13) has questioned this conclusion] the carcinogenic nature of the process materials and document a substantial increase in skin cancer rate for employees in spite of protective measures taken.

Table 2.8: Epidemiological Basis for Synfuels/Carcinogenesis Concern

Reference	Exposure/Experiment	Cancer
Pott, 1775 (118)	English chimney sweeps	Scrotum
Volkmann, 1875 (119)	Handling tar and paraffin from carbonization of lignite	Scrotum
Rehn, 1895 (120)	Production of synthetic dyes from coal tar	Bladder
Lueke, 1907 (121)	Carbon workers	Skin
Henry, 1931 (122)	Coal tar workers	Bladder
Kuroda, 1936 (123)	Coal gas workers	Lung
Kennaway, 1936 (124)	Coal carbonization workers	Lung, larynx
Kennaway, 1947 (125)	Street pavers, asphalters, coal gas workers, chimney sweeps	Lung
Doll, 1952 (126)	Gas retort workers	Lung
Anon, 1958	Coal gas and coke workers	Stomach, kidney, and pancreas
Bruugaard, 1959 (127)	Gas retort workers	Bladder
Sexton, et al, 1960 (12)(87)-(90)	Coal liquefaction	Skin
Doll, 1965 (11)	Gas works	Lung, bladder
Kawai, 1967 (128)	Gas works	Lung
Lloyd, Redmond, et al 1969-1974 (7)(8)(129)-(132)	Steelworkers, coke producers	Lung
Doll, 1972 (10)	Gas works	Lung, bladder, scrotum
Mazumdar, 1975 (9)	Coke oven workers	Lung

Source: Reference (67) as cited in NTIS CONF-770130-2

Concern about an increase in environmental carcinogenesis is less well-founded but cannot be ignored until it is demonstrated that carcinogenic materials can be contained within plant boundaries and the transportation and end-use of the products are not accompanied by an increased cancer risk. Sauter (91) has strongly urged that both the occupational and environmental cancer risk be given maximum attention in national policy decisions affecting the development of a fossil fuels conversion industry.

Polycyclic aromatic hydrocarbons are implicated as primary carcinogens in coal derived materials by virtue of their carcinogenicity and common occurrence in coal tars. The experimental carcinogenicity of coal derived materials increases with increasing boiling point and is correlated with process conditions which favor the formation of PAH.

The Bartlesville Energy Research Center is engaged in a systematic study of coal-derived liquids using modifications of the API-60 method for characterizing petroleum crudes. Reports are available on Syncrude from the COED process using two coals (92)(93), on two oils from the H-Coal process (94), and the product of the Synthoil process (95). Distillates (92)-(95) prepared from the crudes are predominantly hydrocarbon in nature but also contain acidic, basic, and heteroaromatic constituents. Rubin et al (96) has reported as much as 15% removal of constituents from COED Syncrude by acid-base extraction. Jones et al (97), using a newly developed Sephadex LH-20 procedure, reports 3% of the Syncrude is hydrophilic with an additional 8% being capable of strong hydrogen bonding.

Table 2.9 summarizes the distribution (95) of compound types in distillates from two coal conversion processes. Polyaromatics 3 rings and larger are seen to constitute from 5 to 40 wt % of the products depending on process type, coal feed, and distillate boiling range. Diaromatics are present at twice the concentration of polyaromatics in the lower boiling distillates and at equal concentrations for two of three higher boiling distillates considered. The coal used influences the distribution of the products but has a smaller effect on polyaromatic content than does the process.

Table 2.9: Distribution of Constituents in Synthetic Crude Oil Distillates from Coal

	Process					
	Synthoil	COED	COED	Synthoil	COED	COED
	Coal					
	W. Virginia	W. Kentucky	Utah	W. Virginia	W. Kentucky	Utah
	Distillate					
	207°-363°C	205°-380°C	204°-381°C	363°-531°C	>380°C	>381°C
Saturate	16	25	27.8	9.7	23.8	25.8
Monoaromatic	27.3	42	25.1	4.7	25.1	14.4
Diaromatic	21.6	13	17.5	22.6	24.3	18.4
Polyaromatic	7.9	5.4	7.1	41.1	20	25.1
Heteroaromatic	22.2	4.4	15.2	15.6	4.5	7.4
Distillate % of crude	42.6	54.2	45.4	27.3	24.2	40.3

Source: Reference (95) as cited in NTIS CONF-770130-2

The crudes examined contain from approximately 7.5 to 14 wt % polyaromatics if it is assumed that all of the polyaromatic constituents are recovered in the two distillates. Separation at Oak Ridge Laboratory (97) of a COED Syncrude from a Western Kentucky coal by Sephadex LH-20 chromatography produced a polycyclic aromatic fraction constituting 5 wt % of the Syncrude. Preparative scale polycyclic aromatic isolates recently generated (98) by solvent extraction and adsorption chromatography for mouse dermal testing (31) corresponded to 6.0 and 5.1 wt % of the experimental COED Syncrude and Synthoil products, respectively, under study at Oak Ridge Laboratory. Synthetic crude oils thus appear to contain from two to five times the quantity of PAH normally found in petroleum crudes. Tars and other high boiling residues produced as by-products of the conversion process may contain ten times more PAH than do the synthetic crude oils.

The Pittsburgh and Bartlesville Energy Research Centers are examining (92)-(95) (99)-(102) the polyaromatic component of syncrudes and related products by mass spectrometric and nuclear magnetic resonance techniques designed to identify and quantitate classes of constituents. Results indicate the presence of a wide variety of hydrocarbon and heteroatomic constituents and the presence of many alkyl substituted derivatives. Data, some of which appear in Tables 2.10 and 2.11, have been reported (103)-(107) on the quantities of individual polycyclic aromatic hydrocarbons in coal conversion related materials. Results confirm the presence of a variety of known tumorigenic constituents, possible tumor promoters, and cocarcinogens.

Table 2.10: Polycyclic Aromatic Hydrocarbons in Coal Derived Oils and Related Materials

	SRC Process ...Materials...		. Coal Derived Crude Oils .		La.-Miss.	Shale
	Wash Solvent	Process Solvent	Catalytic Hydrogenation	Pyrolysis	Petroleum Crude	Derived Crude Oil
	.... .milligrams per gram.					
Naphthalene	32	0.1	1.6	1.7	0.9	1.4
Methylnaphthalenes	44	5	3.8	4.6	1.6	1.6
Dimethylnaphthalenes	18	18	3.7	6.3	2.1	2.2
Isopropylnaphthalenes	0.2	1.5	–	–	–	–
Biphenyl	10	6	0.2	0.4	–	<0.1
Dimethylbiphenyls	<0.1	2.6	–	–	–	–
Fluorene	0.2	6.6	0.3	0.5	0.2	0.9
9,10-Dihydroanthracene	–	–	1.7	–	–	<0.1
9-Methylfluorene	0.1	3.1	0.1	0.1	–	0.4
1-Methylfluorene	<0.1	3	1.9	–	0.1	1.0
Anthracene plus Phenanthrene	0.1	23	1.4	–	0.3	0.6
Fluoranthene	<0.1	10.5	0.4	–	0.1	0.4
Pyrene	<0.1	11.2	4.3	–	–	0.2
1,2-Benzofluorene	–	–	–	–	<0.1	0.1
2,3-Benzofluorene	–	–	–	–	<0.1	0.1
1-Methylpyrene	–	–	0.6	0.1	<0.1	0.1
Chrysene plus Triphenylene	–	–	0.1	<0.1	–	–
Picene	–	–	0.4	–	–	–
1,12-Benzoperylene	–	–	1.5	–	–	–
Anthanthrene	–	–	0.1	–	–	–

Source: References (103), (105) and (106) as cited in NTIS CONF-770130-2

Table 2.11: Naphthalenes in Synthetic and Natural Crude Oils

	. . . Coal Derived				
	Cat. Hydrogen	Pyrolysis	Shale Oil	LMS Crude	Alaskan Crude
	. . . .	. . . mg/g. . .	. . . .	. . . .	. . . .
Naphthalene	1.6	1.7	1.4	0.9	1.4
2-Methylnaphthalene	2.9	3.5	0.9	0.9	1.8
1-Methylnaphthalene	0.9	1.1	0.7	0.7	1.4
Biphenyl	0.2	0.4	0.1	–	0.4
2,6-Dimethylnaphthalene	0.4	0.8	0.1	0.1	0.5
1,3- + 1,6-Dimethylnaphthalene	1.8	3.0	1.6	1.3	2.3
2,3-Dimethylnaphthalene	1.0	1.5	0.3	0.2	0.4
1,5-Dimethylnaphthalene	0.2	0.7	–	0.1	0.2
1,2-Dimethylnaphthalene	0.3	0.2	0.2	0.4	0.5
Acenaphthylene	1.3	2.2	0.2	0.2	0.6
Acenaphthene	0.5	0.3	–	–	–
Total	11.1	15.4	5.5	4.8	10.5

Source: Reference (111) as cited in NTIS CONF-770130-2

Essentially no data is available on the unambiguous identification and subsequent quantitation of PAH in synthetic fuel materials at the isomeric level. The potent carcinogens 3-methyl cholanthrene and 7,12-dimethylbenzanthracene have been reported (108) in shale oils and cochromatographic and mass spectrometric evidence for the presence of 3-methyl cholanthrene in a variety of synthetic crude oils has been noted but their presence has not been positively identified. Synfuels and related materials have been analyzed for benzo[a]pyrene (BaP) content by isotope dilution/fluorescence spectrometry/gas chromatography following extensive purification by solvent extraction and adsorption or cellulose acetate column, paper, or thin layer chromatography and from approximately 40 ppm BaP to less than 2 ppm BaP were found for the same sample type.

Application (104) of Hecht's procedure (109) to the analyses of one syncrude for methyl chrysene content indicates that chrysene and its 2-methyl and 3-methyl derivatives are present at concentrations of approximately 100 ppm but that the highly carcinogenic (109)(110) 5-methyl derivative constituted less than 20 ppm, if any, of the syncrude. Limitations of existing analytical methodologies and/or resources for detailed chemical studies have impeded the identification and quantitation of known tumorigenic agents at the isomeric level.

Phenols (107)(112)(113) are the major organic contaminants of by-product and process waters from coal conversion. As is illustrated in Table 2.12 polycyclic aromatic hydrocarbons are also found in the waters. The condensate from Synthane gasification is seen to contain predominantly naphthalenes with the 2-methyl derivative being present at a greater concentration than all the other isomers combined. The raw process water from the Solvent Refined Coal process is seen to contain a larger variety of PAH at much higher concentrations than is found for the gasification water or for by-product water from oil shale retorting. Concentrations are in the sub-ppm range for most constituents. As is the case for synthetic oils, these identifications are based primarily on cochromatography or, at best, mass spectrometry and require confirmation.

Table 2.12: Polycyclic Aromatic Hydrocarbons in Coal and Shale Conversion Waters

Constituent	Synthane Gasification Condensate	Solvent Refined Coal Raw Process Water	Simulated In-Situ Shale Retort By-Product Water
	. . . . milligrams per liter (ppm).		
Naphthalene	0.2	5	0.1
2-Methylnaphthalene	1.3	2	0.3
1-Methylnaphthalene	<0.1	*	0.1
2,6-Dimethylnaphthalene	<0.1	*	<0.1
1,3- + 1,6-Dimethylnaphthalene	0.1	*	0.2
1,5- + 2,3-Dimethylnaphthalene	<0.1	*	0.1
1,2-Dimethylnaphthalene	<0.1	*	<0.1
Dimethylnaphthalenes	*	2.3	*
2-Isopropylnaphthalene	*	0.7	*
1-Isopropylnaphthalene	*	2	*
Biphenyl	<0.1	0.2	<0.1
Acenaphthylene	*	<0.1	*
Acenaphthene	**	*	<0.1
Dimethylbiphenyls	*	0.7	*
Fluorene	<0.1	0.3	0.2
9,10-Dihydroanthracene	<0.1	*	<0.1
9-Methylfluorene	*	0.3	*
1-Methylfluorene	ND	0.2	0.2
Anthracene/phenanthrene	ND	1.1	0.3
2-Methylanthracene	ND	*	0.2
1-Methylphenanthrene	ND	0.2	0.1
9-Methylanthracene	ND	*	0.4
Methylphenanthrene	*	0.3	*
Fluoranthene	ND	0.4	<0.1
Dihydropyrene	*	<0.1	*
Pyrene	ND	0.6	0.1
1,2-Benzofluorene	ND	*	<0.1
2,3-Benzofluorene	ND	*	<0.1
1-Methylpyrene	ND	*	0.1
1,2-Benzanthracene	ND	*	<0.1
Chrysene/triphenylene	ND	*	<0.1

*Not reported
**Incomplete resolution
Note: ND = Not detected

Source: References (103) and (105) as cited in NTIS CONF-770130-2

Oak Ridge Laboratory has been collaborating with J.L. Epler (114) and his staff on a study of the Ames Test (115) as a short-term screening bioassay of coal-derived materials. Oils and organic constituents of process-derived waters are subjected to the acid-base solvent partition and adsorption chromatographic fraction procedure (96) found useful in biological studies of tobacco smoke condensates and the resulting fractions are biotested. Early quantitative results have been reported (105). These and more recent studies show that two coal-derived oils are more active in the Ames system than are a shale-derived oil or a petroleum crude oil. Biological activity is concentrated in the basic fractions and neutral fractions thought to contain polyaromatic constituents. A study at Oak

Ridge Laboratory (116) has indicated the presence of azapyrene and possibly other azaarenes in a biologically active subfraction of one basic fraction. The utility of the Ames test for predicting carcinogenic activity is presently under study (31) through a mouse dermal carcinogenicity bioassay of materials previously studied by Epler.

It is not possible to directly extrapolate the results of analysis summarized here to the production or environmental release of PAH by an eventual coal conversion industry. Samples which have been analyzed come from very small scale operations without emission control modules and often from engineering experiments where process conditions are being manipulated to extremes. Materials from more realistic operations, those processing up to 100 tons of coal/day (still far from the 25,000 ton/day commercial plant projected), are only now becoming available. Product, process streams, waste streams, and the working environment of the 30 ton/day Solvent Refined Coal Plant in Fort Lewis, Washington are presently undergoing (86) a systematic chemical and biological study.

The quantities and variety of polycyclic aromatic hydrocarbons found in synfuel-related materials clearly indicate the existence of a cancer threat. There is some evidence (117) that the conversion process can be modified to reduce the production of undesirable constituents without lessening the quality of the final product. Industrial hygiene practices will have to be carefully followed to limit occupational carcinogenesis. Recycling hazardous waste streams as is presently planned in several processes and the development of control technologies optimized to remove PAH may be required to avoid the environmental release of potential carcinogens.

Coal Combustion

Most of the coal mined in the United States is eventually consumed by combustion to produce process, residential, or institutional heating or electric power. An additional quantity is consumed by destructive distillation to produce coke. As is illustrated in Table 2.13, these sources combined with the open burning of coal refuse, account for approximately 90% of the national inventory of airborne benzo[a]pyrene. Of the 600 to 800 metric tons of BaP estimated to be emitted in 1972, controlled combustion and open burning contributed approximately 280 metric tons each.

Table 2.13: Estimated Annual Airborne Emission of Benzo[a]pyrene in the United States

	1968	1972
	. . (metric tons per year) . .	
Coal		
Hand-stoked and underfeed-stoked residential furnaces	380	270
Intermediate size furnaces	9	6
Steam power plants	1	<1
Open burning, coal refuse	308	281
Coke production	0.05-153	0.05-153
Total	~700-850	~550-700
Petroleum and Natural Gas		
Oil, residential through steam power plants	2	2
Gas, residential through steam power plants	2	2
Petroleum catalytic cracking	5	6

(continued)

Table 2.13: (continued)

	1968	1972
	. . (metric tons per year) . .	
Asphalt air-blowing	<1	<1
Asphalt hot-road-mix plant	<1	–
Gasoline-powered, automobiles and trucks	20	10
Diesel-powered, trucks and buses	<1	<1
Rubber tire degradation	9	10
Aircraft	–	–
Lawn mowers and motorcycles	–	–
Total	~41	~32
Other Materials		
Wood, home fireplaces	36	23
Enclosed incinerator, apartment through municipal	31	3
Open burning, domestic and municipal waste	22	–
Open burning, vehicle disposal	45	5
Open burning, forest and agriculture	127	10
Open burning, other	–	9
Total	~261	~50
U.S. Annual Total	100–1,150	640–780

Source: Reference (4) as cited in NTIS CONF-770130-2

Analytical results (4) on samples acquired from the National Air Sampling Network (NASN) for BaP content over the period 1966 to 1970 from urban and nonurban areas further emphasize the importance of coal combustion as a source of atmospheric polycyclic aromatic hydrocarbons. Average seasonal concentrations of airborne BaP ranged from 0.1 to 50 ng/m^3 for rural and urban areas respectively. Concentrations are consistently higher in the winter months regardless of the region sampled in most cases and are always highest for heavily coal-based industrialized regions.

The Pittsburgh and Birmingham areas, both heavy coal consumers, were found to contain maximum annual average concentrations of 14 and 18 ng/m^3 of BaP, respectively, over the five sampling years reported. Perhaps most importantly, the carcinogenicity of organic extract of particulates collected in Birmingham is reported (133) to be greater than that of comparable extracts from Los Angeles area particulates. While well-noted for its "smog," very little of the airborne particulates in the Los Angeles area are coal-derived.

Diehl et al (134) have demonstrated a dependence of emission factors for benzo[a]pyrene, benzo[e]pyrene, and benz[a]anthracene on the type of coal but conclude, as was previously noted for the combustion of petroleum products, that combustion efficiency is the primary factor influencing PAH emissions.

The annual "BaP Inventories" (Table 2.13) identify small residential furnaces as the overwhelming source of heat-generation related BaP. Process- or institutional-heat generation combustors are identified as more important to the inventory than large scale power plants. The studies summarized by Hangebrauck et al (2) provide the basis for these conclusions and illustrate that PAH other than BaP follow the same trend. The data summarized in Table 2.14 further illustrate that the type, and presumably the efficiency, of combustion is a more important factor than is the size or purpose of the combustor.

Table 2.14: Distribution of PAH in Coal Combustion Effluents

Heating Plant	Fuel Rate	Benzo[a]-pyrene	Benzo[a]-pyrene	Pyrene	Benzo[e]-pyrene	Perylene	Benzo[ghi]-perylene	Anthanthrene	Coronene	Anthracene	Phenanthrene	Fluoranthene
		μg/lb	. . . .	milligrams per million Btu heat input	. . . .							
Power Plant												
A	~60*	0.25	0.020	0.160	–	–	–	–	–	–	–	0.20
B	~60*	3.1	0.220	1.025	0.395	0.017	0.200	–	0.005	–	–	0.075
Process												
A	~5*	0.43	0.032	0.24	0.092	–	–	–	–	0.370	–	0.55
B	~300**	140	10	16	7.9	1.6	4.5	0.29	0.33	0.85	10	38
School B	~200**	1.6	0.12	1.7	0.23	–	–	–	–	–	1.0	3.2
Residential												
A	~5**	52	3.8	7.7	5.4	–	0.58	–	1.2	–	29	47
B	~6**	25	1,700	2,700	870	220	1,400	270	49	1,100	2,300	4,300

*Tons per hour. **Pounds per hour.

Source: Reference (2) as cited in NTIS CONF-770130-2

Coal fired power plants contribute relatively little to the natural particulate BaP burden. A high combustion efficiency combined with effective emission control methods reduce the annual emission to one ton or less of BaP. A low annual emission does not eliminate this source as a contributor to the PAH related cancers. Local concentrations may be high.

Diehl et al (134), for example, report the release of up to 50 mg/hr of BaP from power plants rated from 1 x 10^6 to 2 x 10^6 lb steam/hr. A release of 46.6 mg BaP/hr may be calculated from the more recently summarized (1) results of Hangebrauck (2) for a power plant consuming 59 metric tons of coal/hr. While only 840 μg of BaP are released per metric ton of coal consumed, the process releases large quantities of BaP and presumably larger quantities of PAH into a localized environment.

Emissions estimates are dependent on accurate sampling and analysis methods. PAH are thought (4) to be released from the combustor flame in the vapor state later to condense on inorganic particles or to coalesce to form particles of pure condensate. The recent study of sampling techniques by Jones et al (135) addressed the problem of vapor state PAH by designing a sampling system which incorporated an organic adsorbent in the sampling train. The concentrations of anthracene, pyrene, fluoranthene, chrysene, benz[a]anthracene, and benzopyrenes were determined in effluents from the small scale combustion of natural gas and residual oils and from effluents of a carbon black manufacturing facility using the adsorbent sampling system and standard sampling systems.

Expressed as total polycyclic organic matter (POM), Jones et al report from two to ten times as much POM in the effluents using the adsorbent sampler. Enhancements by a factor of ten seem to be associated with the combustion of sulfur-containing fuels. It should be noted that the most volatile PAH, the naphthalenes, were not studied. The logical conclusion is that emissions from coal-fired power plants may be as much as an order of magnitude greater than has been predicted using traditional sampling methods.

Methodological limitations, combined with changes (4) in combustion techniques designed to meet NO_X emission standards, may result in a significant underestimate of the importance of coal-fired power plants to the national particulate BaP inventory. Regional effects, particularly during periods of atmospheric conditions which localize the dispersion of emitted particulates, may be much greater than would be predicted by annual national inventories of BaP.

Differences (133) in carcinogenicity for airborne particulates from different geographic locations suggest that the composition of adsorbed organic matter may be important. The data summarized in Table 2.14 illustrate that pyrene and fluoranthene are consistently present at concentrations equal to or greater than BaP in coal combustion emissions as is also noted (Table 2.7) for residual oil combustion. Benzo[e]pyrene (BeP) is reported present at concentrations approximating that of BaP for coal combustion while BeP was generally present at levels below the detection limit for oil combustors studied (2). Diehl reports (134) that concentrations of benz[a]anthracene are within a factor of two of those for BaP for the coal-fired power plants sampled.

Griest et al (136) have recently surveyed the hydrocarbon content of an ash taken from the stack of a Tennessee Valley Authority coal-fired power plant. The ash was extracted with benzene after adding carbon-14-BaP to the ash to determine recoveries, and the extract was analyzed by fluorescence spectrometry and gas chromatography. The extract was further purified by Florisil and alumina column chromatography to remove paraffins after these components were found to predominate on the ash. Results of these analyses for PAH are summarized in Table 2.15.

Table 2.15: Polycyclic Aromatic Hydrocarbons on Fly Ash from a Coal Fired Power Plant

PAH	Concentration (ng/g)
Naphthalene	8.3
2-Methylnaphthalene	5.0
1-Methylnaphthalene	5.2
Biphenyl	10.3
1,6- and/or 1,3-Dimethylnaphthalene	trace
2,6-Dimethylnaphthalene	trace
1,5- and/or 2,3-Dimethylnaphthalene	trace
9,10-Dihydroanthracene	12.6
Phenanthrene	17.6
2-Methylanthracene	9.1
1-Methylphenanthrene	<24.8
Fluoranthene	<13.4
Pyrene	<19.0
1,2-Benzofluorene	36.8
2,3-Benzofluorene	11.8
1-Methylpyrene	trace
Picene	trace
Total, ppm	0.2

Source: Reference (136) as cited in NTIS CONF-770130-2

Normal paraffins 16- through 34-carbons in chain-length inclusively constituted 8.6 ppm of the ash while the PAH visualized constituted less than 0.2 ppm.

Benzofluorenes and alkyl-substituted naphthalenes, pyrenes, phenanthrenes, and anthracenes were found in addition to the commonly reported pyrene and fluoranthene. The separate analysis for BaP by fluorescence spectrometry and carbon-14 isotope dilution indicated that BaP constituted 35 ppb of the ash. The results cannot be converted to quantitative emission factors because the plant operating characteristics and the sampling protocol were not specified. The data clearly demonstrate, however, that the composition of adsorbed organic matter is considerably more complex than would be indicated by results of analyses for a small number of preselected constituents.

The contribution, if any, of coal fired power plants directly to the aquatic or terrestrial environment is less well-known. Mass flow levels reported (137) for a trace element balance study within a power plant consuming approximately 88 metric tons of coal/hr show that 8.43 x 10^{14} g of fly ash are collected by the precipitator per minute while only 0.037 x 10^4 g/min of fly ash are released into the environment through the stack. Slag and precipitator ash, 394.5 tons per ton of released ash, are often disposed of in ash ponds. It is not known how efficiently the spectrum of PAH contained on the ash are leached or acted upon by microorganisms and subsequently leached into the aquatic environment.

Present technologies are effectively controlling the annual emissions of particulate BaP (and presumably other PAH) from coal-fired power plants. The projected increasing national reliance on coal-based electric power and the possible regional contribution to environmental carcinogenesis from a high point-source emission of PAH, require a continued surveillance of this source of energy-related polycyclic aromatic hydrocarbons.

OIL SHALE

Certain shales, particularly those in the Green River Basin of Colorado are sufficiently rich in organic matter (kerogen) to allow the production of petroleum substitutes by retorting. With shale, polycyclic aromatic hydrocarbon carcinogenesis problems are conceptually similar (but less severe) than those which have already been discussed as accompanying the liquefaction of coal. High grade shales yield only 25 gallons of oil per ton as compared to 1 to 3 barrels per ton of coal.

The high grade oil shale deposits in the Green River formation of Colorado, Utah and Wyoming are so massive, however, as to allow estimates (138) of some 600 billion barrels of recoverable oil from the 600 square miles containing the shale. This quantity of oil is reported (138) to be equivalent to the known reserves of petroleum oil including the 10 billion barrels in the Prudhoe Bay area of Alaska.

Oil is presently produced (139) by heating ("destructive distillation," "retorting") shale extracted by surface or underground mining to approximately 1000°F in above ground retorts. Research is underway at the Laramie Energy Research Center, Laramie, Wyoming, on methods to allow the in-situ conversion of shale to oil.

The primary environmental problems facing the shale oil industry appear to be locating adequate supplies of water for processing, avoiding contaminating the limited western water supply, and disposing of solid wastes. Approximately two tons of spent shale accompanies the production of each barrel of oil. Because of its decreased density following processing, it is not possible to return all of the spent shale to the mine. Almost 600 million tons of spent shale would have to be disposed of annually from an oil shale industry producing one million barrels of oil per day (138).

A report by Schmidt-Collerus (140) suggested that a carcinogenic hazard could be associated with spent shale. Coomes (13) has responded to this and the Sauter (91) paper. He reports that hairless mice living in a bedding of spent shale exhibit no greater incidence of skin cancer than those living in normal bedding and suggest that this is evidence of the noncarcinogenic nature of spent shale. Coomes also reports results from three laboratories which show that the benzo[a]pyrene content of processed shale is no greater than 40 ppb, well within the range of concentration reported for coconut oil, oysters, and oak leaves. Bonomo has reported (108) concentrations of BaP of from 15 to 115 ppb in weathered spent shales.

Crude shale oil contains (13) approximately 3 ppm of BaP, one hundred times that found in spent shale and approximately that found in petroleum crude oils. A shale oil produced by an above-ground simulated in-situ retort has been subjected to solvent extraction and adsorption chromatographic separations to prepare a polycyclic aromatic hydrocarbon enriched isolate for biological study. The isolate composed approximately 2.0 wt % of the shale oil. This is less than the quantity of isolate prepared (3.2 wt %) from a mixture of petroleum crude oils using the same procedure.

The product of simulated in-situ retorting contains the same variety (Table 2.10) of polycyclic aromatic hydrocarbons as is observed in petroleum and coal-derived oils. The highly carcinogenic isomers 3-methyl cholanthrene (MCA) and 7,12-dimethylbenzanthracene have been reported (91)(108) in both raw and processed shale but identifications are based principally on chromatographic retention. Gas chromatographic and mass spectrometric evidence for the presence of trace quantities of MCA in a crude shale oil produced by simulated in-situ retorting has been observed but the data are not considered adequate for positive identification.

Mouse dermal bioassay remains the most widely accepted measure of carcinogenicity. Coomes (13) has published the results summarized in Table 2.16 from a report by the Colony Development Operation (141). The data suggest that crude shale oil is no more carcinogenic than is commonly used fuel oil. It also suggests that subsequent upgrading (by hydrogenation) further reduces the carcinogenicity of the oil.

Table 2.16: Mouse Dermal Carcinogenicity of Shale Oils and Related Materials

Material	P_{mc}*	Material	P_{mc}*
Whole shale oil	0.1	Coke oven coal tar	0.54
Upgraded shale	0.03	Cracked sidestream	0.26
Industrial fuel oil	0.17	Naphthenic distillate	0.06

*Carcinogenicity relative to MCA (142)

Source: Reference (13) as cited in NTIS CONF-770130-2

Reports of polycyclic aromatic hydrocarbons in shale retort-derived waters cannot be extrapolated directly to environmental releases. Data in Table 2.12 were generated using the water recovered by centrifuging the oil-water product of a simulated in-situ retort. The distribution of paraffins in the water was identical to that found in the oil. The polycyclic aromatics may be dissolved in the oil remaining in the water due to incomplete separation. Control technologies optimized to remove "hydrocarbons" may reduce the polyaromatic content of the water to acceptable levels.

The commercial scale conversion of shales to oil will involve processing large quantities of polycyclic aromatic hydrocarbons. The possibility of PAH-related occupational and environmental carcinogenesis cannot be ruled out based on the limited experimental data and process experience available.

CONCLUSIONS

Fossil fuels provide more than 90% of the nation's energy and are the primary technological source of polycyclic aromatic hydrocarbons. Combustion, coal coking, and petroleum catalytic cracking are the primary fossil energy sources of airborne PAH. Transportation and processing of petroleum are the primary technological sources of PAH contaminating the water environment. Energy related operations accompanied by the greatest hazard of occupational carcinogenesis are those involving contact with petroleum- and coal-residuals (e.g., coal tars) and respiration of heavily PAH-contaminated environments (e.g., coking).

Coal and shale conversion will constitute a new source of PAH. A single commercial coal liquefaction plant will produce (assuming 25,000 tons of coal/day, 2 bbl oil/ton of coal, 140 kg oil/bbl, 3 wt % PAH) 210,000 kg of PAH and (assuming 1 ppm BaP) 7 kg of BaP daily as a component of the product. An industry processing 10^6 tons of coal per day would produce approximately 100 metric tons of BaP annually. The quantities, if any, of PAH generated as wastes or emitted during processing are unknown.

Technology appears to exist to handle the increased PAH burden. The petroleum industry, for example, processed 780 metric tons of BaP (assuming 0.78 x 10^9 metric tons petroleum crude oil consumed, 1 ppm BaP) in 1972. Secondly, much of the coal-derived oil is expected to be consumed by combustion and it is well documented that combustion efficiency is more important than is the aromaticity of the fuel in determining final PAH emissions. Conversion technologies remain a potential occupational health and local environmental health threat because it is not known how efficiently the hazardous constituents can be contained in a commercial scale plant.

Based on the very limited data (Table 2.17) available, coal- and shale-derived crude oils contain only 2 to 3 times more BaP than do petroleum crudes. Some processes, particularly those optimized to produce chemical feedstocks, produce products containing 10 to 100 times the BaP and/or PAH content of petroleum crudes. Tars and high boiling products produced in the conversion processes can consist of 40 wt % PAH but these must be compared to their petroleum counterpart, e.g., pitches, rather than crude oils in judging the added impact of the conversion industry on the environmental PAH burden. In general, coal- and shale-derived oils may be expected to contain larger quantities of alkylated PAH, heteroatomic constituents, and hydroaromatics than does petroleum.

Table 2.17: Benzo[a]pyrene Content of Energy Related Materials

Materials, BaP (ppm)	Average BaP (~ppm)
Crude oils	
Petroleum	1
Libya (33), 1.3	
Venezuela (33), 1.6	
Persian Gulf (33), 0.04	
Arabian (143), 1.5	
Shale derived	3
Crude oil (13), 3.1	
Hydrotreated (13), 0.7	
Coal derived	3
Catalytic hydrogenation (107)(143), 1.6, 4.1	
Pyrolysis (143), 4.1	
Petroleum products	
Gasoline	0.4
Commercial (32), 0.46, 0.03, 0.55	
Catalytically reformed (32), 6.2	
Nationwide composite premium (36), 0.48	
Nationwide composite regular (36), 0.21	
Motor oil, new (33)	0.03
Motor oil, used (33)(143), 2.4, 6.0	4
Diesel fuel (33)(143), 0.03, 0.07	0.05
No. 2 heating oil (143), 0.01-0.05	0.03
Asphalt (14), 0.1-27	2
Petroleum pitch (39)	2,000
Miscellaneous	
Creosote (144)	200
Coal tar (42)(144), 1,800, 5,000	3,000
Coal tar pitch (14)	10,000
Raw shale (13)	0.015
Processed shale (13)	0.03
Bituminous coal (42)	<0.001
Reference environmental materials	
Waters	
Uncontaminated ground water (21)	<0.00001
Drinking water (21)	<0.00003
Contaminated by industrial effluent (21)	0.01
Heavily contaminated by coking, oil shale, oil-gas processing (21)	0.5-1.0
Miscellaneous	
Foods (42)	<0.001
Earth's upper crust (21)	0.1-1.0
Biosynthesized per weight dried material, wheat and rye (21)	0.01-0.02

Source: NTIS CONF-770130-2

The experimental carcinogenicity of shale oil is reported to be about equal to that of petroleum crude oil. Coal tars are well-known to be highly carcinogenic. Mutagenicity screening shows two coal-derived oils to be more active than a shale oil which, in turn, is approximately equal in activity to a petroleum crude oil. In all cases, the mutagenic activity is concentrated in the polycyclic aromatic enriched fractions and the basic fractions.

REFERENCES

(1) *A National Energy Plan for Energy Research, Development, and Demonstration: Creating Energy Choices for the Future,* Vol 1, *The Plan,* U.S. Energy Research and Development Administration, ERDA-48, Washington, DC (June 28, 1975).

(2) Hangebrauck, R.P., von Lehmden, D.J., and Meeker, J.E., *Sources of Polynuclear Hydrocarbons in the Atmosphere,* U.S. Department of Health, Education, and Welfare, Public Health Service, Environmental Health Service, National Air Pollution Control Administration, Durham, NC (1967).

(3) *Particulate Polycyclic Organic Matter: Biologic Effects of Atmospheric Pollutants,* National Academy of Sciences, Washington, DC (1972).

(4) *Scientific and Technical Assessment Report on Particulate Polycyclic Organic Matter (PPOM),* U.S. Environmental Protection Agency, EPA-600/5-74-001, Washington, DC (March 1975).

(5) *Particulate Polycyclic Organic Matter,* Prepared by the Committee on Biologic Effects of Atmospheric Pollutants, Division of Medical Sciences, National Research Council (1972).

(6) Surprenant, N., Hall, R. et al, *Preliminary Emissions Assessment of Conventional Stationary Combustion Systems,* Vol II, GCA Report Draft to EPA; GCA-TR-75-26-G(1) (1976).

(7) Lerer, T.J. et al, "Long Term Mortality Study of Steelworkers: vii Mortality Patterns Among Crane Operators," *Jour. Occup. Med.,* 16, p 606 (1974).

(8) Lloyd, J.W., "Long Term Mortality Study of Steelworkers: v Respiratory Cancer in Coke Plant Workers," *Jour. Occup. Med.,* 13, p 53 (1971).

(9) Mazumdar, S., Redmond, C., Sollecito, W., and Sussman, N., "An Epidemiological Study of Exposure to Coal Tar Pitch Volatiles Among Coke Oven Workers," *J. of the Air Pollution Control Association,* 25(4), 382-389 (1975).

(10) Doll, R. et al, "Mortality of Gas-Workers - Final Report of a Prospective Study," *Brit. J. Ind. Mcd.,* 29:394-406 (1972).

(11) Doll, R., Fisher, R.E., Gammon, E.J. et al, "Mortality of Gas-Workers with Special Reference to Cancers of the Lung and Bladder, Chronic Bronchitis, and Pneumoconiosis," *Brit. J. Ind. Med.,* 22, 1-12 (1965).

(12) Sexton, Richard J., "The Hazards to Health in the Hydrogenation of Coal, iv The Control Program and the Clinical Effects," *Archives of Environmental Health,* 1, 208-231 (1960).

(13) Coomes, R. Merril, "Health Effects of Oil Shale Processing," *Quarterly of the Colorado School of Mines,* Proceedings of the Ninth Oil Shale Symposium, James H. Gary, Ed., Vol 71, No 4, pp 101-123 (October 1976).

(14) Wallcave, L., Garcia, H., Feldman, R., Lijinsky, W., and Shubik, P., "Skin Tumorigenesis in Mice by Petroleum Asphalts and Coal Tar Pitches of Known Polynuclear Aromatic Hydrocarbon Content," *Toxicol. Appl. Pharm.,* 18, 41-52 (1971).

(15) Sawicki, E., Elbert, W.C., Hauser, T.R., Fox, F.T., and Stanley, T.W., "Benzo[a]pyrene Content of the Air of American Communities," *Amer. Ind. Hyg. Assoc. J.* 21:443-451 (1960).

(16) Sawicki, E., Hauser, T.R., Elbert, W.C., Fox, F.T., and Meeker, J.E., "Polynuclear Aromatic Hydrocarbon Composition of the Atmosphere in Some Large American Cities," *Amer. Ind. Hyg. Assoc. J.* 23:137-144 (March-April, 1962).

(17) Cautreels, W., and van Cauwenberghe, K., "Fast Quantitative Analysis of Organic Compounds in Airborne Particulate Matter by Gas Chromatography with Selective Mass Spectrometric Detection," *J. of Chromatography,* 131, 253-264 (1977).

(18) Gordon, Robert J., "Distribution of Airborne Polycyclic Aromatic Hydrocarbons Throughout Los Angeles," *Environmental Science & Technology,* 10(4), 370-373 (1976).

(19) Lao, R.C., Thomas, R.S., Oja, H., and Dubois, L., "Application of a Gas Chromatograph–Mass Spectrometric–Data Processor Combination to the Analysis of the Polycyclic Aromatic Hydrocarbon Content of Airborne Pollutants," *Anal. Chem.,* 45, 908 (1973).

(20) Lee, M.L., Novotny, M., and Bartle, K.D., "Gas Chromatography/Mass Spectrometric and Nuclear Magnetic Resonance Determination of Polynuclear Aromatic Hydrocarbons in Airborne Particulates," *Anal. Chem.,* 48(11), 1566-1572 (1976).
(21) Andelman, J.B., and Suess, J.J., "Polynuclear Aromatic Hydrocarbons in the Water Environment," *Bull. Wld. Hlth. Org.,* 43, 479-508 (1970).
(22) Andelman, J.B., and Snodgrass, J.E., "Incidence and Significance of Polynuclear Aromatic Hydrocarbons in the Water Environment," *CRC Critical Reviews in Environmental Control,* 4(1), 69-83 (1974).
(23) Harrison, R.M., Perry, R., and Wellings, R.A., "Polynuclear Aromatic Hydrocarbons in Raw, Potable, and Waste Waters," *Water Research,* 9, 331-346 (1975).
(24) Simon, J.A., and Malhorta, R., "Place of Coal in the Total Energy Needs of the United States," *California Geology,* pp 195-201 (September 1976).
(25) Travers, W.B., and Luney, P.R., "Drilling, Tankers, and Oil Spills on the Atlantic Outer Continental Shelf," *Science,* 194(4267), 791-796 (November 1976).
(26) Smith, H.M., and Hale, J.H., *Crude Oil Characterizations Based on Bureau of Mines Routine Analyses,* Bureau of Mines Report of Investigations 6846, U.S. Department of the Interior (1966).
(27) Smith, N.A.C., Smith, H.M., Blade, O.C., and Garton, E.L., *The Bureau of Mines Routine Method for the Analyses of Crude Petroleum, i The Analytical Method,* Bulletin 490, Bureau of Mines, U.S. Department of the Interior (1951).
(28) Haines, W.E., and Thompson, C.J., *Separating and Characterizing High-Boiling Petroleum Distillates: the USBM-API Procedure,* USERDA, July, LERC/RI-75/5, BERC/RI-75/2 (1975).
(29) Coleman, H.J., Dooley, J.E., Hirsch, D.E., and Thompson, C.J., "Compositional Studies of a High Boiling 370°-535°C Distillate from Prudhoe Bay, Alaska, Crude Oil," presented before the Division of Petroleum Chemistry, Inc., American Chemical Society, Boston Meeting (April 9-14, 1972).
(30) Dooley, J.E., Hirsch, D.E., Thompson, C.J., Vogh, J.W., and Ward, C.C., "Analyzing Heavy Ends of Crude," U.S. Energy Research and Development Administration, Bartlesville Energy Research Center, Bartlesville, OK; reprinted from *Hydrocarbon Processing* (1974).
(31) Holland, J.M., "Dose Response Relationships in Chemical Carcinogenesis," *Coal Technology Program Quarterly Progress Report for the Period Ending December 31, 1976,* ORNL-5252, Oak Ridge National Laboratory, Oak Ridge, TN (1977).
(32) Norris, M.S., and Hill, E.D., "Polynuclear Aromatic Hydrocarbons in Petroleum Products," presented at Fossil Fuel Chemistry and Energy Workshop, University of Wyoming Science Summer Camp, Laramie, WY (July 23-27, 1974).
(33) Graf, W., and Winter, C., "3,4-Benzpyren in Erdöl," *Archiv. für Hygiene und Bakteriologie,* Vol 152, pp 289-293 (1968).
(34) Lebowitz, H., Tam, S., Smithson, G., Jr., Nack, H., Oxley, J., Cavenaugh, G., Burklin, C., and Dickerman, J., *Potentially Hazardous Emissions from the Extraction and Processing of Coal and Oil,* April, EPA-650/2-75-078, NTIS PB241-803, Washington, DC (1975).
(35) Reid, George W., and Streebin, Leale, E., *Evaluation of Waste Waters from Petroleum and Coal Processing,* prepared for Office of Research and Monitoring, U.S. Environmental Protection Agency, EPA-R2-72-001, December, Washington, DC (1972).
(36) Sullivan, James B., *Marine Pollution by Carcinogenic Hydrocarbons,* Proceedings of a Symposium and Workshop held at NBS, May 13-17, Gaithersburg, MD (1974).
(37) Kireeva, I.S., Yanysheve, N.Ya., "Loading of Atmospheric Air with Carcinogenic Polycyclic Aromatic Hydrocarbons from Petroleum Processing Installations," translated from *Gig. Naseiennykh Mest,* No. 9, 113-117 (1970).
(38) Lee, M.L., Bartle, K.D., and Novotny, M.V., "Profiles of the Polynuclear Aromatic Fraction from Engine Oils Obtained by Capillary-Column Gas-Liquid Chromatography and Nitrogen-Selective Detection," *Anal. Chem.,* 47(3), 540-542 (1975).
(39) Grienke, R.A., and Lewis, I.C., "Development of a Gas Chromatographic-Ultraviolet Absorption Spectrometric Method for Monitoring Petroleum Pitch Volatiles in the Environment," *Anal. Chem.,* 47(13), 2151-2155 (1975).

(40) Eckardt, Robert, "Cancer Prevention in the Petroleum Industry," *Int. J. Cancer,* 2, 656-661 (1967).

(41) Sawicki, E., "The Chemical Composition and Potential Genotoxic Aspects of Polluted Atmospheres," presented at the Workshop for the Investigations on the Carcinogenic Burden by Air Pollution in Man in Hanover, Germany on October 22-24, 1975.

(42) Sawicki, E., "Analysis of Atmospheric Carcinogens and Their Cofactors," presented at the Symposium on Environmental Pollution and Carcinogenic Risks at Lyon, France on November 3-5, 1975.

(43) Gross, G.P., *First Annual Report on Gasoline Composition and Vehicle Exhaust Gas Polynuclear Aromatic Content,* (CRC-APRAC Project CAPE-6-68) NTIS Publication PB-200-266.

(44) Gross, G.P., *Second Annual Report on Gasoline Composition and Vehicle Exhaust Gas Polynuclear Aromatic Content,* (CRC-APRAC Project CAPE-6-68) NTIS Publication PB-209-955.

(45) Gross, G.P., *Third Annual Report on Gasoline Composition and Vehicle Exhaust Gas Polynuclear Aromatic Content,* (CRC-APRAC Project CAPE-6-68) NTIS Publication PB-215-527.

(46) Boubel, R.W., and Ripperton, L.A., "Benzo[a]pyrene Production During Controlled Combustion," *J. Air Pollut. Control Assoc.,* 13:553-557 (1963).

(47) Hoffmann, D., and Wynder, E.L., "Studies on Gasoline Engine Exhaust," *J. Air Pollut. Control Assoc.,* 13:322-327 (1963).

(48) Hoffmann, C.S., Jr., Willis, R.L., Patterson, G.H., and Jacobs, E.S., "Polynuclear Aromatic Hydrocarbon Emission from Vehicles," presented to the 160th National Meeting of the American Chemical Society, March, Los Angeles, CA (1971).

(49) Begeman, C.R., "Carcinogenic Aromatic Hydrocarbons in Automobile Effluents," *Vehicle Emissions* (Selected SAE Papers), (Paper 440C presented January 1962 at the SAE Automotive Engineering Congress) New York:Society of Automotive Engineers, Inc. (1964).

(50) Bartle, K.D., Lee, M.L., and Novotny, M., "High-Resolution GC Profiles of Urban Air Pollutant Polynuclear Aromatic Hydrocarbons," *Intern. J. Anal. Chem.,* 3, 349-356 (1974).

(51) Hill, H.H., Jr., Chan, K.W., and Karasek, F.W., "Extraction of Organic Compounds from Airborne Particulate Matter for Gas Chromatographic Analysis," *J. of Chromatography,* 131, 245-252 (1977).

(52) Fox, Marye Anne, and Staley, Stuart W., "Determination of Polycyclic Aromatic Hydrocarbons in Atmospheric Particulate Matter by High Pressure Liquid Chromatography Coupled with Fluorescence Techniques," *Anal. Chem.,* 48(7), 992-998 (1976).

(53) Karasek, T.W., Smythe, R.J., and Laub, R.J., "A Gas Chromatographic – Mass Spectrometric Study of Organic Compounds Adsorbed on Particulate Matter from Diesel Exhaust," *J. Chromatography,* 101(1), 125-136 (1974).

(54) Mentser, M., and Sharkey, A.G., Jr., *Chemical Characterization of Diesel Exhaust Particulates,* USERDA, March, PERC/RI-77/5, Pittsburgh Energy Research Center, Pittsburgh, PA (1977).

(55) Boyer, Kenneth W., and Laitenen, Herbert A., "Automobile Exhaust Particulates - Properties of Environmental Significance," *Environmental Science & Technology,* 9(5), 457-469 (1975).

(56) Candeli, A., Mastrandrea, V., Morozzi, G., and Toccaceli, S., "Carcinogenic Air Pollutants in the Exhaust from a European Car Operating on Various Fuels," *Atmospheric Environment,* 6, 693-705 (1974).

(57) Doran, T., and McTaggart, N.G., "The Combined Use of High Efficiency Liquid and Capillary Gas Chromatography for the Determination of Polycyclic Aromatic Hydrocarbons in Automotive Exhaust Condensates and Other Hydrocarbon Mixtures," *J. Chromatographic Science,* 12, 715-721 (1974).

(58) Jentoff, R.E., and Gouw, T.H., "Analysis of Polynuclear Aromatic Hydrocarbons in Automobile Exhaust by Supercritical Fluid Chromatography," *Anal. Chem.,* 48(14), 2195-2199 (1976).

(59) *Vapor-Phase Organic Pollutants, Medical and Biologic Effects of Environmental Pollutants,* National Academy of Sciences, Washington, DC (1976).
(60) VanDuuren, B.L., "Tumor Promoting and Cocarcinogenic Agents in Chemical Carcinogenesis," pp 34-52, *Chemical Carcinogens,* ACS Monograph 173, C.E. Searle, Ed., American Chemical Society, Washington, DC (1976).
(61) Albagli, A., Oja, H., and Dubois, L., "Size-Distribution Pattern of Polycyclic Aromatic Hydrocarbons in Airborne Particulates," *Environmental Letters,* 6(4), 241-251 (1974).
(62) Kertesz-Saringer et al, "On the Size Distribution of Benzo[a]pyrene Containing Particles in Urban Air," *Atmospheric Environment,* 5, 429-431 (1971).
(63) Pierce, Ronald C., and Katz, Morris, "Dependency of Polynuclear Aromatic Hydrocarbon Content on Size Distribution of Atmospheric Aerosols," *Environmental Science & Technology,* 9(4), 347-353 (1975).
(64) Stanton, M.F., Miller, E., Wrench, C., et al, "Experimental Induction of Epidermoid Carcinoma in the Lungs of Rats by Cigarette Smoke Condensate," *J. National Cancer Institute,* 49:867-873 (1972).
(65) Anon, "How to Use Coal: Gas Judged Superior All Round to Electricity," *The Energy Daily,* Vol 5, No 87, pp 2-3, May 4, Llewellyn King, Ed., Washington, DC (1977).
(66) Wiser, Wendell H., "Some Chemical Aspects of Coal Liquefaction," presented at the Short Course on Coal Characterization and Coal Conversion Processes, Penn State University, University Park, PA, October 29 - November 2, 1973.
(67) Freudenthal, R.I., Lutz, G.A., and Mitchell, R.I., *Carcinogenic Potential of Coal and Coal Conversion Products,* A Battelle Energy Program Report (February 25, 1975).
(68) Redmond, Carol K., Strobino, Barbara Reiber, and Cypess, Raymond H., "Cancer Experience Among Coke By-Product Workers," *Occupational Carcinogenesis, Annals of the New York Academy of Sciences,* 271, 102-115 (1976).
(69) Lawther, P.J., Commins, B.T., and Waller, R.E., "A Study of the Concentrations of Polycyclic Aromatic Hydrocarbons in Gas Works Retort Houses," *Brit. J. Ind. Med.,* 22, 13 (1965).
(70) Jackson, James O., Warner, Peter O., and Mooney, Thomas F., "Profiles of Benzo[a]pyrene and Coal Tar Pitch Volatiles at and in the Immediate Vicinity of a Coke Oven Battery," *American Ind. Hygiene Assoc. J.,* May, pp 276-281 (1974).
(71) Cochran, Neal P., "Oil and Gas from Coal," *Scientific American,* 234(5) (1976).
(72) Hammond, Ogden H., and Baron, Robert E., "Synthetic Fuels: Prices, Prospects, and Prior Art," *American Scientist,* 64, 407-417 (1976).
(73) Klass, Donald L., "Synthetic Crude Oil from Shale and Coal," *Chemtech,* 499-510 (1975).
(74) Howard-Smith, I., and Werner, G.J., *Coal Conversion Technology: A Review,* Millmerran Coal Pty. Ltd., May, Brisbane, Australia (1975).
(75) Glazer, F., Hershaft, A., and Shaw, R., *Emissions from Processes Producing Clean Fuels,* Booz-Allen and Hamilton, Inc., Prepared for Environmental Protection Agency, March (1974).
(76) Katz, D.L., Briggs, D.E., Lady, E.R. et al, *Evaluation of Coal Conversion Processes to Provide Clean Fuels (Fluidized Bed Combustion, Coal Beneficiation; Pyrolysis; Coal Gasification; Coal Liquefaction),* parts I and II, prepared under Research Project EPRI 206-0-0 by the University of Michigan College of Engineering for Electric Power Research Institute, Palo Alto, CA (February 1974).
(77) Massey, Michael J., Dunlap, Robert W., and Luthy, Richard G., *Environmental Assessment in the ERDA Coal Gasification Development Program,* Progress Report for the period July 1976 - December 1976, Carnegie-Mellon University (March 1977).
(78) Jahnig, C.E., and Magee, E.M., *Evaluation of Pollution Control in Fossil Fuel Conversion Processes. Gasification*: Section I: "CO_2 Acceptor Process," Exxon Research and Engineering Company, Linden, NJ, EPA-650/2-74-009d, NTIS PB-241-141 (December 1974), and Section 6: "Hygas Process," EPA-650/2-74-009h, NTIS PB-247-225 (August 1975).

(79) Massey, Michael J., Dunlap, R.W., McMichael, F.C., and Nakles, D.V., *Characterization of Effluents from the Hygas and CO_2-Acceptor Pilot Plants. Fossil Energy,* USERDA, work performed under Contract No E(49-18)-2496, Carnegie-Mellon University, Interim Report, July-September, Pittsburgh, PA (1976).

(80) Kalfadelis, C.D., and Magee, E.M., *Evaluation of Pollution Control in Fossil Fuel Conversion Processes. Gasification,* Section I, "Synthane Process," EPA-650/2-74-009b, NTIS PB 237-113, (June 1974); and *Liquefaction,* Section I, "COED Process," EPA-650/2-74-009e, NTIS PB 240-371, Exxon Research and Engineering Company (January 1975).

(81) Jahnig, C.E., *Evaluation of Pollution Control in Fossil Fuel Conversion Processes. Liquefaction,* Section 2, "SRC Process," Exxon Research and Engineering Company, EPA 650/2-74-009f, NTIS PB-241-792 (March 1975); and Section 5, "Bi-Gas Process," EPA 650/2-74-009g, NTIS PB-243-694 (May 1975).

(82) Shaw, H., and Magee, E.M., *Evaluation of Pollution Control in Fossil Fuel Conversion Processes; Gasification,* Section i, "Lurgi Process," Exxon Research and Engineering Company, EPA-650/2-74-009c, NTIS PB-237-694 (1974).

(83) Magee, E.M., Jahnig, C.E., and Shaw, H., *Evaluation of Pollution Control in Fossil Fuel Conversion Processes; Gasification,* Section 1, "Koppers-Totzek Process," Esso Research and Engineering Company, EPA-650/2-74-009a, NTIS PB-221-675 (January 1974).

(84) Braunstein, H.M., Copenhaver, E.D., and Pfuderer, H.A. (Editors), *Environmental, Health, and Control Aspects of Coal Conversion: An Information Overview,* Volumes I and II, ORNL EIS-94 and EIS-95, Oak Ridge National Laboratory, Oak Ridge, TN (August 1976).

(85) *Environmental Development Plan* (Draft): *Oil Shale Program,* USERDA (by Booz-Allen & Hamilton, Bethesda, MD), Washington, DC (March 8, 1977).

(86) *Carcinogens Relating to Coal Conversion Processes,* USERDA-FE, FE-2213-1, TRW Systems and Energy, Washington, DC (June 14, 1976).

(87) Eckardt, Robert E., "Hydrogenation of Coal," *Archives of Environmental Health,* 1, 232-233 (1960).

(88) Ketcham, N.H., Norton, R.W., "The Hazards to Health in the Hydrogenation of Coal, iii the Industrial Hygiene Studies," *Archives of Environmental Health,* 1, 194-207 (1960).

(89) Sexton, Richard J., "The Hazards to Health in the Hydrogenation of Coal, i an Introductory Statement on General Information, Process Description, and a Definition of the Problem," *Archives of Environmental Health,* 1, 181-186 (1960).

(90) Weil, Carrol S., and Condra, Naomi I., "The Hazards to Health in the Hydrogenation of Coal, ii Carcinogenic Effect of Materials on the Skin of Mice," *Archives of Environmental Health,* 1, 187-193 (1960).

(91) Sauter, D.Y., "Synthetic Fuel Hazard," *The Elements,* 3-4 (April 1977).

(92) Dooley, J.E., Sturm, G.P., Jr., Woodward, P.W., Vogh, J.W., and Thompson, C.J., *Analyzing Syncrude from Utah Coal,* BERC/RI-75/7, August, Bartlesville Energy Research Center, Bartlesville, OK (1975).

(93) Strum, G.P., Jr., Woodward, P.W., Vogh, J.W., Holmes, S.A., and Dooley, J.E., *Analyzing Syncrude from Western Kentucky Coal,* BERC/RI 75/12, Bartlesville Energy Research Center, Bartlesville, OK (November 1975).

(94) Holmes, S.A., Woodward, P.W., Strum, G.P., Jr., Vogh, J.W., and Dooley, J.E., *Characterization of Coal Liquids Derived from the H-Coal Process,* BERC/RI-76/10, Bartlesville Energy Research Center, Bartlesville, OK (November 1976).

(95) Woodward, P.W., Strum, G.P., Jr., Vogh, J.W., Holmes, S.A., and Dooley, J.E., *Compositional Analyses of Synthoil from West Virginia Coal,* BERC/RI-76/2, Bartlesville Energy Research Center, Bartlesville, OK (January 1976).

(96) Rubin, I.B., Guerin, M.R., Hardigree, A.A., and Epler, J.L., "Fractionation of Synthetic Crude Oils from Coal for Biological Testing," *Environmental Research,* 12, 358-365 (1976).

(97) Jones, A. Russel, Guerin, M.R., and Clark, B.R., "Preparative Scale Liquid Chromatographic Fractionation of Crude Oils Derived from Coal and Shale," submitted to *Anal. Chem.,* (May 1977).

(98) Clark, B.R., Ho, C.-h., and Jones, A. Russell, "Approaches to the Chemical Class Analyses of Fossil Derived Materials," preprints of the *Symposium on Analytical Chemistry of Tar Sands and Oil Shale,* Division of Petroleum Chemistry, American Chemical Society New Orleans Meeting, March 20-25, pp 811-822 (1977).

(99) White, Curt M., and Newman, John O.H., *Chromatographic and Spectrometric Investigation of a Light Oil Produced by the Synthoil Process,* PERC/RI-76/3, September, Pittsburgh Energy Research Center, Pittsburgh, PA (1976).

(100) Karn, Fred S., and Brown, Fred R., *Aging Studies of the Synthoil Product,* Part II: *Effects of Time, Temperature, Atmosphere,* PERC/TRI-76/2, September, Pittsburgh Energy Research Center, Pittsburgh, PA (1976).

(101) Sharkey, A.G., "Mass Spectrometric Analysis of Process Streams for Coal-Derived Fuels," *Polynuclear Aromatic Hydrocarbons: Chemistry, Metabolism, and Carcinogenesis,* Vol 1, Ralph Freudenthal and Peter W. Jones, ed, pp 341-347, Raven Press, New York, NY (1976).

(102) Sharkey, A.G., Shultz, J.L., Schmidt, C.E., and Friedel, R.A., "Mass Spectrometric Analysis of Coal Derived Fuel," *Proceedings of the Third ERDA Environmental Protection Conference,* September 23-26, CONF 750967, ERDA-92, pp 687-701, Chicago, IL (1975).

(103) Fruchter, J.S., Laul, J.C., Petersen, M.R., and Ryan, P.W., "High Precision Trace Element and Organic Constituent Analysis of Oil Shale and Solvent-Refined Coal Materials," preprints of the *Symposium on Analytical Chemistry of Tar Sands and Oil Shale,* Division of Petroleum Chemistry, American Chemical Society New Orleans Meeting, March 20-25, pp 793-807 (1977).

(104) Goeckner, N.A., and Griest, W.H., "Determination of Methyl Chrysenes in a Coal Liquefaction Product," *Science of the Total Environment,* in press.

(105) Guerin, M.R., Epler, J.L., Ho, C.-h., and Clark, B.R., "Determining Fugitive Emissions Measurements Needs for an Emerging Industry – Advanced Fossil Fuels Utilization," *Proceedings of Symposium on Fugitive Emissions Measurement and Control,* May, pp 275-304, Hartford, CT (1976).

(106) Kubota, H., Griest, W.H., and Guerin, M.R., "Determination of Carcinogens in Tobacco Smoke and Coal-Derived Samples - Trace Polynuclear Aromatic Hydrocarbons," reprint from *Trace Substances in Environmental Health-IX,* D.D. Hemphill, ed, University of Missouri, Columbia (1975).

(107) Shults, W.D., Ed, *Preliminary Results: Chemical and Biological Examination of Coal-Derived Materials,* March, ORNL/NSF/EATC-18, Oak Ridge National Laboratory, Oak Ridge, TN (1976).

(108) Bonomo, F., "Environmental Effects of Carbonaceous Solid Wastes from Commercial Oil Shale Operations," proceedings of the workshop on *Analytical Chemistry Pertaining to Oil Shale and Shale Oil,* National Science Foundation, June 24-25, pp 133-186, Washington, DC (1974).

(109) Hecht, S.S., Bondinell, W.E., Hoffmann, D., "Chrysene and Methylchrysenes: Presence in Tobacco Smoke and Carcinogenicity," *J. Nat'l. Cancer Inst.,* 53(4), 1121-1133 (1974).

(110) Hecht, S.S., Loy, M., Maronpot, R.R., and Hoffmann, D., "A Study of Chemical Carcinogenesis; Comparative Carcinogenicity of 5-Methylchrysene, Benzo[a] pyrene, and Modified Chrysenes," *Cancer Letters,* 1, 147-154 (1976).

(111) Griest, W.H., and Olerich, G., *Naphthalene Distribution in Synthetic and Natural Crude Oils,* Coal Technology Program Quarterly Progress Report for the Period Ending June 30, Oak Ridge National Laboratory, Oak Ridge, TN (1977).

(112) Ho, C.-h., Clark, B.R., and Guerin, M.R., "Direct Analysis of Organic Compounds in Aqueous By-Products from Fossil Fuel Conversion Processes: Oil Shale Retorting, Synthane Coal Gasification and COED Coal Liquefaction," *J. Environ. Sci. Health,* A11(7), 481-489 (1976).

(113) Schmidt, C.E., Sharkey, A.G., and Friedel, R.A., *Mass Spectrometric Analysis of Product Water from Coal Gasification,* December, NTIS PB 240-835, Pittsburgh Energy Research Center, Pittsburgh, PA (1974).

(114) Epler, J.L., Rao, T.K., and Guerin, M.R., "Evaluation of Feasibility of Mutagenic Testing of Shale Oil Products and Effluents," presented at the *US - Soviet Workshop on Health Effects of Shale Oil Development,* Denver, CO, May 18-20, 1977.

(115) Ames, B.N., McCann, J., Yamasaki, E., "Methods for Detecting Carcinogens and Mutagens with the Salmonella/Microsome Test," *Mutat. Res.,* 31:347-364 (1975).

(116) Ho, C.-h., *Chemical Composition of the Mutagenic Basic Fraction of Syncrudes,* Oak Ridge National Laboratory, Oak Ridge, TN (May 1977).

(117) Massey, Michael J., Nakles, David V., Forney, Albert J., and Haynes, William P., "Role of Gasifier Process Variables in Effluent and Product Gas Production in the Synthane Process," presented at the Environmental Aspects of Fuel Conversion Technology II Symposium sponsored by the Environmental Protection Agency, Hollywood, FL, December 15-18, 1975.

(118) Pott, P., "Cancer Scroti," *Chirurgical Observations,* Hawes Clark and Collings, London, pp 63-68 (1775).

(119) Volkman, R., *Beitrage Zur Chrirurgie,* Druck und Verlag von Breikepf und Hartel (1875).

(120) Rehn, L., *Arch. Klin. Chir.,* 50, p 88 (1895).

(121) Leuke, A.W., "Epithelioma in Carbon Workers," *Cleveland Med. J.,* 6, pp 199-202 (1907).

(122) Henry, S.A., Kennaway, N.M., Kennaway, E.L., "The Incidence of Cancer of the Bladder and Prostate in Certain Occupations," *Jour. Hyg. (Camb.),* 31, pp 199-202 (1931).

(123) Kuroda, S., and Kawahata, K., "Uber die Gewerbliche Entstehung des Lungen Krebses die Generatorgasarbeitern," *Z. Krebsforsch,* 45, pp 36-39 (1936).

(124) Kennaway, N.M., and Kennaway, E.L., "A Study of the Incidence of Cancer of the Lung and Larynx," *Jour. Hyg. (Camb.),* 36, pp 236-267 (1936).

(125) Kennaway, N.M., and Kennaway, E.L., "A Further Study of the Incidence of Cancer of the Lung and Larynx," *Brit. J. Cancer,* 1, pp 260-298 (1947).

(126) Doll, R., *Jour. Ind. Med.,* 9, p 180 (1952).

(127) Bruusgaard, A., *T. Norske, Laegeforen.,* 79, 755 (1959).

(128) Kawai, M., Amamoto, H., Harada, K., "Epidemiologic Study of Occupational Lung Cancer," *Arch. Environ. Health,* 14:859-64 (1967).

(129) Lloyd, J.W., and Clocco, A., "Long-Term Mortality Study of Steelworkers: i Methodology," *Jour. Occup. Med.,* 11, 299 (1969).

(130) Lloyd, J.W. et al, "Long-Term Mortality Study of Steelworkers: iv Mortality by Work Area," *Jour. Occup. Med.,* 12, 151 (1970).

(131) Redmond, C.K. et al, "Long-Term Mortality Study of Steelworkers: iii Follow-up," *Jour. Occup. Med.,* 11, p 513 (1969).

(132) Redmond, C.K. et al, "Long-Term Mortality Study of Steelworkers: vi Mortality from Malignant Neoplasms Among Coke Oven Workers," *Jour. Occup. Med.,* 14, p 621 (1972).

(133) Hueper, W.C., Kotin, P., Tabor, P.C., Payen, W.W., Falk, H., and Sawicki, E., "Carcinogenic Bioassays on Air Pollutants," *Arch. Pathol.,* 74:89-116 (1962).

(134) Diehl, E.K., du Breuil, F., and Glenn, R.A., "Polynuclear Hydrocarbons Emission from Coal-Fired Installations," *Transactions of the ASME Journal of Engineering for Power,* pp 276-282 (April 1967).

(135) Jones, Peter W., Giammar, Robert D., Strup, Paul E., Stanford, Thomas B., "Efficient Collection of Polycyclic Organic Compounds from Combustion Effluents," *Environmental Science & Technology,* 10(8), 806-810 (1976).

(136) Griest, W.H., Kubota, H., Holmberg, R.W., *Analysis of Fly Ash for POM Content,* October, Oak Ridge National Laboratory, Oak Ridge, TN (1976).

(137) Klein, David H., Andren, Anders W., Carter, Joel A., Emery, Jeul F., Feldman, Cyrus, Fulkerson, William, Lyon, William S., Ogle, Jack C., Talmi, Yair, Van Hook, Robert I., and Bolton, Newell, "Pathways of Thirty-Seven Trace Elements Through Coal-Fired Power Plant," *Environmental Science & Technology,* 9, 973 (1975).

(138) Kuhn, E.A., "National Energy Considerations Involving Shale Oil," *Report of the Conference-Workshop Entitled Analytical Chemistry Pertaining to Oil Shale and Shale Oil,* pp 2-7, National Science Foundation, Washington DC, June 23-25, 1974.

(139) Jensen, H., "Historical Survey of the Study of the Qualitative and Quantitative Nature of Shale Oil," *Report of the Conference-Workshop Entitled Analytical Chemistry Pertaining to Oil Shale and Shale Oil,* June 23-25, pp 20-28, National Science Foundation, Washington, DC (1974).

(140) Schmidt-Collerus, J.J., *The Disposal and Environmental Effects of Carbonaceous Solid Wastes from Commercial Oil Shale Operations,* NSF GI 3482X1, Denver Research Institute, Denver, CO (1974).

(141) *Colony Development Operation,* "An Environmental Impact Analysis for a Shale Oil Complex at Parachute Creek, CO," Part I (1974).

(142) Horton, A.W., and Denman, D.T., "Carcinogenesis of the Skin: A Re-Evaluation of Methods for the Quantitative Measurement of the Potencies of Complex Materials," *Cancer Res.,* 15, p 701 (1955).

(143) Kubota, H., Analytical Chemistry Division, Oak Ridge National Laboratory, Oak Ridge, TN, personal communication (1977).

(144) Lijinsky, W., Domsky, I., Mason, G., Ramahi, H.Y., Safavi, T., "The Chromatographic Determination of Trace Amounts of Polynuclear Hydrocarbons in Petrolatum, Mineral Oil, and Coal Tar," *Anal. Chem.,* 35(8), 952-956 (1963).

CHARACTERISTICS OF PNA IN THE ENVIRONMENT

The information in this chapter is based on:

Environmental, Health and Control Aspects of Coal Conversion: Volume 2, An Information Overview, ORNL/EIS-95, April 1977, prepared by H.M. Braunstein, E.D. Copenhaver and H.A. Pfuderer of Oak Ridge National Laboratory for Energy Research and Development under Contract W 7405-eng-26.

Critical Pathways of Polycyclic Aromatic Hydrocarbons in Aquatic Environments, NTIS CONF-790486-3, 1979, prepared by S.E. Herbes, G.R. Southworth, D.L. Shaeffer, W.H. Griest and M.P. Maskarinec of Oak Ridge National Laboratory.

Most of the current information concerning polycyclic aromatic hydrocarbons (PAH) has been generated not from coal conversion processes, but from related technologies. However, as coal conversion pilot plants continue to operate and full-scale plants are built, the literature should begin to address more of the environmental aspects of coal conversion, such as the effects of associated effluents. For example, Herbes, Southworth, and Gehrs (1) have compiled an extensive critical literature review of the environmental hazards of anticipated organic components of aqueous coal liquefaction products, presented at the recent Trace Substances in Environmental Health Conference (University of Missouri 1976).

The environmental role of PAH is of particular concern because some PAH are potential health hazards. PAH are found in small but detectable concentrations in air, water, and soil samples of all types. The concentrations found are typically small, ranging from 0.001 to 10 $\mu g/\ell$, but they accumulate in organic fatty material and can thus concentrate in the food chain (2).

BACKGROUND CONCENTRATIONS

Not all PAH are the result of man's activities. Biosynthesis of PAH by plants and microorganisms is well known; the National Academy of Sciences (3) emphasizes the importance of examining the possible contribution of vegetation to the total carcinogenic burden in the environment.

Andelman and Suess (4) cite the research of Gräf and Diehl (5) and Gräf and Nowak (6) on the synthesis and physiological functions of PAH in plants. Wheat and rye seeds containing only a trace of PAH were grown in PAH-free solutions in both the presence and absence of light. Five- to ten-day-old seedlings contained 10 to 20 μg of benzpyrene (BP) per kilogram of dried material; BP was the only PAH analyzed for in the experiments. The researchers conclude that BP is synthesized by the plants in both the presence and absence of light and thus that PAH likely are synthesized worldwide and have always existed in man's environment. Andelman and Suess agree that the evidence indicates that PAH and other hydrocarbons are naturally ubiquitous. However, Grimmer and Duvel (7) were unable to detect PAH in higher plants grown in PAH-free environments.

A planktonic member of the plant kingdom, the alga *Chlorella vulgaris*, is known to synthesize several PAH (8); algal extractions contained 10 to 50 μg per kg of PAH (9). Phytoplankton play an important role in the annual production of hydrocarbons (cycloparaffin and aromatic); Smith (10) calculates a production rate of 13 tons (87 bbl) of hydrocarbons per square mile of ocean surface. Smith used a planktonic hydrocarbon content of 2,000 ppm of the dry weight (furnished by an analysis of phytoplankton by B.H. Ketchum of the Woods Hole Oceanographic Laboratory) and a value of 6,400 tons per square mile of ocean surface for the annual production rate of phytoplankton to determine the annual hydrocarbon output by oceanic phytoplankton.

Blumer (11) studied the BP content of rural soils in Massachusetts and Connecticut (Table 3.1). It was suggested that the BP in these soils is indigenous because the rural areas sampled are distant from major highways and industries. The natural sources are thought to be pyrolytic decomposition of wood, transformation of plant organic matter to peat and lignite, and activity of soil microorganisms. These studies prompted Blumer to postulate, as did Gräf and coworkers, that hydrocarbons, including carcinogenic species, have been in the environment throughout man's entire history.

Andelman and Suess (4) report the synthesis of BP by soil bacteria in citing the research of Knorr and Schenk (12). Laboratory-cultured bacteria synthesized and accumulated BP in amounts of 2 to 6 μg per kilogram of dried material. Suess (13) reports the existence of bacterially produced carcinogenic PAH in the upper layers of the earth in concentrations of 100 to 1,000 μg/kg; these PAH leach through the soil, giving groundwater PAH concentrations of 0.001 to 0.010 μg/ℓ in the absence of anthropogenic sources.

Mallet (14) studied old sediments and found 19.5 μg/kg of benzo[a]pyrene (BaP) in a fragment from a 50-m depth. It was concluded that this hydrocarbon could be formed from phytoplanktonic lipids under the influence of anaerobic bacteria, particularly *Clostridium putride*. Mallet and Tissier (15) further investigated the biosynthesis of BP by bacteria in a series of laboratory experiments using *C. putride* and *Escherichia coli*. The initial experiment used humus from a forest soil near an industrial area. This soil was sterilized, analyzed for BP, inoculated with *C. putride*, and stored for 4 months at laboratory temperature. The initial concentration, 80 to 100 μg of BP per 100 g of soil, increased to 160 μg at the end of 4 months. A second experiment involved forest soils extracted from rural areas having an initial BP concentration of 1 to 2 μg per 100 μg of soil. One batch of this rural soil was ground, sterilized dry at 170°C, and incubated for 6 months anaerobically after inoculation with *C. putride*. The same experiment was also performed with *E. coli*. Table 3.2 shows that, after 6 months, about 4 μg of BaP was detected in soils containing both *C. putride* and *E. coli*. In another experiment, Mallet and Tissier (15) extracted from the soils what they believe to be the active substances for the formation of BaP, namely, fatty acids, and added them to soft agar inoculated with *C. putride*. After 4 months of incubation at 31°C, no trace of BaP was detected in the control tubes; however, significant but small amounts of the hydrocarbon were found in the samples inoculated with the *C. putride*.

Table 3.1: BaP in Soils

Origin and Type of Soil	Concentration (μg/kg)
Oak forest, West Falmouth, Cape Cod, Mass.	40
Pine forest, West Falmouth, Cape Cod, Mass.	40
Mixed forest, West Falmouth, Cape Cod, Mass.	1,300
Mixed forest, eastern Conn.	240
Garden soil, West Falmouth, Cape Cod, Mass.	90
Plowed field, eastern Conn.	900

Source: Reference (11) as cited in ORNL/EIS-95

Table 3.2: Soil Sterilized in the Pasteur Oven at 170°C

BP Without Addition of Bacteria (μg)	Bacteria Added	BP After Experiment (μg/100 g)
Possible traces	*Clostridium putride*	2.6
Traces	*Clostridium putride*	4.2
1.40	*Clostridium putride*	3.95
Traces	*Clostridium putride*	1.95
2	*Clostridium putride*	2.6
0.75	*Escherichia coli*	2.25
Traces	*Escherichia coli*	2.20
1	*Escherichia coli*	4.75
	Escherichia coli	5.1
	Escherichia coli	5

Source: Reference (15) as cited in ORNL/EIS-95

These investigations corroborate the results observed in marine muds, where the amounts of BP are generally considerable; BP is capable of being synthesized by anaerobic microorganisms or reduced by certain aerobic bacteria of the *Bacillus* type (15).

Other papers by Mallet et al (16)(17) also suggest the biosynthesis of BaP by anaerobic bacteria at the expense of lipids. A coworker of Mallet, Brisou (18) reports that it now seems possible to accept that:

(1) The biosynthesis of polybenzene hydrocarbons, including BaP for certain anaerobic bacteria occurs;

(2) This biosynthesis depends on only certain bacterial strains and is not a property vested in all the anaerobes; and

(3) This biosynthesis has for a starting point not only lipids (purified fatty acids), but other substrates (particularly aliphatic terpenes).

However, in a recent paper, Hase and Hites (19) found no biosynthesis of PAH by anaerobic bacteria present in New England river sediments, but they did find that the bacteria accumulated PAH from the sediments.

The PAH compounds also occur naturally as minerals (20). These relatively rare minerals, curtisite, idrialite and pendletonite, are found in association with mercury ores at scattered locations such as Shaggs Springs, Sonoma County, California and Ordejov, Czechoslovakia.

Geissman, Sim and Murdoch (1967) are cited by Blumer (20) as postulating that these minerals originated from organic matter by pyrolysis and were deposited by distillation or from solution. A pyrolytic origin is consistent with the fact that the minerals were found in association with bituminous substances and inflammable gases. The PAH and sulfur-containing composition of curtisite and idrialite is shown in Table 3.3. In contrast to the variety of PAH in these two minerals, pendletonite is almost pure coronene with only a small alkyl-derivative contribution.

Table 3.3: PAH and Sulfur-Containing Series in Curtisite and Idrialite

First Member Mass (amu)	Occurrence*	Suggested Structure	Chromatographic Position
216	CU	Benzofluorene (+ thienologs ?)	With chrysene
228	CU	Chrysene	(UV evidence)
234	(CU)	Tribenzothiophene	With chrysene
252	(CU)	4,5-10,11-dimethylenechrysene	With chrysene
254	CU	Naphthenochrysene (2 alicyclic carbons)	With alkylchrysenes
260	(CU)	Naphthenotribenzothiophene (2 alicyclic carbons)	With chrysene
266	*CU*, ID	Dibenzofluorene (+ thienologs ?)	With picene
276	(CU)	Anthanthrene	After picene (UV: not benzoperylene)
278	*CU*, ID	Picene	(UV evidence)
284	CU, ID	Tetrabenzothiophene	Before picene
296	CU	Dinaphthenobenzofluorene (3 alicyclic C/ring)	With benzofluorene
300	(CU)	Coronene	With anthanthrene

(continued)

Table 3.3: (continued)

First Member Mass (amu)	Occurrence*	Suggested Structure	Chromatographic Position
302	(CU)	Dibenzofluoranthene, dimethylenepicene	After picene
304	*ID*, CU	Naphthenopicene (2 alicyclic C)	With picene
310	(ID)	Naphthenotetrabenzothiophene (2 alicyclic C)	With tetrabenzothiophene
316	*ID*, (CU)	Tribenzofluorene	After picene
322	ID	Dinaphthenotetrabenzothiophene (3 alicyclic C)	With picene/tetrabenzothiophene
328	ID	Benzopicene	After picene, with tribenzofluorene
334	ID	Pentabenzothiophene	With alkylpicenes
340	(ID)	Tetranaphthenopicene (5 alicyclic C)	With alkylpicenes
354	ID	Naphthenobenzopicene (2 alicyclic C)	With benzopicene
366	(ID)	Tetrabenzofluorene	Last PAH fraction
372	(ID)	Dinaphthenopentabenzothiophene (3 alicyclic C)	After picene
378	(ID)	Dibenzopicene	Last PAH fraction

*CU is curtisite; ID is idrialite. Major series in italics; minor series in parentheses.

Source: Reference (20) as cited in ORNL/EIS-95

More recently, Blumer (21) proposed mechanisms for the geosynthesis of aromatic hydrocarbon minerals such as idrialite, curtisite and pendletonite. They arise from sediments containing organic compounds that are located deeper in the earth than those regions in which petroleum is formed and in which pyrolytic temperatures may reach 500°C. Although Blumer's mineral specimens came from museum collections, it was believed that they were representative of the original composition in situ since no contamination peaks were detected by mass spectrometry (20).

Volcanic activity and forest fires are also sources of background PAH (13). Forest fires especially can be important in determining background levels, particularly in the atmosphere. Also, Blumer and Youngblood (22) and Youngblood and Blumer (23) suggest that PAH compounds in soils and recent sediments are formed largely from natural forest and prairie fires. This suggestion is based on the high degree of similarity in the molecular weight distribution of the many series of alkyl homologs in the PAH compounds, and the fact that this distribution exhibits little variance over a wide range of depositional environments. Table 3.6 (in the next section) indicates that 140 tons of BP are emitted each year as a result of the open burning of forest and agricultural land as compared with the 192 tons emitted per year by coke production.

Hydrocarbons, including PAH, are ubiquitous in the environment, and their presence cannot be attributed exclusively to man's activities.

EXISTING CONCENTRATIONS

Existing concentrations of PAH include those from natural and anthropogenic sources. These sources release PAH compounds into the water environment (24) (Table 3.4) and atmosphere (25) (Table 3.5). Table 3.4 indicates the PAH compounds commonly found in water along with information about their relative carcinogenicity; Table 3.5 lists polycyclic compounds found in air, those that occur in cigarette smoke and exhaust gas and those that are associated with coal.

Table 3.4: PAH Commonly Found in Water

IUPAC Name*	Earlier Name**	Mol Wt	Relative Carcino-genicity***	Abbreviation
Anthracene	Anthracene	178	?	An
Benzo[a] anthracene	1,2-Benzanthracene	228	+	BaA
Benzo[b] fluoranthene	3,4-Benzfluoranthene	252	++	BbF
Benzo[j] fluoranthene	10,11-Benzfluoranthene	252	++	BjF
Benzo[k] fluoranthene	11,12-Benzfluoranthene	252	–	BkF
Benzo[a] pyrene	3,4-Benzpyrene	252	+++	BaP
Benzo[e] pyrene	1,2-Benzpyrene	252	+	BeP
Benzo[ghi] perylene	1,12-Benzperylene	276	–	BghiP
Chrysene	Chrysene	228	+	Ch
Fluoranthene	Fluoranthene	202	–	Fl
Indeno[1,2,3-cd] pyrene	2,3-o-Phenylenepyrene	276	+	IP
Phenanthrene	Phenanthrene	178	?	Ph
Perylene	Perylene	252	–	Per
Pyrene	Pyrene	202	–	Pyr

*IUPAC 1957 rules. 1960. *J. Am. Chem. Soc.* 82:5545-84.
**Pre-1957 naming.
***Relative activity on mouse epidermis: +++, active; ++, moderate; +, weak; ?, unknown; –, inactive.

Source: Reference (24) as cited in ORNL/EIS-95

Table 3.5: Polycyclic Compounds Found in Air, Cigarette Smoke, and Exhaust Gases

Compound*	Occurrence**
Naphthalene	A,T
2-Methylnaphthalene	A,T
Alkylnaphthalenes	A
Azulene	A,T
Acenaphthene	A,T
Acenaphthylene	A,T
Dibenzofuran	A
Carbazole	A
Dibenzothiophene	A
Fluorene	A,T
Anthracene (A)	A,T
Phenanthrene	A,T
2-Methylanthracene	T
11H-Benzo[b] fluorene	A

(continued)

Table 3.5: (continued)

Compound*	Occurrence**
11H-Benzo[a] fluorene	A,T
7H-Benzo[c] fluorene	A
Fluoranthene (Fluor)	A,T
8-Methylfluoranthene	T
Alkyl fluoranthene	T
Naphthacene	G
Benz[a] anthracene (BaA)***,†	A,T
Chrysene (Ch)***,†	A,T
Alkyl chrysene	T
Benzo[c] phenanthrene***	T
Pyrene (P)	A,T,G,D
1-Methylpyrene	A,T
4-Methylpyrene	T
2,7-Dimethylpyrene	A
Naphtho[2,1,8,7-klmn] xanthene	T
10,11-Dihydro-9-H-benzo[a] cyclopent[i] anthracene	T
2,3-Dihydro-1-H-benzo[a] cyclopent[h] anthracene	T
7H-Dibenzo[c,g] carbazole***	T
Benzo[b] fluoranthene***	A,T,G,D
Benzo[ghi] fluoranthene	A,T
Benzo[j] fluoranthene***,†	A,T
Benzo[k] fluoranthene (BkF)	A,T,G,D
2-Methylfluoranthene***	T
Methylfluoranthene	T
Benzo[a] naphthacene	T
Dibenzo[b,h] phenanthrene	T,G,D
Dibenz[a,h] anthracene***,†	T
Benzo[a] pyrene (BaP)***,†	A,T,G,D
Methylbenzo[a] pyrene	T
Hydroxybenzo[a] pyrene	T
Benzo[a] pyrenequinone	A
Benzo[e] pyrene (BeP)***,†	A,T
Perylene (Per)	A,T
Dibenzo[a,l] naphthacene	A,G,D
Dibenzo[a,j] naphthacene	T
Naphtho[2,1,8-qra] naphthacene (Naphtho[2,3-a] pyrene)	T
Phenalen-1-one	A
Dibenzo[a,i] pyrene***,†	A,T
Dibenzo[a,e] pyrene	T,G,D
Dibenzo[cd,jk] pyrene (anthanthrene)	A,T,G,D
Dibenzo[cd,jk] pyrene-6,12-dione (anthanthrone)	A
Benzo[ghi] perylene (BghiP)	A,T,G,D
Dibenzo[b,pqr] perylene	G
Coronene (Cor)	A,T
Dibenzo[a,h] pyrene***,†	A,T
Tribenzo[a,i] fluorene	T
13H-Dibenzo[a,i] fluorene	T
Dibenzo[a,c] naphthacene	T
Benzo[h] quinoline	A
Ra-Benzo[h] quinoline††	A
Rb-Benzo[h] quinoline††	A
Benz[c] acridine†††	A,G
Ra-Benz[c] acridine††	A

(continued)

Table 3.5: (continued)

Compound*	Occurrence**
Rb-Benz[c]acridine††	A
Dibenz[a,h]acridine***,†††	A,T
Indeno[1,2,3-ij]isoquinoline	A,G
Phenanthridine	A
11H-Indeno[1,2-b]quinoline	A
Acridine	A
Benzo[f]quinoline††	A
Ra-Benzo[f]quinoline††	A
Rb-Benzo[f]quinoline††	A
Benz[a]acridine††	A
Rb-Benz[a]acridine††	A
Dibenz[a,j]acridine***,†††	A,T
Rb-Dibenz[a,j]acridine	A
7H-Benz[de]anthracen-7-one	A
Indeno[1,2,3-cd]pyrene***,†	A
Dibenz[e,l]pyrene***	A
Xanthene-9-one	A
Dibenz[a,i]acridine	A

*Abbreviations used for most common compounds in parentheses
**A, air; T, tobacco smoke; G, gasoline exhaust; D, diesel exhaust
***Reported by Public Health Services to have carcinogenic activity
†Coal-associated PAH
††R, alkyl groups, substituted alkyl groups; Ra and Rb, various substitutes
†††Coal-associated heterocyclics

Source: Reference (25) as cited in ORNL/EIS-95

Air

The yearly global emission of BP, from data for 1966–1969, is estimated by Suess (13) to be about 5,000 tons, the greatest contribution coming from coal combustion (Table 3.6); U.S. yearly emission of BP was estimated to be about 25% of the worldwide total, or 1,283 tons. In comparison, Olsen and Haynes (26) estimate the U.S. annual BaP emissions to be 481 tons (Table 3.7) by assuming an annual consumption and emission rate that was derived from a large number of sources.

Measured average annual airborne BaP concentrations for 1966–1970 in urban areas in the United States are reported in Table 3.8. Altoona, Pennsylvania appears to have the highest recorded value (29.5 ng/m^3 in 1967), followed by Chattanooga, Tennessee (22.9 ng/m^3 in 1967). The U.S. average BaP concentration for the five-year period is about 2.0 ng/m^3.

Table 3.9 lists nonurban annual average BaP concentrations for 1966–1970. As expected, these values are generally an order of magnitude less than those for the urban concentration in Table 3.8. Clarion County, Pennsylvania, which had the highest values for each of the 5 years (1.5, 2.1, 1.0, 1.2 and 1.2 ng/m^3), compares favorably with most urban sites, but the values are only a fraction of the highest values recorded at some of the Pennsylvania urban sampling stations (e.g., Pittsburgh, 13.8 ng/m^3 in 1969).

Table 3.6: Estimated BaP Emissions to the Atmosphere

Sources of Formation	BaP Emission (tons/yr) U.S.A.	Worldwide*	Worldwide
Heating and power generation, using			
Coal	431	1,945	2,376
Oil	2	3	5
Gas	2	1	3
Wood	40	180	220
Subtotal	475	2,129	2,604
Industrial processes			
Coke production	192	841	1,033
Catalytic cracking	6	6	12
Subtotal	198	847	1,045
Refuse and open burning			
Enclosed incineration			
Commercial and industrial	23	46	69
Other	11	22	33
Open burning			
Coal refuse fires	340	340	680
Forest and agriculture	140	280	420
Other	74	74	148
Subtotal	588	762	1,350
Vehicles			
Trucks and buses	12	17	29
Automobiles	10	6	16
Subtotal	22	23	45
Grand total	1,283	3,761	5,044

*Excluding U.S.A.

Source: Reference (13) as cited in ORNL/EIS-95

Table 3.7: Estimated Annual BaP Emissions for the United States

Source	Estimated BaP Emission Rate	Estimated Annual Consumption or Production	Estimated Annual BaP Emission
Heat Generation			
	(μg/10^6 Btu)	(10^{15} Btu)	(tons)
Coal			
Residential			
Hand-stoked	1,400,000	0.26	400
Underfeed	44,000	0.20	9.7
Commercial	5,000	0.51	2.8
Industrial	2,700	1.95	5.8
Electrical	90	6.19	0.6
Oil	200	6.79	1.5
Gas	100	10.57	1.2
Total			421.6

(continued)

Table 3.7: (continued)

Source	Estimated BaP Emission Rate	Estimated Annual Consumption or Production	Estimated Annual BaP Emission
Refuse Burning			
	(μg/ton)	(10^6 tons)	(tons)
Incineration			
Municipal	5,300	18	0.1
Commercial	310,000	14	4.8
Open burning			
Municipal refuse	310,000	14	4.8
Grass, leaves	310,000	14	4.8
Auto components	26,000,000	0.20	5.7
Total			20.2
Industry			
	(μg/bbl)	(10^6 bbl)	(tons)
Petroleum catalytic cracking*			
Fluid catalytic cracker			
No CO boiler**	240	790	0.21
With CO boiler	14	790	0.012
Houdriflow catalytic cracker			
No CO boiler	218,000	23.3	5.6
With CO boiler	45	43.3	0.0024
Thermo for catalytic cracker***			
No CO boiler	90,000	131	13.0
With CO boiler	<45	59	<0.0029
Thermo for catalytic cracker†			
No CO boiler	–	119	0.0041
With CO boiler	<31	0	0
Asphalt road mix	50 μg/ton	187,000 tons	0.000010
Asphalt air blowing	<10,000 μg/ton	4,400 tons	<0.000048
Carbon-black manufacturing	††	††	††
Steel and coke manufacturing	††	††	††
Chemical complex	††	††	††
Total			18.8
Motor Vehicles			
	(μg/gal)	(10^{10} gal)	(tons)
Gasoline			
Automobiles	170	4.61	8.6
Trucks	>460	2.01	>10
Diesel	690	0.257	2.0
Total			>20.6
Total (all sources tested)			481

*Catalyst regeneration.
**CO boiler—carbon monoxide waste heat boiler.
***Air lift.
†Bucket lift.
††Atmospheric samples indicate that BaP emissions from these processes are not extremely high.

Table 3.8: Annual Average Ambient BaP Concentrations at National Air Sampling Network Urban Stations (ng/m^3)

Station	1966	1967	1968	1969	1970
Alabama					
Birmingham	18.5	–	–	–	–
Gadsden	3.5	–	2.4	1.8	2.5
Huntsville	–	3.1	2.7	1.8	1.6
Mobile	6.5	–	4.2	2.6	–
Montgomery	–	2.3	2.9	2.0	1.3
Alaska					
Anchorage	2.3	1.9	1.7	1.3	0.8
Arizona					
Phoenix	1.7	2.5	2.1	2.2	–
Tucson	0.6	0.7	0.7	0.5	0.4
Arkansas					
Little Rock	1.2	0.9	0.9	1.1	0.7
West Memphis	1.1	2.2	2.2	2.4	0.6
California					
Burbank	2.5	–	–	2.9	1.9
Glendale	–	1.0	1.6	1.6	1.0
Long Beach	–	2.1	2.1	2.3	1.0
Los Angeles	2.1	1.3	1.8	1.9	1.2
Oakland	2.7	1.7	1.6	1.6	1.0
Ontario	–	–	0.9	0.6	0.6
Pasadena	1.8	–	2.3	–	0.7
Riverside	–	–	1.3	0.8	0.7
Sacramento	–	–	1.4	1.8	0.7
San Bernardino	–	–	1.0	0.9	0.8
San Diego	1.7	1.6	1.2	1.4	0.7
San Francisco	1.1	1.5	1.8	1.2	0.6
Colorado					
Denver	2.3	2.4	2.3	2.5	2.2
Connecticut					
Hartford	2.3	2.1	1.4	2.0	1.4
New Haven	3.5	1.9	1.4	2.1	1.2
Delaware					
Newark	1.0	1.4	0.9	–	0.4
Wilmington	2.2	2.7	1.9	1.7	1.1
District of Columbia	2.4	1.9	1.9	4.3	–
Florida					
Jacksonville	–	–	2.9	2.3	1.4
Tampa	–	–	1.5	1.0	0.5
Georgia					
Atlanta	1.4	3.0	1.8	1.9	0.9
Hawaii					
Honolulu	0.2	0.5	0.6	0.6	0.2
Idaho					
Boise City	3.5	2.4	2.0	6.0	1.1
Illinois					
Chicago	3.3	3.0	3.1	3.9	2.0
Springfield	–	–	1.1	1.3	0.9
Indiana					
East Chicago	6.8	5.7	1.9	6.8	5.3
Hammond	3.9	2.5	2.1	3.3	1.7
Indianapolis	10.4	5.7	4.1	5.2	2.3
Muncie	2.4	1.6	–	–	–
New Albany	5.4	–	–	4.3	3.7
South Bend	2.2	2.6	3.7	3.7	2.4
Terre Haute	–	3.7	–	4.0	2.8
Iowa					
Davenport	3.2	–	–	1.7	0.9
Des Moines	2.5	2.7	1.1	0.9	0.7
Cedar Rapids	–	0.8	0.7	–	0.3
Kansas					
Kansas City	1.2	–	1.2	1.1	2.4
Topeka	–	0.5	0.7	0.4	0.3
Wichita	0.8	0.5	1.0	0.7	0.5

(continued)

Table 3.8: (continued)

Station	1966	1967	1968	1969	1970
Kentucky					
Ashland	10.5	–	9.3	10.9	6.7
Covington	3.1	1.9	3.6	4.1	4.4
Lexington	–	1.8	3.0	–	1.6
Louisville	2.5	2.1	2.7	1.9	–
Louisiana					
New Orleans	2.3	1.8	1.6	1.5	1.1
Maine					
Portland	–	–	2.3	–	1.1
Maryland					
Baltimore	2.8	3.8	2.3	2.8	2.1
Massachusetts					
Worcester	–	–	1.7	1.5	1.6
Michigan					
Detroit	4.7	5.4	5.1	3.9	2.6
Flint	–	1.4	0.8	1.7	1.5
Grand Rapids	–	2.8	3.4	1.7	0.9
Trenton	–	–	1.4	1.6	0.8
Minnesota					
Duluth	2.2	–	2.7	2.1	1.1
Minneapolis	1.6	1.3	1.1	1.4	0.6
Moorhead	0.7	–	0.9	1.0	1.6
St. Paul	1.8	2.3	1.8	1.8	1.0
Missouri					
Kansas City	–	–	1.8	1.6	1.1
St. Louis	–	2.3	–	3.3	–
Montana					
Helena	–	0.8	0.9	0.5	–
Nebraska					
Omaha	2.7	1.3	1.9	1.6	1.0
Nevada					
Las Vegas	1.3	1.1	1.4	–	–
Reno	–	4.6	3.1	–	–
New Hampshire					
Concord	0.6	1.5	1.0	0.7	0.6
New Jersey					
Camden	3.0	–	1.6	2.4	1.9
Glassboro	0.7	0.8	1.2	1.1	1.2
Jersey City	4.2	3.5	2.3	2.7	4.7
Marlton	1.2	1.6	1.3	–	1.4
Newark	2.1	3.3	2.1	1.8	1.5
Paterson	–	1.9	2.0	1.2	1.2
Perth Amboy	2.1	2.1	1.2	1.2	1.0
Trenton	2.2	–	1.0	1.5	1.1
New Mexico					
Albuquerque	2.0	1.9	1.8	1.1	1.1
New York					
New York City	4.1	3.9	–	3.6	3.0
North Carolina					
Charlotte	5.7	6.3	5.6	4.9	1.9
Durham	–	–	8.0	3.4	3.9
North Dakota					
Bismarck	–	–	0.9	1.0	0.4
Ohio					
Akron	4.1	3.7	3.0	–	–
Cincinnati	3.6	1.9	1.8	2.9	2.6
Cleveland	3.1	2.9	3.0	3.8	2.8
Columbus	2.9	1.7	2.2	2.7	1.6
Dayton	2.7	3.7	2.4	1.9	1.5
Toledo	1.8	1.9	1.8	1.5	1.4
Youngstown	7.3	8.2	5.6	9.9	7.1
Oklahoma					
Oklahoma City	1.5	0.7	0.7	0.7	0.9
Tulsa	0.7	0.6	0.8	0.5	0.8

(continued)

Table 3.8: (continued)

Station	1966	1967	1968	1969	1970
Oregon					
Eugene	–	2.4	–	–	–
Medford	–	4.8	8.2	4.1	–
Portland	3.3	3.5	4.1	2.6	2.3
Pennsylvania					
Allentown	–	1.8	1.2	1.9	2.4
Altoona	–	29.5	18.0	22.3	19.3
Bethlehem	–	2.9	2.1	2.0	2.7
Harrisburg	–	–	1.3	1.5	1.5
Lancaster	2.3	–	–	–	–
Philadelphia	3.8	5.9	2.9	4.0	2.4
Pittsburgh	4.9	7.0	6.3	13.8	5.9
Reading	2.3	2.9	2.4	1.8	1.6
Scranton	–	5.2	6.1	7.7	2.9
Warminster	0.9	2.2	0.9	1.0	–
West Chester	–	1.1	1.0	1.3	–
Wilkes Barre	–	–	1.6	1.5	1.3
York	–	1.8	1.9	2.0	1.2
Rhode Island					
East Providence	–	1.6	1.2	1.2	1.2
Providence	3.6	2.8	2.0	2.2	2.1
South Carolina					
Columbia	–	4.2	6.2	1.3	–
Greenville	5.0	–	18.6	7.0	3.4
Tennessee					
Chattanooga	8.4	22.9	7.4	4.2	5.5
Knoxville	–	7.0	9.8	4.7	–
Memphis	1.7	1.6	1.3	0.7	1.4
Nashville	5.5	7.0	6.0	2.8	2.6
Texas					
Dallas	1.4	–	–	2.0	1.9
Houston	0.9	–	–	–	1.2
San Antonio	0.6	1.4	0.9	0.6	1.0
Utah					
Ogden	0.5	–	0.8	0.7	2.5
Salt Lake City	1.2	0.7	1.0	0.7	1.4
Vermont					
Burlington	0.8	–	0.7	0.5	0.7
Virginia					
Danville	3.2	–	2.5	1.8	2.7
Hampton	–	2.2	1.5	0.9	1.1
Lynchburg	–	9.2	8.7	6.3	4.5
Norfolk	2.8	3.5	4.9	3.9	1.8
Portsmouth	–	7.7	10.2	3.4	4.9
Richmond	–	5.2	–	2.2	2.1
Roanoke	–	7.5	7.7	5.3	6.2
Washington					
Seattle	2.7	1.8	2.0	1.6	1.5
West Virginia					
Charleston	3.4	–	4.6	2.6	2.1
Wisconsin					
Kenosha	–	–	1.4	1.7	1.3
Madison	–	–	1.3	–	1.1
Milwaukee	4.1	–	4.7	4.0	2.5
Superior	–	–	3.3	1.6	1.5
Wyoming					
Casper	–	–	0.9	0.6	0.4
Cheyenne	0.5	–	0.6	0.5	0.4

Source: Reference (25) as cited in ORNL/EIS-95

Table 3.9: Annual Average Ambient BaP Concentrations at National Air Sampling Network Nonurban Stations (ng/m^3)

Station	1966	1967	1968	1969	1970
Arizona					
Grand Canyon	0.3	0.2	0.2	0.2	0.1
Maricopa County	-	0.2	0.5	0.3	0.3
Arkansas					
Montgomery County	0.3	0.1	0.2	0.2	0.1
California					
Humboldt County	0.4	0.4	0.3	0.5	0.1
Idaho					
Butte County	-	-	0.2	0.1	0.1
Indiana					
Monroe County	0.5	-	0.5	0.3	0.2
Parke County	0.9	-	0.4	0.3	0.4
Maine					
Acadia National Park	0.2	-	0.3	0.1	0.2
Missouri					
Shannon County	-	0.2	0.2	0.2	0.2
Montana					
Glacier National Park	-	0.3	0.4	0.4	-
Nebraska					
Thomas County	0.2	-	0.2	0.1	0.1
Nevada					
White Pine County	0.1	-	0.1	0.1	0.1
New Hampshire					
Coos County	0.2	0.2	0.2	0.1	0.1
New York					
Jefferson County	0.2	-	0.2	0.3	0.2
North Carolina					
Cape Hatteras	0.2	-	0.2	0.1	0.2
Oklahoma					
Cherokee County	0.2	0.2	0.2	0.2	0.2
Oregon					
Curry County	0.1	1.1	0.1	0.1	0.1
Pennsylvania					
Clarion County	1.5	2.1	1.0	1.2	1.2
Texas					
Matagorda County	0.3	0.1	0.2	0.1	0.3
Vermont					
Orange County	0.9	-	0.3	0.3	0.2
Virginia					
Shenandoah National Park	0.9	0.3	0.3	0.3	0.2

Source: Reference (25) as cited in ORNL/EIS-95

Table 3.10 demonstrates the seasonal effect on BaP concentrations in urban air; concentrations are higher in winter than in summer, the difference often being an order of magnitude.

Shabad (27) cites Yanysheva and Balenko (1966) in reporting the safe lifetime BP dose for the human lungs as 4.3 mg. On the basis of this value, the concentration of atmospheric BP should not exceed 120 ng/m^3. A recent standard of BaP concentration (28) for industrial workers was determined to be 200 ng/m^3, about twice as great as the suggested atmospheric value. The highest recorded BaP value in the United States for 1966-1970 (29.5 ng/m^3 in 1967 for Altoona, Pennsylvania, Table 3.8) is almost an order of magnitude less than the proposed industrial standard and fourfold less than the atmospheric standard.

Table 3.10: Seasonal Effect on the BaP Concentrations of Various Urban Atmospheres

Location	Year	Concentrations (μg/m^3 of air) Summer (low)	Winter (high)
Atlanta, GA	1958	0.0016	0.015
	1960	0.0009	0.014
Birmingham, AL	1958	0.006	0.074
	1960	0.003	0.062
Cincinnati, OH	1958	0.002	0.026
	1960	0.0012	0.018
Detroit, MI	1958	0.0034	0.031
Location I	1961	0.0072	0.017
Location II	1961	0.0036	0.0137
Location III	1963	0.0002	0.0018
Los Angeles, CA	1958	0.0004	0.013
Nashville, TN	1958	0.0014	0.055
New Orleans, LA	1958	0.0020	0.006
	1960	0.0006	0.007
New York, NY			
Herald Square	1963-64	0.0005	0.0094
Columbus Circle	1964	0.0007	0.0026
Philadelphia, PA	1958	0.0025	0.019
San Francisco, CA	1958	0.0003	0.0075
South Charleston, WV	1960	0.0006	0.012

Source: Reference (26) as cited in ORNL/EIS-95

Water

According to the NAS report (29), about 6 million tons of petroleum hydrocarbons enter the oceans annually; the major contributors are marine transportation and runoff (urban and river), yielding 2.1 and 1.9 million tons per year respectively (Table 3.11). Coastal refineries and industrial and domestic waste together were estimated to contribute 0.8 million tons per year; natural seeps and atmospheric fallout each add another 0.6 million tons per year.

Table 3.12 shows the concentration of BP in marine sediments, primarily off the Mediterranean coast. These sediments are muds, sands, and shells; there is no easily discernible correlation between sediment, depth of sample, or BP concentration.

Concentrations of PAH in surface waters, mainly German rivers, are compiled in Table 3.13 (multiple values correspond to different sources of the same information). The average total PAH concentration is about 1.2 μg/ℓ; the highest values (12 μg/ℓ) are recorded in samples taken at the discharge site of shale oil and coke by-product effluent. However, 500 to 3,500 m downstream of the effluent discharge, concentrations are reduced to 1-2 μg/ℓ. The PAH levels in Thames

River water are comparable with those in the German rivers (24).

Table 3.11: Comparison of Estimates for Petroleum Hydrocarbons Annually Entering the Ocean, 1969-1971

	Authority (millions of tons per year)		
Source	MIT SCEP Report* (1970)	USCG Impact Statement** (1973)	NAS Workshop (1973)
Marine transportation	1.13	1.72	2.133
Offshore oil production	0.20	0.12	0.08
Coastal oil refineries	0.30		0.2
Industrial waste		1.98	0.3
Municipal waste	0.45		0.3
Urban runoff			0.3
River runoff***			1.6
Subtotal	2.08	3.82	4.913
Natural seeps			0.6
Atmospheric rainout	9.0†		0.6
Total	11.08		6.113

*From Study of Critical Environmental Probe (SCEP). 1970. *Man's impact on the global environment. Assessment and recommendations for action*. Cambridge, MA. MIT Press.

**From U.S. Coast Guard. 1973. *Draft environmental impact statement*. For International Convention for the Prevention of Pollution from Ships, 1973.

***Input from recreational boating assumed to be incorporated in the river runoff value.

†Based upon assumed 10% return from the atmosphere.

Source: Reference (29) as cited in ORNL/EIS-95

Table 3.12: Concentration of BP in Marine Sediments

Source	Sample	Depth (m)	BP Concentration (μg/kg of dry sample)
Greenland, West Coast	Sand	0.20	5
Italy, Bay of Naples	Mud, sand, shell*	15-45	1,000-3,000
(6 locations)	Sand, shell	13	7.5
	Mud, sand	2-65	10-530
	Mud, sand, shell**	55	260-960
	Muddy sand	120	1.4
	Mud, sand***		100-560
French Mediterranean Coast		0-0.03	1,800
		0.03-0.08	3,600
		0.08-0.13	5,000
		0.13-0.18	2,500
		0.23-0.28	2,200
		0.33-0.38	730
		0.48-0.53	420
		1.00	26
		2.00	16
	Sand	14	400
	Black mud	16	1,500
	Sand	48	75
	Sand	53	Traces
	Beige mud	82	400

(continued)

Table 3.12: (continued)

Source	Sample	Depth (m)	BP Concentration (μg/kg of dry sample)
Estuary, French Mediterranean	Sand		34
	Sand	1	20
	Sand	4	15
	Sand	5	25
	Mud	102	Not detected
French Mediterranean Coast			50
French Mediterranean Coast			20
French Channel Coast	Mud		15,000
French Channel and Atlantic Coasts (11 locations)	Mud, sand		Not detected to 1,700

*Five samples were collected 300 m from shore, area is highly industrialized.
**Four samples were collected in the vicinity of volcanic pollution.
***Four samples; this island is affected by pollution.

Source: Reference (4) as cited in ORNL/EIS-95

Table 3.13: Concentration of PAH in Surface Waters

Source	Concentration (μg/ℓ) BP	Carcinogenic PAH	Total PAH
Bodensee		0.003	
		0.0004*	
	0.0013	0.030	0.065
Alprhine		0.005*	
River Rhine		0.050-0.500**	
River Rhine at Mainz		0.080*	
River Rhine at Mainz			
(March 1964)	0.049	0.240	0.73
(March 1964)	0.114	0.730	1.50
River Main, at Seligenstadt	0.0024	0.155	0.48
River Danube, at Ulm			
(April 1964)	0.0006	0.055	0.28
(May 1964)		0.078	0.20
River Gersprenz, at Munster			
(January 1964)	0.0096	0.055	0.16
(April 1964)		0.038	0.12
River Aach, at Stockach*		0.050	
River Aach, at Stockach*			
Sample I	0.043	1.30	3.0
Sample II	0.016	0.90	2.5
Sample III	0.004	0.50	1.4
Sample IV	0.005	1.10	3.1
River Schussen (Bodensee)		0.50*	
River Schussen	0.01	0.20	1.0
River Argen (Bodensee)		0.07*	
River Seine	***		
River Plyussa			
At discharge site of shale-oil effluent	12		

(continued)

Table 3.13: (continued)

Source	BP	Concentration (µg/ℓ) Carcinogenic PAH	Total PAH
3,500 m downstream	1		
At the water intake of Narva	0.1		
A river			
15 m below discharge of coke by-product effluent	8-12		
500 m downstream	2-3		
Peat (turf) water	0.05		

*Extrapolated from centrifugate fractions.
**Extrapolated from activated-carbon adsorption analyses.
***Considerable amount.

Source: Reference (4) as cited in ORNL/EIS-95

Concentrations of PAH have been detected in wastewater samples (Table 3.14). In many instances, domestic effluents have higher PAH concentrations than do factory effluents. The PAH levels in domestic sewage increase by 50- to 1,000-fold during periods of heavy rain (Table 3.15) due to increased runoff, primarily from highways (9). Table 3.16 shows the PAH content of road dust and a comparison of it with the atmospheric dust of urban air in Dortmund, a large German city. Many of the PAH concentrations in road dust are more than 1,000-fold those levels found in soil (Table 3.17). Also, the PAH concentrations in road dust are shown to be several times greater than those in the atmospheric dust of Dortmund. Therefore, it is evident why runoff from highways causes a dramatic increase in the PAH concentration found in sewage during a heavy rain as opposed to PAH sewage levels during dry weather (Table 3.15).

Table 3.14: PAH in Various Wastewater Samples (ng/ℓ)

PAH	Effluents Domestic	Effluents Factory	Domestic	Sewage*	
Fluoranthene	2,416	2,198	273	3,420	2,660
Pyrene	1,763	1,957	–	3,120	2,560
Benzo[a]anthracene	319	167	191	1,360	343
Benzo[b]fluoranthene	202	114	36	870	525
Benzo[j]fluoranthene	205	45	37	1,740	1,100
Benzo[k]fluoranthene	193	32	31	460	336
Benzo[a]pyrene	74	100	38	100	368
Benzo[ghi]perylene	219	73	40	480	120
Indeno[1,2,3-cd]pyrene	238	57	22	930	476

*High percentage industry.

Source: References (9)(24) as cited in ORNL/EIS-95

Table 3.15: PAH in Domestic Sewage During Dry and Wet Weather (ng/ℓ)

PAH	Concentration in Sewage During Dry Weather	Concentration in Sewage During Heavy Rain
Fluoranthene	352	16,350
Pyrene	254	16,050
Benzo[a]anthracene	25	10,360
Benzo[b]fluoranthene	39	9,910
Benzo[j]fluoranthene	57	10,790
Benzo[k]fluoranthene	22	4,180
Benzo[a]pyrene	1	1,840
Benzo[ghi]perylene	4	3,840
Indeno[1,2,3-cd]pyrene	17	4,980

Source: Reference (9)(24) as cited in ORNL/EIS-95

Table 3.16: PAH in Road Dust (mg/kg)

Substances Detected	Dust from Federal Road 31: 1	2	3	4	5	Atmospheric Dust Dortmund
Fluoranthene	126	166	78	37	32	3.8
Pyrene	114	147	62	–	27	2.77
Benzo[a]anthracene	21	59	19	17	16	0.53
Benzo[j]fluoranthene	60	97	36	14	6	0.62
Benzo[b]fluoranthene	26	62	19	18	15	0.16
Chrysene	22	64	–	–	–	–
Benzo[a]pyrene	9	17	12	2	3	0.05
Benzo[ghi]perylene	22	47	21	6	12	0.51
Benzo[k]fluoranthene	13	26	13	6	8	0.48
Indeno[1,2,3-cd]pyrene	24	61	20	8	9	0.96
Known compounds	437	746	280	108	128	9.88
Unidentified compounds	400	800	22	17	18	0.20
PAH total	837	1,546	302	125	146	10.08
Carcinogenic PAH	162	360	106	59	49	2.31
% fraction of carcinogens of the total amount	19	23	35	47	33	23
Number of substances isolated	32	32	22	20	20	18

Source: Reference (9)(24) as cited in ORNL/EIS-95

Table 3.17: Concentration of PAH Found in Forest Soil (μg/kg)

PAH	South of Darmstadt: Mixed Woods	Beech Woods	Oak Woods	Near Lake Constance: Spruce Woods	Mixed Woods	Beech Woods
Benzo[a]pyrene	2.5	4.0	1.5	1.5	2.5	1.5
Benzo[ghi]perylene	10	70	20	10	20	10
Benzo[b]fluoranthene and benzo[j]fluoranthene	30	110	50	35	25	25

Source: Reference (9) as cited in ORNL/EIS-95

Soil

Concentrations of PAH in forest soil are listed in Table 3.17. The soil samples were taken from beech, oak, and mixed woods; in one location, soils from beech woods had the highest PAH concentrations, but in the other, soils from mixed woods had the highest levels. Thus, no apparent correlation between PAH concentration in soil and ground cover can be detected from this table.

FATE IN NATURAL WATER

The processes which affect PAH in the aquatic environment may be considered as two general types: those which result in transport, and those which result in transformation. Transport processes are those in which only the physicochemical form is altered, while transformational processes result in alteration to a different chemical species. As such, transport processes are driven by chemical equilibrium considerations, while transformational processes are essentially nonequilibrium. A second general distinction among transport and transformation processes is whether or not they are mediated by living organisms (biotic vs abiotic). Major transport processes in the aquatic environment are summarized in Table 3.18.

Table 3.18: Major Processes Influencing PAH Behavior in Natural Waters

	Transport	Transformation
Abiotic	Dispersion Sorption Volatilization	Oxidation Photolysis
Biotic	Bioaccumulation	Metabolism Microbial degradation

Source: NTIS CONF-790486-3

In natural stream systems, PAH will be transported to the ocean, decomposed, or removed. Removal can consist of consumption by biota, adsorption on biological material, and/or adsorption on suspended or bottom mineral matter (2). Decomposition depends on biological metabolism or chemical oxidation, whereas physical transport usually occurs in the absence of the other mechanisms. Transient storage compartments for hydrocarbons include water and organisms, whereas sediments probably constitute the most significant long-term storage site (Figure 3.1). Losses from an aqueous system include evaporation, tidal and riverine flushing, and microbiological decomposition (30).

Marine

In marine environments, pollutants are transported and distributed by currents, waves, and turbulent mixing. However, present information is insufficient to predict accurately either the complete physical behavior or the entire biological fate of a pollutant in oceanic deep waters (31). For example, the fate of a pollutant may be determined as often by the microbial environments as by the chemical and physical surroundings. Many pollutants, including the components of crude oil, may be degraded by the action of bacterial yeasts and other microflora in either water or bottom sediments. In the event that a benthic population is altered or destroyed, the loss of microbial degradation may cause harmful chemicals to be released (these same chemicals ordinarily could be degraded by a healthy benthic population).

Figure 3.1: Conceptual Model of Probable Flows of Hydrocarbons Through Major Compartments of an Estuarine Ecosystem

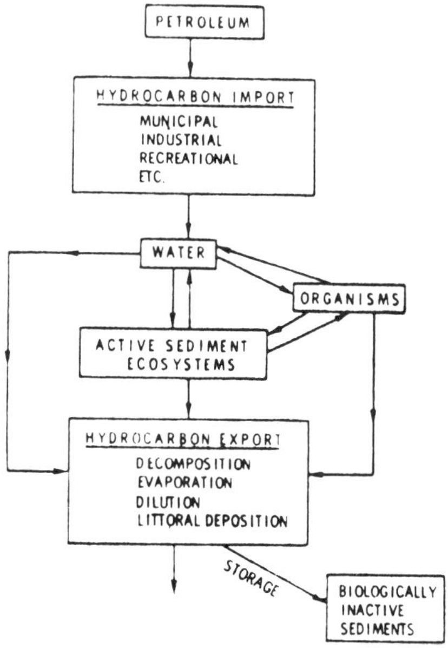

Source: Reference (30) as cited in ORNL/EIS-95

Most chemical reactions occur in the ocean at the air-sea interface, the sediment-water interface, or the boundaries between the water and the particles that are suspended in the water (32). Materials that concentrate at the air-water interface are subject to oxidation, ultraviolet radiation, evaporation of volatile constituents, and polymerization (32).

When considering the rate of sediment supply to the oceans and the apparent rate of accumulation in deep ocean basins, it is likely that 90% or more of river-borne particles are deposited on continental shelves. The particles that are greater than 0.5 μ in diameter have residence times in the oceanic water column of less than 100 years. For smaller particles the residence may be between 200 and 600 years. There is also rapid removal of particles by filter feeders that concentrate the particles in fecal pellets, which drop to the ocean bottom (32).

Accumulation in Biota

Upon entering the aquatic environment, hydrocarbons are available for uptake by the biota. Farrington and Quinn (33), in studying marine mollusks, report that marine organisms can synthesize some low-boiling hydrocarbons, but state that synthesis of naphthenic and aromatic hydrocarbons is too complex for those marine organisms. Thus, when hydrocarbons were extracted from clams collected in Narragansett Bay, they were identified as petroleum hydrocarbons because they included aromatic species. Marine mussels have also been recognized as good indicators of petroleum hydrocarbons. They appear to take up and release the hydrocarbons in proportion to the hydrocarbon concentrations in surrounding seawater (30). Lee, Sauerheber, and Benson (34) also report the presence of hydrocarbons in mussels. The experiments were performed in the laboratory on the marine mussel *Mytilus edulis*. The mussel rapidly filtered hydrocarbons across its gills, but seemed to lack the ability to metabolize any of the hydrocarbons. The nontoxic paraffinic hydrocarbons were taken up to a greater extent (10 mg per mussel) than were the aromatics (2 to 20 μg per mussel). When the polluted mussels were placed in uncontaminated seawater, they discharged the hydrocarbons: 90% of the heptadecane, 80% of the Tetralin (1,2,3,4-tetrahydronaphthalene), and most of the BaP; naphthalene and toluene were also discharged.

Other experiments by Lee and coworkers (35)(36) indicate that unlike mussels, marine copepods and the blue crab, *Callinectes sapidus*, have the ability to metabolize paraffinic and aromatic hydrocarbons. When transferred to fresh seawater, both copepod and blue crab released most of the accumulated radiolabeled hydrocarbons. After 17 days, the temperate water copepod *Calanus plumchrus* discharged all but 1 x 10^{-5} μg of benzypyrene of the original ingested amount, 22 x 10^{-4} μg. The hepatopancreas, assumed to be the main site of hydrocarbon metabolism, was the only organ in the blue crab where an appreciable buildup of radioactivity occurred. Based on later work, Lee and coworkers report that similar rates of hydrocarbon uptake, metabolism, and discharge have been observed in the California spiny lobster *Panulirus interruptus*, the spider crab *Libinia emarginata*, and the American lobster *Homarus Americanus*, leading to speculation that all marine crustaceans have the ability to metabolize hydrocarbons.

The oyster, a marine counterpart of the mussel, also has the ability to take up hydrocarbons. In one study (37), oysters from Galveston Bay were found to contain hydrocarbons, 56% of which were aromatic. Ehrhardt (37) considers the stability of hydrocarbons, as they pass through the food chain, responsible for uptake of the chemical by the oyster. In a study by Blumer (21), hydrocarbons in the liver of the basking shark were found to reflect the hydrocarbon content of its planktonic food. On the basis of this supporting evidence, Ehrhardt concludes that even highly organized animals accumulate the entire range of hydrocarbons to which they are exposed.

From laboratory studies of North Sea plankton, Whittle (38) proposed that plankton concentrate hydrocarbons from seawater, possibly via surface absorption; this further supports the accumulation of hydrocarbons by lower members of the aquatic food chain. Another indication that highly organized animals can take up hydrocarbons is from research by Ogata and Miyake (39). They demonstrated that aromatic hydrocarbons (benzene, xylene, and toluene) rapidly infiltrate fish and possibly impart an offensive odor. They conclude that aromatic hydrocarbons can accumulate in fish and that the hydrocarbons might be transferred to man through his consumption of fish.

Stegeman and Teal (40) also found hydrocarbon uptake in their laboratory study of the oyster, *Crassostrea virginica*. They note that the hydrocarbon concentration in the seawater, as well as the lipid content of the oyster, apparently affects the rate and extent of accumulation. The oysters having a higher lipid content accumulated more hydrocarbons. When oysters were transferred to uncontaminated seawater, there was a rapid discharge of about 90% of the hydrocarbons taken in. However, the amount retained was over 30 times that found in the oysters prior to exposure to the hydrocarbons. The hydrocarbons discharged by the oysters were greater in aromatic content and seemed to be somewhat degraded relative to the hydrocarbons present at the time of exposure; this prompted Stegeman and Teal to suggest the possibility that oysters modify the hydrocarbons.

Accumulation in Sediments

Shelton and Hunter (41) document the existence of hydrocarbons in bottom sediments. They cite the American Petroleum Institute's estimate of the national pollution potential from oil as about 450 million gallons annually, some of which reaches aquatic environments. Shelton and Hunter state that a significant portion of the aquatic input is incorporated into bottom sediments of receiving waters. Much of the PAH found in bottom sediments may be the result of natural forest and prairie fires. The PAH, while adsorbed on particles or in solution, may eventually reach the beds of the various water bodies and be taken up by bottom organisms or, in the absence of light and under anaerobic conditions, remain in a stable condition for long periods (13). Additional discussion of hydrocarbons in sediments is presented in the following sections.

Solubility

The fate of hydrocarbons in natural water systems is influenced by their solubilities. The solubility of a hydrocarbon determines whether it will go into solution, become suspended in the water column, or adsorb onto the sediments.

In discussing hydrocarbon solubility, it is important to consider the possible incorporation of hydrocarbons into micelles, a process thought to be responsible for some hydrocarbon solubilization in natural systems (42)(43). A micelle is composed of an aggregate of single-surface-active molecules (surfactants), each possessing a hydrophobic hydrocarbon chain and an ionizable or water-soluble group, hydrophilic in nature (e.g., sodium lauryl sulfate) (42):

$$\underbrace{CH_3CH_2CH_2CH_2CH_2CH_2CH_2CH_2CH_2CH_2CH_2CH_2}_{\text{Hydrophobic}}\underbrace{-SO_4^-Na^+}_{\text{Hydrophilic}}$$

The solubilization of a hydrocarbon is thought to occur by association of the hydrocarbon (1) with the hydrocarbon core of a micelle, (2) with the polar surface of a micelle, or (3) with both the interior and the surface; a hydrocarbon having low solubility in pure water (e.g., decane) can be incorporated into the micellar structure and become solubilized.

Boehm and Quinn (43) list the following natural sources for surfactant humic materials in seawater:

(1) Dissolved organic matter in seawater;
(2) Fulvic acid extracted from a marine sediment;
(3) Organic matter extracted by seawater from a marine sediment;
(4) Material isolated from seawater at a chloroform-seawater interface; and
(5) Organic matter contributed by a municipal sewage effluent.

Removal of this dissolved organic matter from natural samples resulted in a 50 to 99% decrease in the amounts of hydrocarbons (primarily n-alkanes) solubilized. Solubilities of phenanthrene and anthracene were unaffected by the removal. However, high concentrations of surfactants (10 to 50 mg/ℓ) can increase the solubility of BP by a factor of 2 to 10 (44).

Another source of surfactants is microbiological. Microorganisms are thought to produce surfactants as metabolites to solubilize hydrocarbon droplets prior to assimilation (45)(46).

McAuliffe (47) has performed extensive determinations of the solubility of hydrocarbons. His studies indicate that branching increases the solubility of paraffin, olefin, and acetylene hydrocarbons, but not of cycloparaffin, cycloolefin, and aromatic hydrocarbons. For each homologous series of hydrocarbons, the logarithm of the solubility in water was found to be a linear function of the hydrocarbon molar volume. McAuliffe determined his solubilities in distilled water with measurements by a gas-liquid partition chromatographic technique. The chromatograph indicated slight impurity peaks, but McAuliffe assumed that these impurities did not interact with the hydrocarbon to alter the solubility.

Solubility of PAH in natural water systems seems to be poorly defined. Boehm and Quinn (43) report the PAH, phenanthrene, to be quite water-soluble. Three concentrations of phenanthrene (1.0, 0.5, and 0.1 mg/100 ml) added to Narragansett Bay water yielded solubility values of 60 to 70 μg/100 ml for all three concentrations. In two samples of Providence River water, Boehm and Quinn found phenanthrene solubilities of 24.6 and 15.5 μg/100 ml. The authors give no explanation for the higher solubility of phenanthrene in Narragansett Bay water than in Providence River water. They define soluble hydrocarbons as those hydrocarbons passing a 0.5-μm filter and solubilization as the process resulting in passage of the hydrocarbons by this filter. These researchers also found that neither salinity nor dissolved organic carbon had an effect on the solubility of phenanthrene, implying that phenanthrene is not solubilized by micelles.

McGinnes and Snoeyink (2) report that PAH are not soluble in water, but are present either as particulate material or as material adsorbed onto solid surfaces in natural water systems. Andelman and Suess (4) report that the solubility of PAH is very low in pure water, but that the increase of solubility of PAH by the addition of water-soluble organic compounds is a phenomenon that is important in the passage of these compounds into and through environmental waters. Thus, the solubilization of PAH is enhanced by the presence of micelles; this conclusion seems to contradict the results of Boehm and Quinn (43), who found dissolved organic carbon to have no influence on phenanthrene and anthracene solubility. However, since no two PAH compounds can be assumed to behave similarly with regard to solubility or any other property without sufficient experimental data, the results of Boehm and Quinn cannot definitely be reported as contradictory to the work of Andelman and Suess (4), who do not identify the specific PAH compounds about which they are writing. Grasselli (48) records phenanthrene as insoluble, but conditions for determination were not stated.

Clearly, the literature indicates that the solubilities of hydrocarbons, particularly PAH, in natural waters are poorly understood, presumably because of insufficient experimental data.

Degradation–Photooxidation

Photooxidation of hydrocarbons, particularly PAH, is the most important process in hydrocarbon degradation in air and water (13). Suess refers to results from his other published data and predicts that BP degradation in natural waters will be higher in the upper strata because the factors that promote degradation (illumination, temperature, and oxygen concentration) are high; degradation will decrease with depth due to a reduction in water clarity, illumination, temperature, and oxygen concentration. He also predicts that degradation by photooxidation of PAH in river, lake, and sea deposits will be minimal due to the lack of penetrating radiation and oxygen.

Feldman (49) found that toluene and related aromatics having side chains can be photooxidized in the presence of a variety of materials such as H_2O, H_2O_2, and $H_2O + FeCl_3$; tetrahydronaphthalene, for example, photooxidized in the presence of H_2O and O_2, gives polymeric materials. Feldman suggests that hydrocarbon photooxidation is a precursor to the formation of tar in marine waters.

McGinnes and Snoeyink (2) studied the photodecomposition of particulate PAH in water using benzpyrene (BP) and 1,2-benzanthracene (BA). The BP particles, 1.5 μm in diameter, irradiated for 4 days with ultraviolet light of varying intensities, decomposed (as shown in Figure 3.2) from 1 mg/ℓ to a nominal concentration of 0.3 to 0.45 mg/ℓ; 55 to 65% of the original BP in suspension was photodecomposed. The residual BP did not further decompose, regardless of the length of irradiation, because the decomposed outer layers (about 0.2 μm thick) functioned as a shield, protecting the inner BP from photodecomposition. This implies that a residual BP core will remain in natural systems for all photodecomposed BP particles having a diameter greater than 0.4 μm.

The primary BP photodecomposition products are identified as benzpyrene quinones, probably benzpyrene-6,12-quinone and benzpyrene-1,6-quinone:

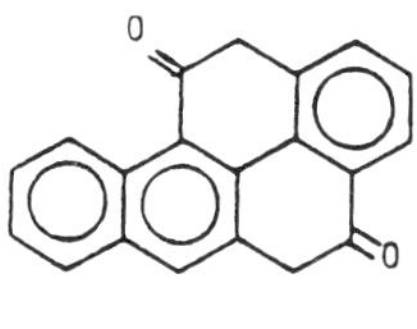

BP-1,6-Quinone

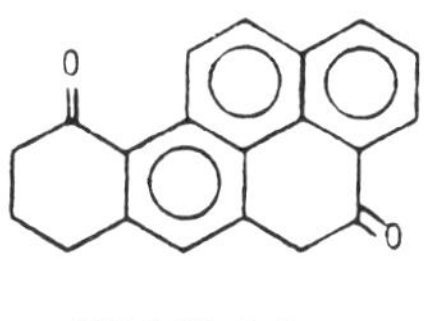

BP-6,12-Quinone

The decomposition products are unstable in the presence of ultraviolet light and are assumed by McGinnes and Snoeyink to decompose rapidly to secondary and tertiary decomposition products. McGinnes and Snoeyink (2) also irradiated BA particles having a diameter of 1.25 μm in a manner similar to that used for BP. Figure 3.3 shows that, in contrast to BP, BA continues to decompose with time. BA photocomposes to a benzanthracene-7,12-quinone:

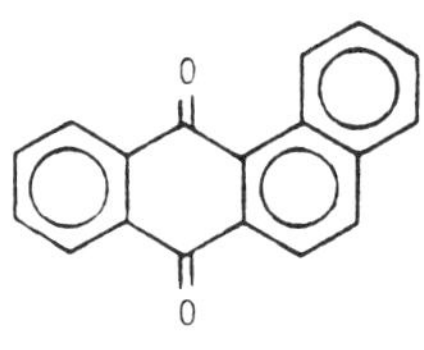

BA-7,12-Quinone

Figure 3.2: Decomposition of Particulate Suspensions of BP Under Ultraviolet Light of Various Intensities

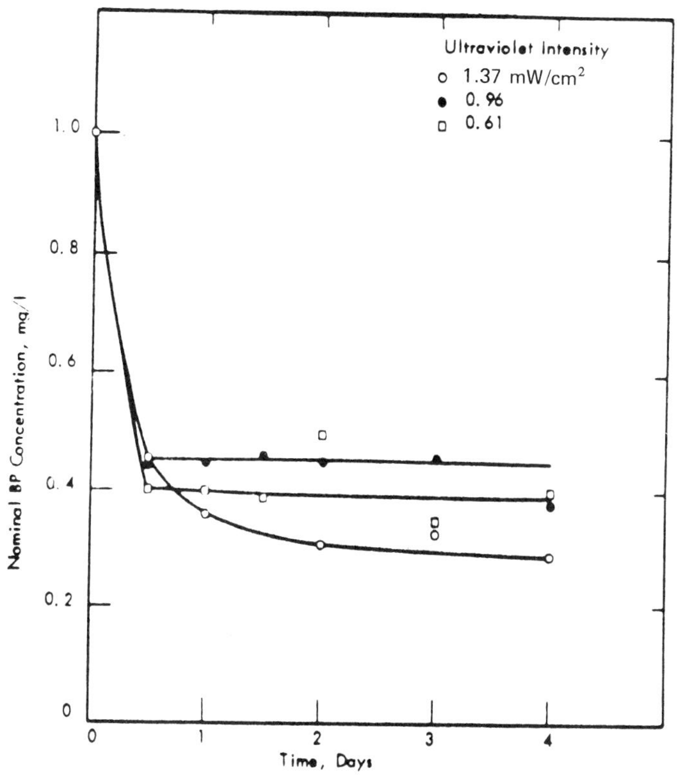

Source: Reference (2) as cited in ORNL/EIS-95

A secondary decomposition product is speculated. The BA photodecomposition products were assumed to be either soluble in water or transparent to ultraviolet radiation, thus explaining the complete decomposition of BA. Although these are laboratory test results, the researchers feel confident that the results can be extended to natural water systems.

The effect of particle size on decomposition of particulate PAH was also investigated by McGinnes and Snoeyink. Samples were irradiated at an ultraviolet intensity of 0.96 mW/cm^2; BP samples underwent irradiation until a steady state was reached while BA samples were irradiated for a 24-hour period. This investigation showed an increase in percent decomposition for all samples when particle size was decreased, thereby increasing total surface area available for photodecomposition. This implies surface decomposition of both BA and BP.

Figure 3.3: Decomposition of Particulate Suspensions of BA Under Ultraviolet Light of Various Intensities

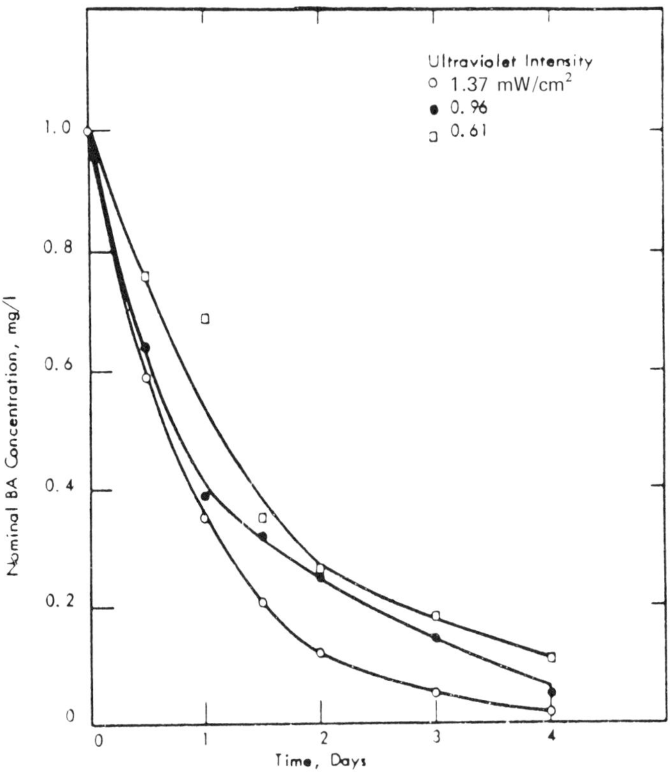

Source: Reference (2) as cited in ORNL/EIS-95

Freshwater Biodegradation

Several researchers have proposed biodegradation as a means for decomposition of hydrocarbons in the environment. Feldman (49) states that the interactions or pathways of such materials as trace or heavy metals, vitamins, amino acids, EDTA, nutrient forms of phosphorus and nitrogen, DDT, and PCB influence the biodegradation of hydrocarbons. These trace materials and compounds tend to concentrate in the surface layer and can interfere with the succession of biota, each of which uses a particular petroleum fraction, or an intermediate product.

Ludzack and Ettinger (50) measured the biodegradability of naphthalene, ethylbenzene, and Tetralin; the data were expressed in terms of the cumulative carbon dioxide recovered as a percent of theoretical recovery vs time in days. Theoretical carbon dioxide was obtained from the stoichiometric ratio of chemical to carbon dioxide after complete oxidation. The results indicate that naphthalene, Tetralin, and ethylbenzene were biodegraded at a moderate rate (more than 15 days for 50% CO_2 recovery); microbiological adaption to these compounds was poor.

Kostyaev (51) states that the activity of phenol-oxidizing bacteria, which exist in air, water, and soil, depends primarily on the temperature of the surrounding medium. He conducted experiments at 10°C and found that 5% of the original phenol decomposed, whereas at 30°C, 65% of the original phenol decomposed. Jeffery (52) found that the rate of phenol oxidation in his studies increased with increased air flow through the medium.

Marine Biodegradation

Suess (13) points out the existence of some controversy concerning the ability of marine organisms to degrade PAH. He cites the works of Reichart et al (1971) and Malaney et al (1968) as two examples of biological treatment processes having little or no efficiency in removing PAH and the work of Poglazova et al (1972) as an example of successful degradation of PAH by a mixed microbial culture. According to an NAS report (29),

> Microorganisms capable of oxidizing constituents under the right conditions have been found in virtually all parts of the marine environment that have been examined.... Both laboratory experiments and some field observations have shown that microorganisms consume the least toxic fraction of petroleum (normal alkanes) in a few days or months, depending on temperature and nutrient supply. The fraction containing aromatics and naphthenes is more toxic than the alkanes and also degrades more slowly.

The degradation of PAH may also be assisted by more highly developed aquatic fauna (13).

Sedimentary Biodegradation

A significant portion of the oil pollutants discharged into aquatic environments is incorporated into bottom sediments of receiving waters. Shelton and Hunter (41) cite Tauson (1934) in discussing one mechanism of the anaerobic decomposition of hydrocarbons in freshwater sediments. Tauson's calculations indicate that the lower-molecular-weight saturated hydrocarbons may be oxidized microbially to CO_2 and H_2O at the expense of sulfate oxygen, whereas higher members of the series, starting with nonane (C_9H_{20}), may be transformed into naphthenes and polynaphthenes that are of no use to sulfur-reducing bacteria due to the negative heat effect accompanying their oxidation with sulfate.

Shelton and Hunter (41) report that there will be a gradual loss of the low-boiling petroleum fractions and an accumulation of naphthenic and polynaphthenic acids in sediments if the transformation of oil under natural anaerobic conditions occurs as Tauson indicates. These acids can be expelled to the surface and completely oxidized if, for example, there is gas formation on a river bottom to act as a carrier.

Baker (53) suggests that the fate of hydrocarbons in marine sediments is to form crude oil. The oil-forming mechanism is as follows:

(1) Recent sediments and crude oils contain some of the same kinds of hydrocarbons that appear to be derived from the organic compounds produced by living organisms.

(2) Because the majority of the hydrocarbons which may later become collected to form crude oil deposits appear to be present in the sediments relatively soon after decomposition, no geologic time factor seems to be involved in hydrocarbon formation.

(3) Formation waters may contain natural solubilizers, which would enhance the solubility of hydrocarbons. Such waters could selectively dissolve and release sediment hydrocarbons in proportions that correspond to hydrocarbon occurrence in crude oils.

(4) The solubilizing ability of the formation waters is thought to be sensitive to both salt concentration and dilution. Consequently, hydrocarbons could be abruptly released as the interstitial waters pass from shale to sand or as the micelles diffuse throughout the sand.

(5) The hydrocarbons that are released would appear for the first time as tiny oil droplets (oil shows). They could coalesce and collect in the highest parts of the reservoir rock to form regions of high oil saturation (oil pools). The principles involved here are the classical ones of buoyancy and capillary-pressure-entry-pressure mechanisms.

In summary, the ultimate fate of hydrocarbons in the water environment is (1) dispersal in the water column, (2) incorporation into sediments, or (3) oxidation by chemical or biological means to CO_2 (29).

FATE IN SOILS

Soils exhibit differences in pH, texture, structure, and chemical composition (54). Thus, movement of additives within the soils is affected by the interrelation of the macrofeatures (texture, structure, pore space, and density), the humic nature of the surface horizons, and the four components of soil (solid, liquid, gaseous, and living). The solid component has a surface microcharacter that adsorbs, reacts with, and catalyzes reactions of both added and indigenous substances. This microcharacter is generated by crystalline mineral colloids, amorphous mineral colloids, organic colloids, and their surface character. The soil liquid component has special acidity-alkalinity properties due to the variable salt solution concentration, which may be buffered and which change with changing soil moisture.

Since the water phase is strongly adsorbed, it competes for solid surfaces. The soil atmosphere (gaseous component) is a modification of normal air; microbial activity and limited rates of diffusion cause a lower oxygen content and a higher carbon dioxide content than in air. The living component is the most complex. Organisms introduce into the soil many enzymes capable of metabolizing both natural organic substances and synthetic organic additives.

Leaching

Contamination of groundwater by leaching of organic chemicals through soil is an environmental concern (55). The ability of a chemical to reach groundwater depends upon not only its movement through, but also its disappearance from, the soil; that is, if the rate of degradation is sufficiently rapid, as compared with the rate of leaching, the chemical will disappear before it can reach the groundwater. Determination of soil leaching rates indicates how long a chemical is retained in the top soil, where it is most subject to degradation or dissipation. Four major elements of leaching (55) are:

(1) Soil adsorption, which determines an underlying pattern;

(2) Porous flow and diffusion, which disperse the chemical;

(3) Adsorption dynamics, which introduce a factor of hysteresis; and

(4) Water infiltration and evaporation, which determine the actual amount of water movement and, hence, the observed amount of chemical movement.

Adsorption

Many different intermolecular interactions are involved in adsorption of organic chemicals such as PNA by soils (Table 3.19) (56). Clay minerals (mainly kaolinite and montmorillonite)), organic matter (humin, humic acids, and fulvic acids), and hydrated metal oxides (aluminum and iron oxides), the three major soil components for adsorption, can each contribute to or compete for the adsorption of a soil contaminant, depending on the interactions available to the system. Hamaker and Thompson (56) conclude that adsorption in soil is largely due to small molecules interacting with macromolecules such as the soil organic matter.

Table 3.19: Intermolecular Interactions

Type of Interaction	Forces Involved
Van der Waals-London	Electrostatic
Hydrophobic bonding	Entropy generation
Charge transfer	Electrostatic
Hydrogen bonding	Electrostatic
Ligand exchange	Electrostatic
Ion exchange	Electrostatic
Direct and induced ion-dipole and dipole-dipole forces	Electrostatic
Chemisorption	Electrostatic
Magnetic bonding	Magnetic

Source: Reference (56) as cited in ORNL/EIS-95

Baker and Luh (57) attribute pyridine sorption onto sodium kaolinite and sodium montmorillonite from aqueous solution to cationic exchange of pyridinium ions with sodium ions. The extent of sorption depends upon the pH of the solution. Maximum pyridine sorption for sodium kaolinite occurs at a pH of 5.5, and maximum pyridine sorption for sodium montmorillonite occurs at a pH of 4.0. Sorption does not occur if the pH is greater than 6.0.

Roberts, Street and White (58) describe the physical processes of phenol adsorption by amine-activated montmorillonite. Wetting with water usually results in a considerable basal spacing expansion, whereas the adsorption of the phenol causes a progressive contraction of the mineral lattice. Formation of a quaternary ammonium chloride montmorillonite results in a much greater basal expansion and a correspondingly greater adsorption of phenol. The amount of phenol adsorbed increases as the number of milliequivalents of quaternary ammonium base in the clay increases.

Porous Flow, Adsorption Dynamics, and Water Infiltration

A chemical flowing through a porous medium is either dispersed or retarded, depending on the porosity of the medium (55). This is a dynamic effect, which depends upon the movement of the liquid through pores. Diffusion allows the chemical to flow into stagnant pores to be released slowly when the main body of chemical has passed. A frictional effect caused by adsorption results in slow release of a chemical as soil readjustment takes place.

The pattern of rainfall and evapotranspiration determines the infiltration and movement of water into and through soil. Chemicals are leached downward following a rain, but as evaporation dries out the upper layers of soil, there is a compensating upward movement of water and chemicals. Letey and Oddson (59) observed that energy from the sun causes evaporation at a soil surface, resulting in the soil becoming drier; consequently, soil suction increases and a hydraulic gradient causes upward water flow, which could transport organic chemicals back to the surface. Hamaker (55) concludes, on the basis of published information, that little, if any, organic chemical ever leaches to groundwater.

Decomposition

The decomposition of hydrocarbons by soil microorganisms is the most important process in hydrocarbon degradation. Meikle (60) discusses the decomposition of saturated, unsaturated, alicyclic aliphatic, and aromatic hydrocarbons. Saturated aliphatic hydrocarbons undergo three types of reaction in soil and in cultures of microorganisms: oxidation of the terminal carbon atom, oxidation of the C_2-carbon atom, and dehydrogenation:

(1) $CH_3[CH_2]_nCH_3 \longrightarrow CH_3[CH_2]_nCH_2OH$

(2) $CH_3[CH_2]_nCH_2CH_3 \longrightarrow CH_3[CH_2]_n\underset{}{C}H(OH)CH_3$

(3) $CH_3[CH_2]_nCH_2CH_2[CH_2]_mCH_3 \longrightarrow CH_3[CH_2]_nCH{=}CH[CH_2]_mCH_3$

According to Meikle, the predominant soil reaction is Reaction (1). However, the usual isolated product of the microbiological oxidation of saturated aliphatic hydrocarbon chains is a carboxylic acid (Stewart et al 1959, as cited by Meikle) or glycolipid (Jones and Howe 1968; Makula and Finnerty 1968, as cited by Meikle) rather than the hydroxyl compound implied in Reaction (1). Alcohols are thought to be the intermediates in the sequence that produces the carboxylic acid. Meikle believes Reaction (2) to be of relatively minor importance in the oxidation of saturated aliphatics. As evidence of Reaction (3), Meikle cites the work of Tamura and Manire (1968), in which hexadecane was converted to a mixture of internal monohexadecanes.

Microbial decomposition of alicyclic aliphatic hydrocarbons is discussed by Meikle (60). Reaction (4) illustrates the hydroxylation of an unactivated methylene group, and Reaction (5) shows a steroid aromatization reaction.

(4) 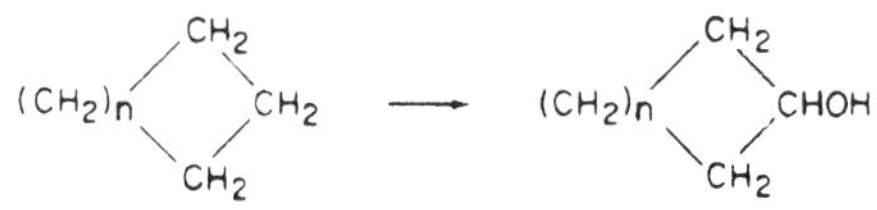

(5) 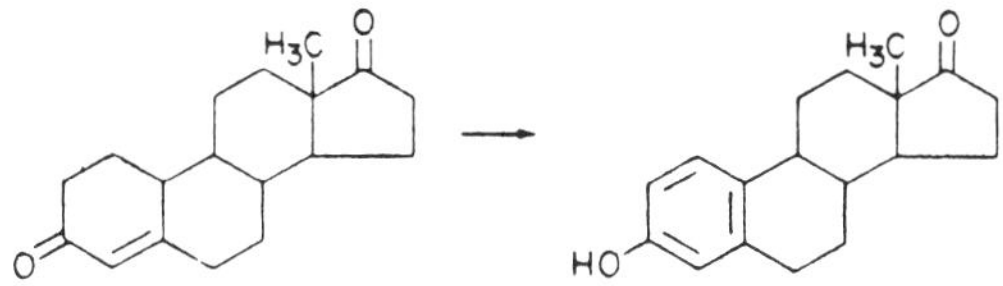

Microbiological degradation of the aromatic nucleus is accomplished with relative ease, resulting in destruction of the ring and leading to the ultimate use of the fragments by microorganisms as a source of carbon for growth.

Fernley and Evans (61) and Davies and Evans (62) both present the degradative pathways of the PAH naphthalene by soil pseudomonads (Figure 3.4 through Figure 3.6). Both groups of researchers agree that naphthalene is oxidatively metabolized by soil pseudomonads through D-trans-1,2-dihydroxynaphthalene to 1,2-dihydroxynaphthalene, which then undergoes ring cleavage. At this point, however, disagreement arises. Fernley and Evans say that the ring fission product is o-carboxy-cis-cinnamic acid (Figure 3.4), whereas Davies and Evans believe that the ring fission product is cis-o-hydroxybenzalpyruvate, isolated as the crystalline perchlorate (Figure 3.6). The two groups of researchers do agree that naphthalene is completely dissimilated by soil pseudomonads.

Phenanthrene and anthracene can be completely dissimilated by soil pseudomonads (63) (Figure 3.7). As with naphthalene metabolism, the ring fission mechanism with ring cleavage occurs at the 3,4-dihydroxyphenanthrene (Figure 3.7). The final metabolism is accomplished through the naphthalene pathway (Figure 3.6). Anthracene metabolism by ring fission (Figure 3.8) is similar to that for phenanthrene, although the final metabolism is conducted by unknown pathways. Evans, Fernley and Griffiths (63) state that PAH, occurring as various derivatives, can be completely dissimilated by soil pseudomonads.

Muller and Korte (64) studied the microbial degradation of BaP. The chemical was sprayed on ground waste, and after three weeks of composting, the digested material was extracted by organic solvents. They found that BaP withstood degradation by at least 99.5%; however, the researchers indicate that Lorbacher, Püls, and Schlipköter (1971) found that BaP can be accumulated and converted by special strains of microorganisms.

Hussien, Tewfik and Hamdi (65) demonstrated the ability of 22 strains of *Rhizobium* to degrade the aromatic compounds, catechol, protocatechuic acid, p-hydroxybenzoic acid, and salicylic acid. At 1 mM concentration and in the presence of 4.8 mM sodium glutamate, all rhizobia tested degraded catechol (99 to 100%), p-hydroxybenzoic acid (79 to 99%), protocatechuic acid (81 to 97%), and salicylic acid (20 to 83%). Increased glutamate concentration favored degradation of p-hydroxybenzoic and salicylic acids, had little observable effect on catechol, and inhibited the degradation of protocatechuic acid. The researchers found that most strains of rhizobia tested readily used catechol and protocatechuic acid without detectable accumulation of intermediate phenolic compounds, but p-hydroxybenzoic acid apparently was converted to protocatechuic acid before ring cleavage. Also, they found salicylic acid to be slowly degraded by different strains of rhizobia in liquid cultures, being first converted to another compound identified as gentisic acid.

An understanding of simultaneous leaching and degradation is needed to properly understand the movement of chemicals through the soil and into the groundwater (55).

Figure 3.4: Suggested Pathway for the End-Ring Attack on Naphthalene by Soil Pseudomonads

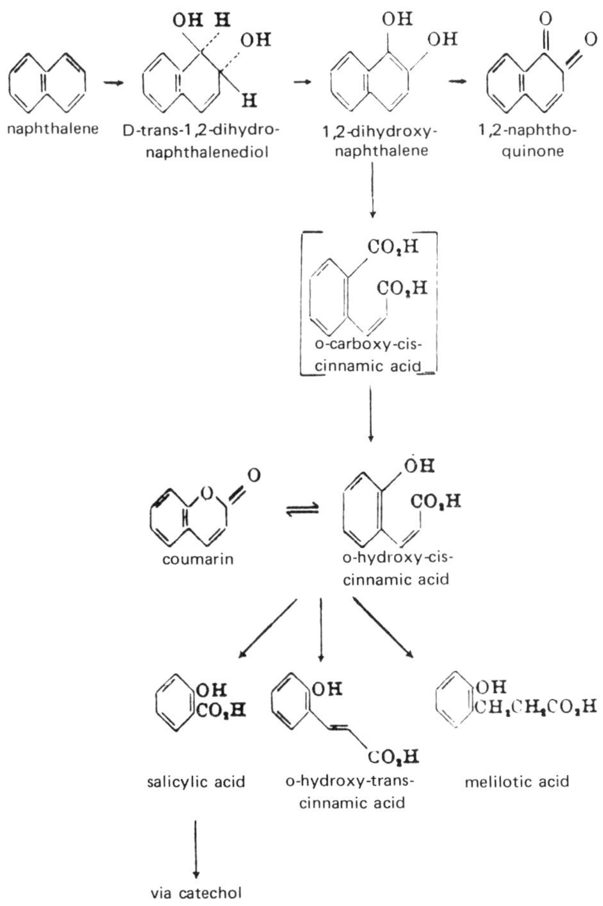

Source: Reference (61) as cited in ORNL/EIS-95

Figure 3.5: Possible Transformations of cis-o-Hydroxybenzalpyruvate

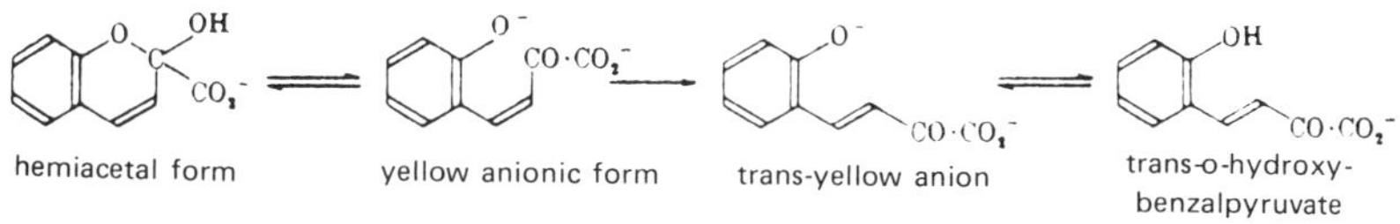

Source: Reference (62) as cited in ORNL/EIS-95

Figure 3.6: Proposed Pathway of Naphthalene Metabolism by Soil Pseudomonads

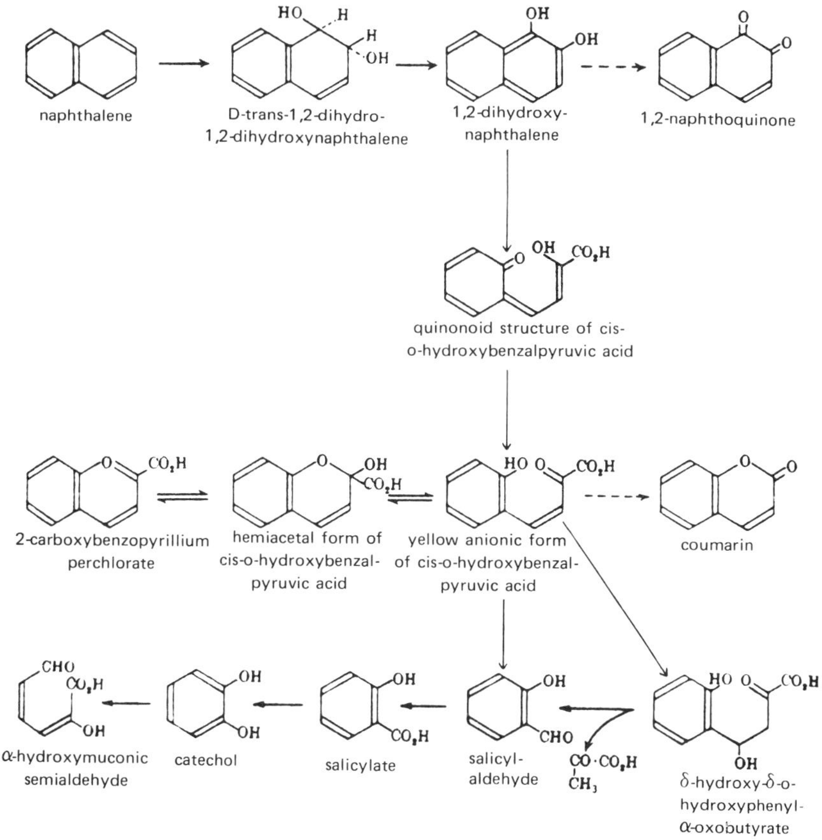

Source: Reference (62) as cited in ORNL/EIS-95

Figure 3.7: Proposed Pathway of Phenanthrene Metabolism by Soil Pseudomonads

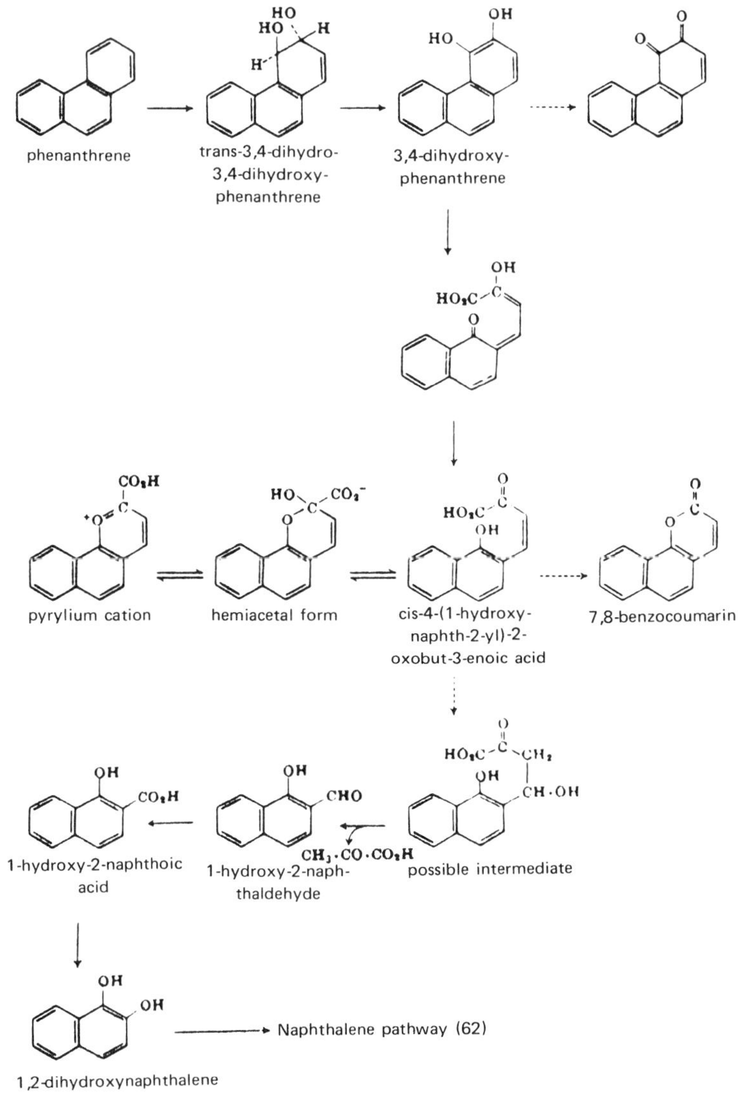

Source: Reference (63) as cited in ORNL/EIS-95

Figure 3.8: Proposed Pathway of Anthracene Metabolism by Soil Pseudomonads

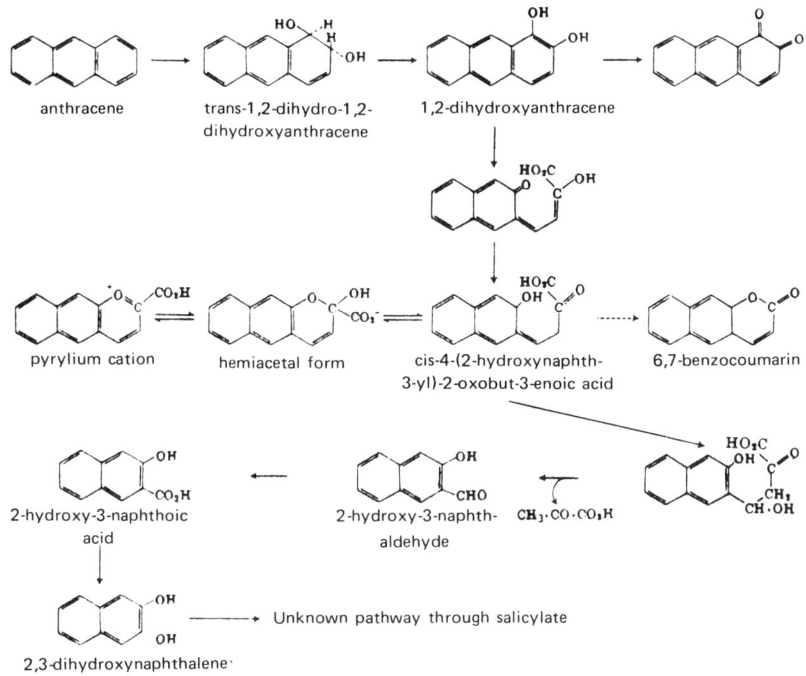

Source: Reference (63) as cited in ORNL/EIS-95

FATE IN AIR

In the atmosphere, polycyclic organic matter is associated with particulate matter, especially soot. Figure 3.9 illustrates some carcinogenic polycyclics that have been identified in urban air.

Collection of PAH on particulate filters is the method generally employed during air pollution surveys. Pupp et al (66) report that the volatility of some PAH compounds hindered their collection on particulate filters; losses occurred from the filter surface. The PAH compounds, pyrene and BaP, which have equilibrium vapor concentrations of 500 $\mu g/10^{-3}\ m^{-3}$ or higher at ambient temperatures, were reported to have considerable losses from filters during air sampling. These losses (plus losses by sublimation) and the fact, reported by Thomas, Mukai, and Tebbens (67), that not all PAH are absorbed even in the presence of excess soot led Pupp et al to question the accuracy of present collection methods. If the present collection methods are not totally representative of ambient PAH concentrations, then the discrepancy would most likely be in such a direction that the reported PAH atmospheric concentrations would be lower than the true value due to losses and unmeasured PAH.

Figure 3.9: Carcinogenic Polycyclics Identified in Urban Air

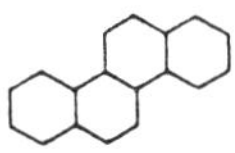

Chrysene

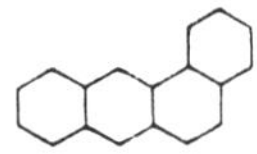

Benz(a)anthracene
(1, 2-benzanthracene)

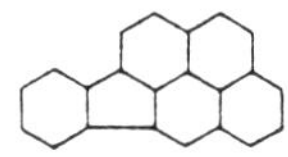

Indeno(1, 2, 3-cd)pyrene
(o-phenylenepyrene)

Benzo(e)pyrene
(1, 2-benzopyrene)

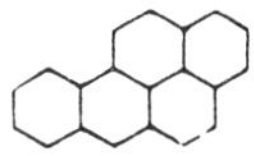

Benzo(a)pyrene
(3, 4-benzopyrene)

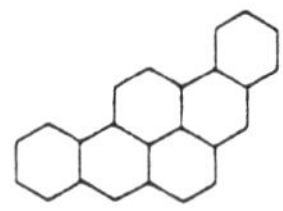

Dibenzo(a, i)pyrene
(3, 4, 9, 10-dibenzopyrene)

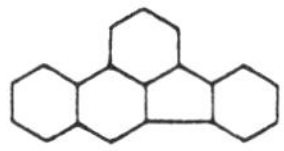

Benz(e)acephenanthrylene
(3, 4-benzofluoranthene)

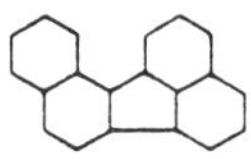

Benzo(j)fluoranthene
(10, 11-benzofluoranthene)

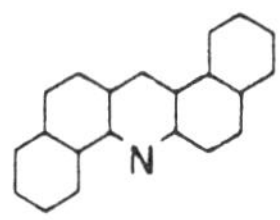

Dibenz(a, h)acridine
(1, 2, 5, 6-dibenzacridine)

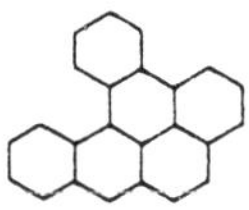

Dibenzo(e, l)pyrene
(1, 2, 3, 4-dibenzopyrene)

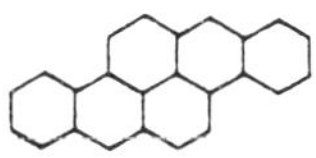

Dibenzo(a, h)pyrene
(3, 4, 8, 9-dibenzopyrene)

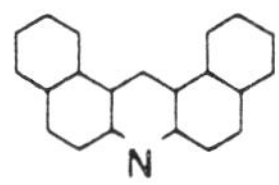

Dibenz(a, i)acridine
(1, 2, 7, 8-dibenzacridine)

Source: Reference (26) as cited in ORNL/EIS-95

Physical Reactivity

Particle size is the physical property having the greatest influence on the bioenvironmental behavior of aerosols containing polycyclic organic matter (3). A seasonal study of PAH association with particulate matter found that most PAH were associated with particles less than 3.0 μm in diameter (68). During the summer months, 6 to 23% of the total atmospheric PAH were associated with particles having diameters $\leqslant$1.0 μm, whereas particles having a diameter $\leqslant$3.0 μm contained 56 to 70% of the total PAH content. During the colder winter months (this study was conducted in Toronto), the PAH content of the smaller respirable particles ($<$1.0 μm) increased.

Pierce and Katz (68) note that particles having a diameter $\leqslant$5.0 μm are in the respirable size range, that is, the size range most likely to be deposited in the pulmonary portion of the respiratory tract. Figure 3.10 shows that 60% of all particulate matter in the size range of 0.5 to 2.0 μm that is inhaled is retained in the lung. The highest retention, 77%, occurs at 1.0 μm. DeMaio and Corn (69) found that aerosol particles $<$5.0 μm in diameter contained more than 75% of the weight of selected polycyclic hydrocarbons.

Figure 3.10: Retention of Particulate Matter in Lung in Relation to Particle Size

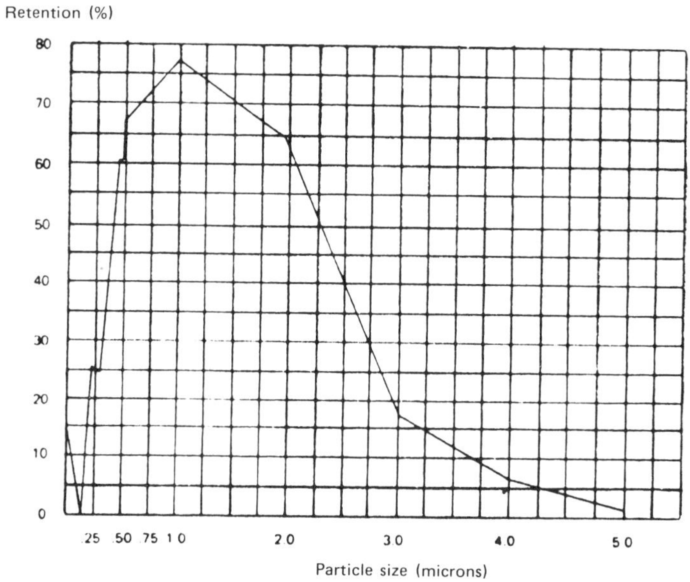

Source: Reference (26) as cited in ORNL/EIS-95

Atmospheric suspended particles evolve by the following mechanisms (3):

(1) Growth or change in particles by homogeneous or heterogeneous chemical reactions of gases on the surface of particles;

(2) Change in particles by attachment and adsorption of trace gases and vapors to aerosol particles;

(3) Net change by collision between particles undergoing Brownian motion or differential gravitational settling;

(4) Net change by collision between particles in the presence of turbulence in the suspending gases;

(5) Gain or loss in concentration by diffusion or convection from neighboring air volumes;

(6) Loss by gravitational settling;

(7) Removal at the earth's surface on obstacles by impaction, interception, Brownian motion, and turbulent diffusion;

(8) Loss or modification by rainout within clouds; and

(9) Loss by washout below cloud level.

These mechanisms can influence aerosol chemical composition, as a function of size, and can affect the observed size distribution of aerosols.

The NAS report states that the lifetime of polycyclic organic matter in air depends on the carrier aerosol and on the chemical alteration of the matter itself. In dry atmospheric conditions, residence times of particles <5 μm in diameter exceed 100 hr, but in the presence of sunlight, chemical reactivity may lead to transition of polycyclic organic matter adsorbed on soot to other material in several hours.

Chemical Reactivity

The major mode of polynuclear compound removal from the atmosphere is through chemical reactions; for example, photooxidation is probably one of the most important processes in atmospheric removal of PAH (3). Thomas, Mukai and Tebbens (67) indicate that BaP and other arenes are primarily adsorbed on the surface of soot through hydrogen bonding. Chemisorption is apparently not feasible because of the rapid rate of photomodification, indicating that BaP must be primarily in an exposed position on the surface.

Tebbens, Thomas and Mukai (70) demonstrated a 15 to 50% loss of BaP in smoke on filters with 6 hr of sunlight exposure. Another study (71) shows a 10% destruction of BaP adsorbed on soot or a filter after a 48-hr exposure to light of unstated intensity and a 50% destruction after a 1-hr exposure to light and synthetic smog.

Products of photooxidation of adsorbed aromatic hydrocarbons (anthracene and pyrene adsorbed on silica gel and alumina) are given in the NAS report (3):

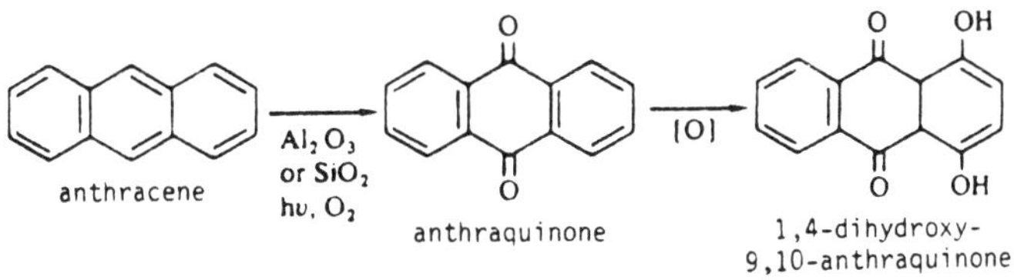

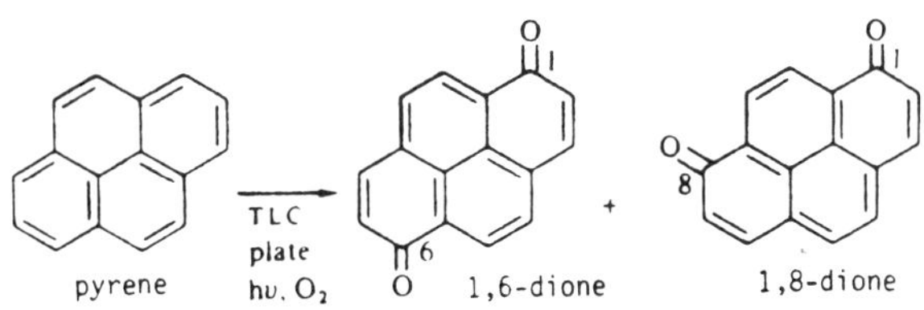

The oxidation of both anthracene and pyrene took place in the presence of ultraviolet light and oxygen, the degree of anthracene oxidation being somewhat dependent on the adsorbent. The pyrene oxidation was conducted on thin-layer chromatography plates.

PAH also react with ozone, peroxides, nitrogen oxides, and sulfur oxides and undergo one-electron oxidation (3). They react with ozone in one of four ways:

(1) Cleavage of phenanthrene-like double bonds, which results in diacid formation under oxidative conditions;

(2) Oxidation at anthracene-9,10-like positions to yield quinones;

(3) A more complex nuclear oxidation; and

(4) A side-chain oxidation.

A typical example is the ozonolysis of benz[a]anthracene, which undergoes cleavage of the phenanthrene-like double bond to yield a cleavage product, a quinone, and a phthalic acid:

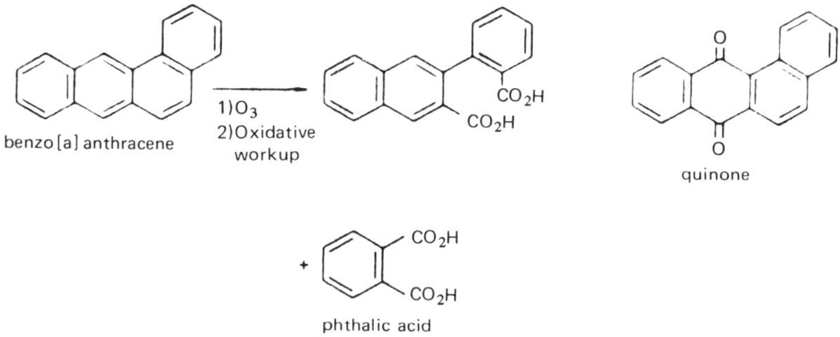

The peroxide reaction involves the interaction of benzoyl peroxide [$(C_6H_5CO_2)_2$] with the PAH, in this case BaP, to yield the 6-benzoyloxy derivative:

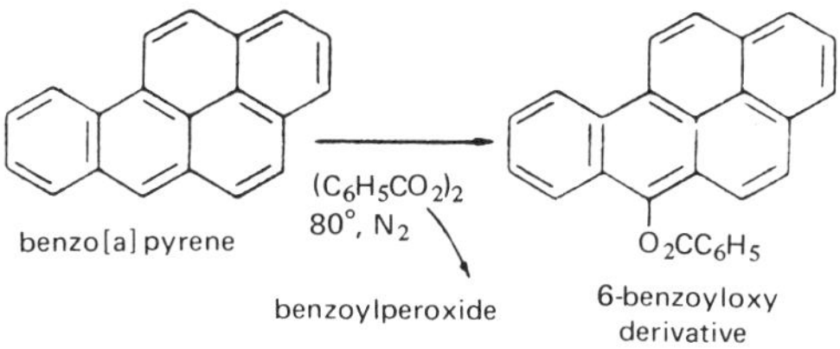

In the presence of oxygen, this peroxide reaction could lead to further oxidation. Nitrogen oxides or dilute aqueous HNO_3 oxidizes anthracene to anthraquinone:

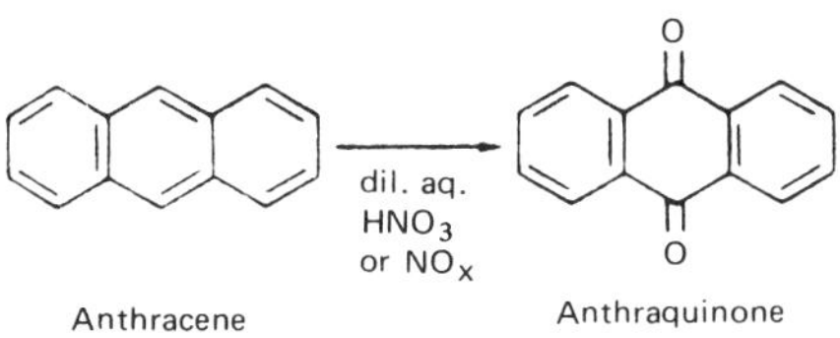

Specifically, the addition and substitution of NO_2 in anthracene yields an oxidation product by loss of HNO_2 from the intermediate compound:

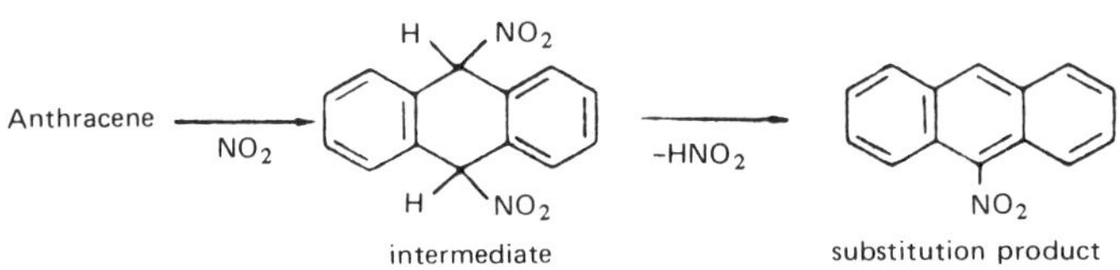

The one-electron oxidation of PAH results in the formation of radical cations as primary products; these react rapidly with water or other nucleophiles or with oxygen. The complete reaction leads to formation of quinones (with water and oxygen), nucleophile adducts, or dimeric hydrocarbons. One-electron oxidation of BaP also illustrates the formation of quinones:

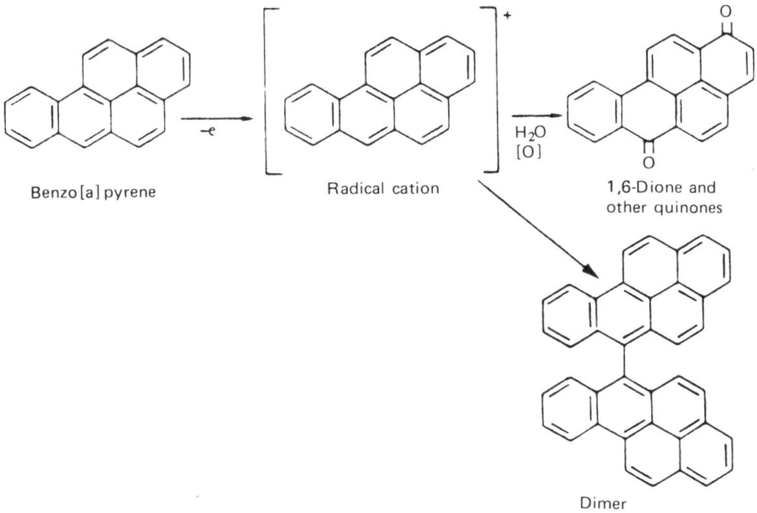

PAH also react with sulfur oxides, yielding sulfuric and sulfonic acids as products; these sulfur compounds will no longer appear in the benzene-soluble fraction because they are water-soluble.

ENVIRONMENTAL TRANSPORT SUMMARY

The environmental transport of hydrocarbons is summarized by Suess (13):

> ...the transport pattern of PAH in the environment appears to be relatively simple. The background PAH, which are formed by biosynthesis, are quite static and, obviously, remain in the plants and microorganisms in which they were formed and, more generally seen, stay within their own ecosystems. However, it appears probable that PAH in ground-water are leached out from the soil.
>
> In contrast, PAH formed by high temperature processes, whether resulting from natural open burning and volcanic eruptions or from man-induced combustion reactions, are all emitted into the atmosphere, and thus are subject to the same dynamic forces which govern the movement, transport and fallout of aerosols generally. Because a significant portion of PAH, adsorbed onto the aerosols, will decompose by photooxidation while still in the atmosphere, either stationary or in motion, their fallout at greater distances from the source (delayed fallout) will be relatively very limited. However, where fallout of PAH occurs, it will contaminate the upper layers of the earth, including vegetation and forests, as well as rivers, reservoirs and lakes, and some of them will also reach the oceans. Runoff and the rivers will carry eventually some of this fallout to the open seas and oceans. As waste treatment plants do not remove all PAH, the coastal

waters will receive an additional load from domestic and industrial waste effluents either directly, or indirectly through the rivers....

The degradation of PAH in the atmosphere by photooxidation will also continue to some extent when they have settled back on earth and water surfaces, for as long as they are exposed to sunlight. However, some PAH will be degraded by soil bacteria and aquatic organisms. PAH, while adsorbed on particles or in solution, may eventually reach the bed of the various water bodies and be taken up by bottom organisms, or remain there stable for extremely long periods, given absence of light and anaerobic conditions.

REFERENCES

(1) Herbes, S.E., Southworth, G.R., and Gehrs, C.W., "Organic Contaminants in Aqueous Coal Conversion Effluents: Environmental Consequences and Research Priorities," *Tenth Annual Conference on Trace Substances in Environmental Health, Univ. of Missouri, June 8-10* (1976).

(2) McGinnes, P.R. and Snoeyink, V.L., *Determination of the Fate of Polynuclear Aromatic Hydrocarbons in Natural Water Systems*, University of Illinois at Urbana-Champaign, Water Resources Center, WRC Res. Rept. 80, UILU-WRC-74-0080, PB-232 168 (1974).

(3) National Academy of Sciences (NAS), *Particulate Polycyclic Organic Matter*, Washington D.C., National Academy of Sciences (1972).

(4) Andelman, J.B. and Suess, M.J., "Polynuclear Aromatic Hydrocarbons in the Water Environment," *Bull. W.H.O.*, 43, pp. 479-508 (1970).

(5) Gräf, W. and Diehl, H., "Concerning the Naturally Caused Normal Level of Carcinogenic Polycyclic Aromatics and Its Causes," *Arch. Hyg. Bakt.*, 150 (4), 49-59 (1966).

(6) Gräf, W. and Nowak, W., "Promotion of Growth in Lower and Higher Plants by Carcinogenic Polycyclic Aromatics," *Arch. Hyg. Bakt.*, 150, 513-528, ORNL/tr-4111 (1966).

(7) Grimmer, G. and Duvel, D., "Investigations of Biosynthetic Formation of Polycyclic Hydrocarbons in Higher Plants," *Z. Naturforsch.*, 25b, 1171-1175 (1970).

(8) Borneff, J., Selenka, F., Kunte, H., and Maximos, A., "Experimental Studies on the Formation of Polycyclic Aromatic Hydrocarbons in Plants," *Environ. Res.*, 2, 22-29 (1968).

(9) Borneff, J. and Kunte, H., "Carcinogenic Substances in Water and Soil. XVII. Concerning the Origin and Estimation of the Polycyclic Aromatic Hydrocarbons in Water," *Arch. Hyg. Bakt.*, 149, 226-243 (1965).

(10) Smith, P.V., Jr., "Studies of the Origin of Petroleum: Occurrence of Hydrocarbons in Recent Sediments," *Bull. Am. Assoc. Petrol. Geol.*, 38 (3), 377-404 (1954).

(11) Blumer, M., "Benzpyrenes in Soil," *Science*, 134, 474-475 (1961).

(12) Knorr, M. and Schenk, D., "The Question of the Synthesis of Polycyclic Aromatics by Bacteria," *Arch. Hyg. Bakt.*, 152 (3), 282-285 (1968).

(13) Suess, M.J., "The Environmental Load and Cycle of Polycyclic Aromatic Hydrocarbons," presented at The International Conference on Environmental Sensing and Assessment, Las Vegas (Sept. 14-19, 1975).

(14) Mallet, L., "Investigation of Polybenzene Hydrocarbons in Oil Sediments," *Pollution des Milieux Vitaux par Les Hydrocarbures Polybenzeniques du Type Benzo-3,4 Pyrene*, ed. L. Mallet, pp. 111-113, Paris, Librairie Maloine (1972).

(15) Mallet, L. and Tissier, M., "Experimental Biosynthesis of Polybenzene Hydrocarbons of the Benzo-3,4-Pyrene Type at the Expense of Forest Soils," *Pollution des Milieux Vitaux par Les Hydrocarbures Polybenzeniques du Type Benzo-3,4 Pyrene*, ed. L. Mallet, pp. 105-107, Paris, Librairie Maloine (1972).

(16) Mallet, L., Heros, M., and Brisou, J., "Biosynthesis and Bioregression of the Carcinogenic Polybenzene Hydrocarbons of the Benzo-3,4-Pyrene Type, Taken as Controls, at the Expense of Lipids," *Pollution des Milieux Vitaux par Les Hydrocarbures Poly-*

benzeniques du Type Benzo-3,4 Pyrene, ed. L. Mallet, pp. 177-180, Paris, Librairie Maloine (1972).
(17) Mallet, L., Zanghi, L., and Brisou, J., "Investigations of the Possibilities of Biosynthesis of Polybenzene Hydrocarbons of the Benzo-3,4-Pyrene Type by a *Clostridium Putride* in the Presence of Marine Plankton Lipids," *Pollution des Milieux Vitaux par Les Hydrocarbures Polybenzeniques du Type Benzo-3,4 Pyrene*, ed. L. Mallet, pp. 153-157, Paris, Librairie Maloine (1972).
(18) Brisou, J., "Lipids, Sterols, Terpenes and Bacterial Biosynthesis of 3,4-Benzopyrene," *Pollution des Milieux Vitaux par Les Hydrocarbures Polybenzeniques du Type Benzo-3,4 Pyrene*, ed. L. Mallet, pp. 181-183, Paris, Librairie Maloine (1972).
(19) Hase, A. and Hites, R.A., "On the Origin of Polycyclic Aromatic Hydrocarbons in Recent Sediments: Biosynthesis by Anaerobic Bacteria," *Geochem. Cosmochim. Acta.* 40, 1141-1143 (1976).
(20) Blumer, M., "Curtisite, Idrialite and Pendletonite, Polycyclic Aromatic Hydrocarbon Minerals: Their Composition and Origin," *Chem. Geol.* 16, 245-256 (1975).
(21) Blumer, M., "Polycyclic Aromatic Hydrocarbons in Nature," *Sci. Am.* 234 (3), 35-45 (1976).
(22) Blumer, M. and Youngblood, W.W., "Polycyclic Aromatic Hydrocarbons in Soils and Recent Sediments," *Science* 188, 53-55 (1975).
(23) Youngblood, W.W. and Blumer, M., "Polycyclic Aromatic Hydrocarbons in the Environment: Homologous Series in Soils and Recent Marine Sediments," *Geochem. Cosmochim. Acta* 39, 1303-1314 (1975).
(24) Harrison, R.M., Perry, R., and Wellings, R.A., "Polynuclear Aromatic Hydrocarbons in Raw, Potable and Waste Waters," *Water Res.* 9, 331-346 (1975).
(25) Environmental Protection Agency (EPA), *Scientific and Technical Assessment Report on Particulate Polycyclic Organic Matter (PPOM)*, EPA-60016-75-001 (1975).
(26) Olsen, D. and Haynes, J.L., *Preliminary Air Pollution Survey of Organic Carcinogens. A Literature Review*, APTD 69-43, National Air Pollution Control Administration, Raleigh, N.C. (1969).
(27) Shabad, L.M., "Concerning the Possibility of Establishing Maximum Permissible Doses and Concentrations (MPDs and MPCs) for Carcinogens," *Hygiene Sanitation* 36 (10-12), 115-120 (1971).
(28) Standards Advisory Committee on Coke Oven Emissions, "Advisory Panel Recommends Adoption of Benzo(a)pyrene as Hazard Indicator," *Occupational Safety and Health Reporter* 4 (50), 1639-1640 (1971).
(29) National Academy of Sciences (NAS), *Petroleum in the Marine Environment*, Washington, D.C., National Academy of Sciences (1975).
(30) DiSalvo, L.H., Guard, H.E., and Hunter, L., "Tissue Hydrocarbon Burden of Mussels as Potential Monitor of Environmental Hydrocarbon Insult," *Environ. Sci. Technol.* 9 (3), 247-251 (1975).
(31) Ruivo, M., "Summary of Discussion," *Marine Pollution and Sea Life, Food and Agriculture Organization of the United Nations Conference, Rome*, ed. M. Ruivo, pp. 113-14, London, Fishing News (Books) Ltd. (1970).
(32) Gross, M.G., "Waste Removal and Recycling by Sedimentary Processes," *Marine Pollution and Sea Life, Food and Agriculture Organization of the United Nations Conference, Rome*, ed. M. Ruivo, pp. 152-158, London, Fishing News (Books) Ltd. (1970).
(33) Farrington, J.W. and Quinn, J.G., "Petroleum Hydrocarbons in Narragansett Bay. I. Survey of Hydrocarbons in Sediments and Clams (*Mercenaria mercenaria*)," *Estuarine Coastal Mar. Sci.* 1, 71-79 (1973).
(34) Lee, R.F., Sauerheber, R., and Benson, A.A., "Petroleum Hydrocarbons: Uptake and Discharge by the Marine Mussel *Mytilus edulis*," *Science* 177, 344-346 (1972).
(35) Lee, R.F., "Fate of Petroleum Hydrocarbons in Marine Zooplankton," *Proceedings of the Conference on Prevention and Control of Oil Pollution*, American Petroleum Institute, Washington, D.C. (1975).
(36) Lee, R.F., Ryan, C., and Neuhauser, M.L., "Fate of Petroleum Hydrocarbons Taken Up from Food and Water by the Blue Crab *Callinectes sapidus*," *Marine Biol.* 37, 363-370 (1976).

(37) Ehrhardt, M., "Petroleum Hydrocarbons in Oysters from Galveston Bay," *Environ. Pollut.* 3, 257-271 (1972).
(38) Whittle, K., Mackie, P.R., and Hardy, R., "Hydrocarbons in the Marine Ecosystem," *S. Aft. J. Sci.* 70 (5), 141-144 (1974).
(39) Ogata, M. and Miyake, Y., "Identification of Substances in Petroleum Causing Objectionable Odour in Fish," *Water Res.* 7, 1493-1504 (1973).
(40) Stegeman, J.J. and Teal, J.M., "Accumulation, Release and Retention of Petroleum Hydrocarbons by the Oyster *Crassostrea virginica*," *Mar. Biol.* 22, 37-44 (1973).
(41) Shelton, T.B. and Hunter, J.V., "Anaerobic Decomposition of Oil in Bottom Sediments," *J. Water Pollut. Control Fed.* 47 (9), 2256-2270 (1975).
(42) Elworthy, P.H., Florence, A.T., and Macfarlane, C.B., *Solubilization by Surface-Active Agents*, London, Chapman and Hall Ltd. (1968).
(43) Boehm, P.D. and Quinn, J.G., "Solubilization of Hydrocarbons by the Dissolved Organic Matter in Sea Water," *Geochim. Cosmochim. Acta* 37, 2459-2477 (1973).
(44) Andelman, J.B. and Snodgrass, J.E., "Incidence and Significance of Polynuclear Aromatic Hydrocarbons in the Water Environment," *CRC Crit. Rev. Environ. Control* 4 (1), 69-83 (1974).
(45) Barnett, S.M., Velankar, S.K., and Houston, C.W., "Mechanism of Hydrocarbon Uptake by Microorganisms," *Biotechnol. Bioeng.* XVI, 863-865 (1974)
(46) Goma, G., Pareilleux, A., and Durand, G., "Kinetics of Degradation of Hydrocarbons by *Candida lipolyfica*," *Arch. Mikrobiol.* 88, 97-109 (1973).
(47) McAuliffe, C., "Solubility in Water of Paraffin, Cycloparaffin, Olefin, Acetylene, Cycloolefin, and Aromatic Hydrocarbons," *J. Phys. Chem.* 70 (4), 1267-1275 (1966).
(48) Grasselli, J.G., *Atlas of Spectral Data and Physical Constants for Organic Compounds*, Cleveland, Ohio, CRC Press (1973).
(49) Feldman, M.H., *Petroleum Weathering: Some Pathways, Fate and Disposition on Marine Waters*, Corvallis, Oregon, Pacific Northwest Environmental Research Laboratory, Program Element 1BA025, EPA 660/3-73-013, PB-227 278 (1973).
(50) Ludzack, F.J. and Ettinger, M.B., "Biodegradability of Organic Chemicals Isolated from Rivers," in *Industrial Waste Conference, Purdue Univ.*, pp. 278-282 (1963).
(51) Kostyaev, V. Ya., "Effect of Phenol on the Hydrochemical Regime, Phytoplankton, and Plant Overgrowth in Artificial Bodies of Water," *Tr. Inst. Biol. Vnutr. Vod. Akad. Nauk. SSR* 24, 119-151 (1973).
(52) Jeffery, J., "Experiments on the Biological Treatment of Phenolic Effluents," *Gas World – Coking* 153 (3994), 23-25 (1961).
(53) Baker, E.G., "A Hypothesis Concerning the Accumulation of Sediment Hydrocarbons to Form Crude Oil," *Geochim. Cosmochim. Acta.* 19, 309-317 (1960).
(54) Ahlrichs, J.L., "The Soil Environment," *Organic Chemicals in the Soil Environment*, ed. C.A.I. Goring and J.W. Hamaker, pp. 3-43, New York, Marcel Dekker, Inc. (1972).
(55) Hamaker, J.W., "The Interpretation of Soil Leaching Experiments," *Environmental Dynamics of Pesticides*, ed. R. Haque and V.H. Freed, pp. 115-133, New York, Plenum Press (1975).
(56) Hamaker, J.W. and Thompson, J.M., "Adsorption," *Organic Chemicals in the Soil Environment*, ed. C.A.I. Goring and J.W. Hamaker, pp. 49-143, New York, Marcel Dekker, Inc. (1972).
(57) Baker, R.A. and Luh, M.-D., "Pyridine Sorption from Aqueous Solution by Montmorillonite and Kaolinite," *Water Res.* 5, 839-848 (1971).
(58) Roberts, A.L., Street, G.B., and White, D., "The Mechanism of Phenol Adsorption by Organo-Clay Derivatives," *J. Appl. Chem.* 14, 261-265 (1964).
(59) Letey, J. and Oddson, J.K., "Mass Transfer," *Organic Chemicals in the Soil Environment*, ed. C.A.I. Goring and J.W. Hamaker, pp. 399-440, New York, Marcel Dekker, Inc. (1972).
(60) Meikle, R.W., "Decomposition: Qualitative Relationships," *Organic Chemicals in the Soil Environment*, ed. C.A.I. Goring and J.W. Hamaker, pp. 145-251, New York, Marcel Dekker, Inc. (1972).
(61) Fernley, H.N. and Evans, W.C., "Oxidative Metabolism of Polycyclic Hydrocarbons by Soil Pseudomonads," *Nature* 182, 373-375 (1958).

(62) Davies, J.I. and Evans, W.C., "Oxidative Metabolism of Naphthalene by Soil Pseudomonads," *Biochem. J.* 91, 251-261 (1964).

(63) Evans, W.C., Fernley, H.N., and Griffiths, E., "Oxidative Metabolism of Phenanthrene and Anthracene by Soil Pseudomonads – the Ring-Fission Mechanism," *Biochem. J.* 95, 819–831 (1965).

(64) Muller, W.P. and Korte, F., "Microbial Degradation of Benzo-[a]-pyrene, Monolinuron, and Dieldrin in Waste Composting," *Chemosphere* 3, 195–198 (1975).

(65) Hussien, Y.A., Tewfik, M.S., and Hamdi, Y.A., "Degradation of Certain Aromatic Compounds by Rhizobia," *Soil Biol. Biochem.* 6, 377-381 (1974).

(66) Pupp, C., Lao, R.C., Murray, J.J., and Pottie, R.F., "Equilibrium Vapour Concentrations of Some Polycyclic Aromatic Hydrocarbons, As_4O_6 and SeO_2 and the Collection Efficiencies of These Air Pollutants," *Atmos. Environ.* 8, 915–925 (1974).

(67) Thomas, J.F., Mukai, M. and Tebbens, B.D., "Fate of Airborne Benzo[a]pyrene," *Curr. Res.* 2 (1), 33-39 (1968).

(68) Pierce, R.C. and Katz, M., "Dependency of Polynuclear Aromatic Hydrocarbon Content on Size Distribution of Atmospheric Aerosols," *Environ. Sci. Technol.* 9 (4), 347–353 (1975).

(69) DeMaio, L. and Corn, M., "Polynuclear Aromatic Hydrocarbons Associated with Particulates in Pittsburgh Air," *J. Air Pollut. Control Assoc.* 16, 67–71 (1966).

(70) Tebbens, B.D., Thomas, J.F., and Mukai, M., "Fate of Arenes Incorporated with Airborne Soot," *J. Am. Ind. Hyg. Assoc.* 27, 415–422 (1966).

(71) Falk, H.L., Markul, I., and Kotin, P., "Aromatic Hydrocarbons. IV. Their Fate Following Emission into the Atmosphere and Experimental Exposure to Washed Air and Synthetic Smog," *A. M. A. Arch. Ind. Health* 13, 13-17 (1956).

MICROBIAL AND PHOTOCHEMICAL EFFECTS ON PNA

The information in this chapter is based on:

Environmental, Health and Control Aspects of Coal Conversion: Volume 2, An Information Overview, ORNL/EIS-95, April 1977, prepared by H.M. Braunstein, E.D. Copenhaver and H.A. Pfuderer of Oak Ridge National Laboratory for Energy Research and Development Administration under Contract W-7405-eng-26

Formation and Transformation of Polycyclic Organic Matter from Coal Combustion, NTIS COO-4347-1, 1978, prepared by D.F.S. Natusch of Colorado State University under Contract EE-77-S-02-4347.

MICROBIAL INTERACTIONS WITH PNA

Industrial waste such as that expected from coal conversion processes contains polycyclic aromatic hydrocarbons (PAH). It is important to draw on related industry experience with PAH to assess the role of microorganisms in determining the fate of these compounds in the environment.

The literature is inconsistent in defining the fate of benzo[a]pyrene (BaP). Authors in both laboratory and field degradation experiments report highly variable information on degradation. Benzpyrene (BP) has been found almost universally to be degraded by soil microorganisms.

Degradation by Soil Microorganisms

Soil microflora can destroy PAH. BP-contaminated soil was collected and analyzed by Khesina et al (1), who showed that soil microflora can destroy up to 70% of the BP contained in the soil. A mutant strain of the soil bacteria *Bacillus megaterium* was isolated and found capable of metabolizing (apparently by an oxidation reaction) BP and several other PAH. The bacteria metabolized the PAH compounds both in culture media and in natural soil contaminated with industrial waste products. The bacteria could degrade 80 to 90% of the BP originally present in 8 days of cultivation (1).

Benzo[a]pyrene: Poglazova et al (2)-(5) reported a series of experiments in which bacteria were isolated from BaP-polluted and nonpolluted soils collected in various areas of the U.S.S.R. Nonspore-forming bacteria were isolated from BaP-contaminated soil (100 mg of BaP/kg of soil) surrounding an oil refinery. Pure cultures were isolated by repeated inoculation of soil samples onto meat-peptone agar. Seventeen bacterial strains of different morphological and cultural characteristics were isolated. All the isolates assimilated BaP to some degree, all were spore-forming, and many belonged to the genus *Megaterium*. Four of the seventeen cultures (PBK No. 2/I, PBK No. 2/II, PBK No. 5, and PBK No. 13) were selected for quantitative investigations. The selected bacteria strains were cultured on nutrient agar to which BaP had been added.

BaP decomposition or modification was determined by measuring the difference between the BaP introduced into the medium and the amount in the medium and microorganisms after incubation. The experiment was run four times, and the results (Table 4.1) were averaged. Cultures of bacterial strains No. 13 and 2/II showed a gradual disappearance of BaP; cultures containing bacteria strains No. 2/I and No. 5 completely retained the BaP in the medium. These results indicate that BaP-polluted soils do contain bacteria capable of modifying the BaP.

Using nonspore-forming bacteria isolated from his 1966 work, Poglazova et al (3) used members of the *Mycobacterium* genus and the mutant *Bacillus megaterium* mutilate for measurement of BaP assimilation. Results (Table 4.2) showed that *Mycobacterium rubrum* and a *M. flavum* strain exhibited the highest BaP degradation–54 and 48% respectively. All the *Mycobacteria* could degrade BaP to at least 14%.

Table 4.3 shows BaP degradation potentials of *Bacillus megaterium* mutilate and isolates PBK No. 5 and PBK No. 13. The cultures were subjected to both solid and aerated liquid media. The bacteria showed a greater BaP dissimilatory ability in the liquid media than on the solid media; this tendency is possibly due to closer contact of the bacteria cells with the hydrocarbon in liquid media (3).

Poglazova et al (4) again used *Bacterium megaterium* mutilate and isolates PBK 5 and PBK 13 in a study of BaP degradation. Test soil contaminated with BaP was moistened with meat-peptone broth and sterilized. Test cultures were inoculated onto the soil and incubated. Results (Table 4.4) show that all cultures could oxidize BaP (up to 85%) in the contaminated and sterilized soil. Continued studies with incubation of unsterilized soil contaminated with BaP showed that natural soil microorganisms could also degrade the hydrocarbon when it was present in high concentrations (Table 4.4).

Poglazova et al (5) again compared BaP degradation between natural soil bacteria and selected species. Natural bacteria cultures were isolated from soils containing varying amounts of BaP. Results (Table 4.5) show that natural soil microorganisms in all the soil samples decomposed significant (20 to 40%) amounts of BaP. Poglazova et al (5) also sterilized the soil samples and inoculated them with either *Bacillus sphaericus, Bacillus megaterium* mutilate, or *Pseudomonas*-146.

Table 4.1: Change in BaP Total Content in Medium and Cells of Soil Microorganisms During Their Culturing

Bacterial Cultures	Duration of Culturing (days): 2 — BaP Extracted (μg), Experiment 1	2: Experiment 2	2: Experiment 3	2: Untreated with Bacteria (%)	3 — BaP Extracted (μg) Experiment 4	3: Treated with Bacteria (%)	4 — BaP Extracted (μg), Experiment 1	4: Experiment 2	4: Experiment 3	4: Experiment 4	4: Treated with Bacteria (%)
PBK No. 13	176	174	175	13	114	43	63	84	117	118	53
PBK No. 2/II	176	192	165	11	154	23	–	132	118	150	33
PBK No. 5	201	201	188	0	204	0	199	195	207	–	0
PBK No. 2/I	198	188	182	0	195	0	210	180	195	–	0
Control	208	201	198	–	203	–	210	198	–	–	–

Source: Reference (2) as cited in ORNL/EIS-95

Table 4.2: BaP Content Change Due to Dissimilation by *Mycobacteria*

Culture	BaP Introduced (μg)	BaP After 4 days of Culturing (μg)	Deficit (%)
Mycobacterium flavum var. *methanicum* strain B_4b	280	200	29
M. flavum var. *methanicum* strain C_1	280	145	48
M. lacticola strain B_3	280	188	33
M. lacticola strain D_5	280	237	15
M. rubrum var. *propanicum* strain B_4a	280	241	14
M. rubrum var. *propanicum* strain D_2	280	219	22
M. rubrum	280	128	54
M. smegmatis	280	180	35

Table 4.3: Comparison of BaP Dissimilation by Microorganisms in Solid and Liquid Media

Culture	Solid Medium: BaP Introduced (μg)	Solid Medium: BaP Remaining After 4 days of Culturing (μg)	Solid Medium: Deficit (%)	Liquid Medium: BaP Introduced (μg)	Liquid Medium: BaP Remaining After 4 days of Culturing (μg)	Liquid Medium: Deficit (%)
B. megaterium mut.	200	170	15	570	370	36
PBK No. 13	200	166	17	570	535	6
PBK No. 5	200	160	20	520	240	54

Source: Reference (3) as cited in ORNL/EIS-95

Table 4.4: Destruction of BaP by Soil Microorganisms*

Culture	BaP** (μg/10g)	Deficit (%)
Sterilized Soil		
No. 13	270	79
No. 5	200	85
B. megaterium mutilate	234	82
Unsterilized Soil		
Natural soil microflora	525	60
Natural microflora + No. 13	370	71
Natural microflora + No. 5	350	73
Natural microflora + *B. megaterium* mutilate	290	78

*Soil naturally contaminated with BaP; original BaP concentration 1,306 μg per 10 g of soil.
**Remaining after eight days of culturing.

Source: Reference (4) as cited in ORNL/EIS-95

Table 4.5: Decomposition of BaP by Microflora of Various Soils

Soil Sample Location (Territory)	Initial BaP (μg/ml)	BaP* (μg/ml)	Deficit (%)
Klyaz'menskii reservoir	5	4	20
MONIKI hospital	5	3	40
"Neftegaz" factory	5	3	40

*After 5 days of cultivation

Source: Reference (5) as cited in ORNL/EIS-95

The bacterial cultures were previously shown to be BaP metabolizers and, after incubation, showed degradation of 70 to 80% in the most heavily polluted soils (Table 4.6). Of the tested soils, those containing a greater quantity of aromatic compounds provide more favorable conditions for promoting the processes of PAH metabolism; however, the mechanism of this phenomenon remains obscure (5).

Table 4.6: Decomposition of BaP in the Soils by Several Microorganisms

Location and Strain (Territory)	Initial BaP* (μg)	BaP** (μg)	Deficit (%)
Klyaz'menskii reservoir			
B. sphaericus		0.007±0.0005	48
B. megaterium mutilate	0.0135±0.0015	0.0055±0.0015	59
Pseudomonas-146		0.006±0.0001	55
MONIKI hospital			
B. sphaericus		0.9±0.2	82
B. megaterium mutilate	5.15±0.55	1.3±0.1	75
Pseudomonas-146		0.7±0.1	86
"Neftegaz" plant			
B. sphaericus		2.4±1.0	76
B. megaterium mutilate	10.0±2.0	2.0±1.0	80
Pseudomonas-146		2.0±1.2	80

*In 10 g of soil
**In 10 g of soil after 5 days of cultivation

Source: Reference (5) as cited in ORNL/EIS-95

Muller and Korte (6) experimented with BaP degradation by natural soil microorganisms. BaP in an acetone-water solution was sprayed on ground waste (10 to 20 μg/g) and composted for three weeks. Results showed that BaP was able to withstand the composting process to at least 99.5%. No degradation by microorganisms was observed in the extractable matter.

Other PAH: Fedoseeva et al (7) cultured *Bacillus megaterium* mutilate with a variety of PAH compounds. The *B. megaterium* mutilate was known to be a BaP-degrading bacteria from the earlier work by Poglazova; the work by Fedoseeva et al, however, examined the ability of the bacteria strain to degrade PAH compounds other than BaP.

The compounds used in this study were 9,10-dimethyl-1,2-benzanthracene (DMBA), 1,2,5,6-dibenzanthracene (DBA), 1,2-benzanthracene (BaA), pyrene (P), 1,2-benzpyrene (BeP), BaP, 1,12-benzperylene (1,12-BPL), and perylene (PL). One of the above hydrocarbons was added to each growth medium to a concentration of 10 mg/ml of medium. Each hydrocarbon-containing medium preparation was then inoculated with the strain of bacteria. Cultures were incubated for four days. After incubation, the remaining undegraded hydrocarbon was removed from the medium along with the microorganisms growing in it by extraction with n-octane accompanied by mechanical shaking for 24 hours.

Results of the Fedoseeva et al work (Table 4.7) shows that *B. megaterium* mutilate could oxidize all the PAH compounds tested and was not BaP-specific. The percentage of the hydrocarbon oxidized was found to be relatively the same for each compound, regardless of the compound's solubility or its ability to act as an enzyme inducer.

Table 4.7: Oxidation of Aromatic Hydrocarbons by Microorganisms

Compound Investigated	. . . . Concentration . . . . (μg/ml nutrients medium) (control)	(μg/ml medium)*	Hydrocarbon Metabolized (% of control)
DMBA	10.5	7.3	
	10.8	6.8	37
	9.5	5.2	
BaP	8.4	5.8	
	8.0	5.2	35
	8.8	5.4	
	9.55	5.8	
DBA	9.45	5.0	42
	9.9	6.1	
	8.6	6.3	
BaA	9.7	6.1	32
	8.6	6.0	
	7.9	4.2	
P	7.8	3.2	55
	9.0	3.8	
	10.0	7.9	
BeP	10.5	6.7	23
	9.5	8.4	

(continued)

Table 4.7: (continued)

Compound Investigated	Concentration (μg/ml nutrients medium) (control)	Concentration (μg/ml medium)*	Hydrocarbon Metabolized (% of control)
1,12-BPL	9.0	5.8	
	10.1	5.9	40
	9.5	5.3	
	12.6	9.4	
PL	13.0	7.4	33
	12.0	9.5	

*After 4 days incubation.

Source: Reference (7) as cited in ORNL/EIS-95

Fedoseeva et al (7) compared the oxidizing ability of *B. megaterium* mutilate with high and low concentrations of perylene, which has a low solubility, and BaP, which has a high solubility. The results (Table 4.8) show that the bacteria could oxidize both hydrocarbons over a wide range of concentrations with practically no percentage change. These results seem to indicate that microorganisms can oxidize PAH compounds in their surroundings to an extent that is practically independent of compound concentration.

Table 4.8: Oxidation of Aromatic Hydrocarbons in Relation to Their Concentration in the Culture Medium

Compound Investigated	Concentration, μg/ml Before Incubation*	Concentration, μg/ml After Incubation	Hydrocarbon Metabolized (% of control)
PL	12.6	9.4	
	13.0	7.4	33
	12.0	9.5	
PL	2.04	1.55	
	1.98	1.43	31
	2.3	1.39	
BP	9.7	4.8	
	10.25	6.3	46
	10.0	5.0	
BP	1.9	1.1	
	2.1	1.05	45
	2.0	1.1	

*Control.

Source: Reference (7) as cited in ORNL/EIS-95

Degradative Pathway: Phenanthrene decomposition was studied by Rogoff and Wender (8) to identify intermediate transformation products in the oxidative pathway used by soil microorganisms. The microorganisms were isolated from mixed cultures grown from soil samples. The isolate was a gram-negative motile bacillus that produced small, circular, granular colonies and liquefaction on gelatin plates, and saccate liquefaction with a whitish sediment in gelatin stabs.

Further biochemical tests showed the isolate to be a member of the genus *Pseudomonas* and resembled the type species *P. aeruginosa.* The isolate also showed some resemblance to the naphthalene-attacking *P. boreopolis.*

The experiment was carried out in 500-ml Erlenmeyer flasks, with 96 hr incubation at 37°C before determinations of residual phenanthrene were taken. Decomposition varied from 46 to 78% with the different isolate cultures tested. The *Pseudomonas* isolate, which consistently decomposed from 65 to 78% of the phenanthrene, was used in the study. Interestingly, after 36 to 48 hr of incubation, the fermentation liquor became a deep buff color and gave a positive bluish-green ferric chloride test, which indicates the presence of a phenolic compound, possibly a hydroxynaphthoic acid (8).

The oxidation of phenanthrene by soil isolates was also studied much earlier by Tausson (9). The soil isolate was able to decompose relatively large amounts of phenanthrene in a fairly short time. Growth experiments were with o-hydroxybenzyl alcohol, o-hydroxybenzaldehyde, salicylic acid, and catechol as substrates. Tausson postulated, on the basis of the rapidity with which the compounds were assimilated, that the primary attack on the phenanthrene molecule occurred at the 9,10-position and resulted in the formation of the abovementioned compounds as intermediates (Pathway I, Figure 4.1).

Figure 4.1: Possible Pathways of Phenanthrene Oxidation by Microorganisms

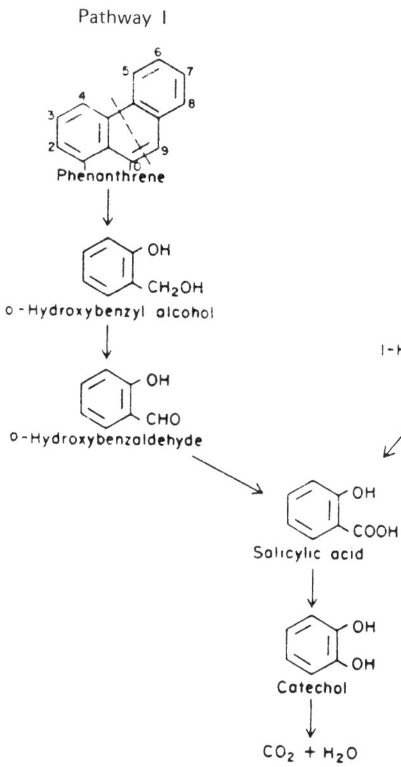

Source: Reference (8) as cited ORNL/EIS-95

Rogoff and Wender (8) isolated 1-hydroxy-2-naphthoic acid from their cultures and found that the phenanthrene-grown cells could oxidize salicylic acid and catechol. Based on their results, Rogoff and Wender postulated that the attack on the phenanthrene molecule occurred via the end-ring (Pathway II, Figure 4.1). The end-ring attack first occurs at the 3,4-bond of the phenanthrene molecule (10). The possibility of differences between points of attack in the Tausson postulate and the Rogoff and Wender postulate due to the metabolic pattern differences of the organisms used cannot be eliminated (8).

Degradation by Marine Organisms

Sisler and ZoBell (11) describe the microbial use of aromatic hydrocarbons (naphthalene, anthracene, phenanthrene, BaA, and DBA) in laboratory cultures. The bacteria were mixed cultures of marine origin. This work involved the dispersal of the aromatic hydrocarbon on ignited sand and the immersion of the sand and substrate into seawater. Seawater is a physiologically balanced solution and contains all the essential salts necessary to satisfy the mineral requirements of marine microorganisms (11).

The amount of hydrocarbon used was measured by the amount of carbon dioxide evolved from oxidation of the substrate. The carbon dioxide produced by the control was subtracted from the total carbon dioxide produced by the culture acting on the substrate to give the net amount of hydrocarbon oxidized by the microorganisms (Table 4.9). The marine bacteria assimilated phenanthrene and anthracene more rapidly than naphthalene, or the higher homologues (BaA and DBA). No intermediates in the oxidation processes were isolated.

Table 4.9: Amount of Carbon Dioxide Produced by Action of Bacteria on 25 mg of Hydrocarbons in Four Days at 32°C

Hydrocarbon	CO_2 Produced (mg)	Amount Oxidized (%)
Naphthalene	44.2	51
Anthracene	53.5	64
Phenanthrene	58.5	68
Diaminobenzene	14.0	23
1,2-Benzanthracene	44.2	47
1,2,5,6-Dibenzanthracene	11.6	13
None (control)*	0	–

*The controls used as checks against adventitious organic matter produced less carbon dioxide than the endogenous controls, hence are condensed as one control shown here.

Source: Reference (11) as cited in ORNL/EIS-95

The degradation of naphthols by marine organisms has received little attention. Sikka, Miyazaki, and Lynch (12) studied the ability of selected species of marine bacteria, yeasts, and filamentous fungi to degrade 1-naphthol. The compound used in the study was 1-naphthol-1-^{14}C having a specific activity of 15.2 mci/mmol.

Organisms isolated from a marine environment and used in this study were *Brevibacterium* sp, *Flavobacterium* sp, *Serratia marina,* and *Spirillum* sp (bacteria); *Candida parapsilosis, Rhodotorula glutinus,* and *Trichosporon fermentans* (yeasts);

Aspergillus fumigatus, Culcitalna achraspora, Halosphaeria mediosetigera, and *Humicola alopallonella* (filamentous fungi). The bacteria were grown on Difco marine broth, whereas the yeasts and filamentous fungi were grown in a medium containing 5 g of glucose, 2.4 g of NH_4NO_3, and 1.0 g of yeast extract per liter of artificial seawater prepared from Rila marine mix (Rila Products). The organisms were grown on their respective nutrient media containing 1.0 ppm of 1-naphthol. The microorganisms showed the ability to degrade 1-naphthol and also produced different types and metabolites of 1-naphthol (Table 4.10).

Table 4.10: Distribution of ^{14}C in the Culture Solution of Various Microorganisms Incubated with ^{14}C-1-Naphthol

	Percent of Initial ^{14}C						
	Ether-Soluble Products						Water-
	Rf-Values						Soluble
	0	0.09	0.31	0.39	0.64	Total	Products*
Brevibacterium sp	–	0	0	77.8	0	77.8	3.6
Flavobacterium sp	10.6	0	0	72.2	0	76.8	4.7
Serratia marina	5.7	0	0	59.4	0	65.1	15.8
Spirillum sp	7.9	0	0	57.1	0	65.0	14.3
Candida parapsilosis	0	0	0	74.7	0	74.7	7.5
Rhodotorula glutinis	0	0	0	73.0	0	73.0	8.0
Trichosporon ferments	0	0	0	71.8	0	71.8	9.6
Aspergillus fumigatus	2.4	2.4	6.9	0	1.1	12.8	74.8
*Culcitalna achraspora***	4.0	2.1	3.2	0.3	3.1	12.7	70.7
*Halosphaeria mediosetigera***	1.7	3.0	11.6	2.8	0	19.1	70.7
*Humicola alopallonella***	0	2.1	2.8	0	0.9	5.8	90.2

*Corrected for nonbiological conversion.

**These organisms were incubated with 1-naphthol for 7 days, whereas the others were incubated for 3 days.

Source: Reference (12) as cited in ORNL/EIS-95

The cultures of *Brevibacterium, Candida, Rhodotorula,* and *Trichosporon* incubated with ^{14}C-1-naphthol produced only small amounts of water-soluble metabolites from 1-naphthol. The remaining organisms were able to convert 1-naphthol to both ether- and water-soluble metabolites. *Culcitalna, Halosphaeria, Humicola,* and *Aspergillus* were more effective than *Flavobacterium, Spirillum,* and *Serratia* in converting 1-naphthol to its water-soluble metabolites.

In the first four genera, ^{14}C in the ether phase had decreased to less than 20% of the original amount. *Humicola* was found to be the most effective in converting 1-naphthol into water-soluble metabolites. It was also stated that soil microorganisms such as *Fusarium* and *Pseudomonas* converted a significant amount of 1-naphthol into unidentified water-soluble metabolites (12).

Resistance to Degradation

Activated sludges from municipal waste treatment systems have been shown to be mostly ineffective in the treatment of aromatic, organic industrial wastes. The inability of treatment systems to degrade the hydrocarbons could result in introduction of the hydrocarbons into local water tables as pollutants.

Oxidated Sludge: Malaney et al (13) studied the resistance of carcinogens to oxidation by oxidated sludge. The compounds studied (Figure 4.2) were mostly aromatic, organic industrial wastes. Sludge samples were taken from municipal treatment plants at Nashville, Franklin, and Ashland City, Tennessee.

Figure 4.2: Carcinogens and Their Molecular Structures That Were Studied for Biologic Oxidation

9,10–Dimethyl-1,2-benzanthracene

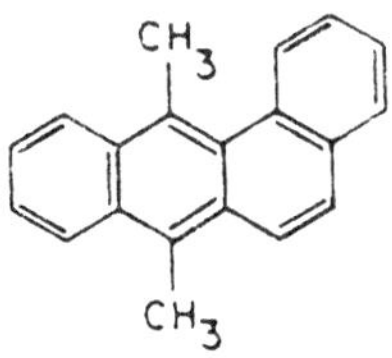

1,2-Benzanthracene

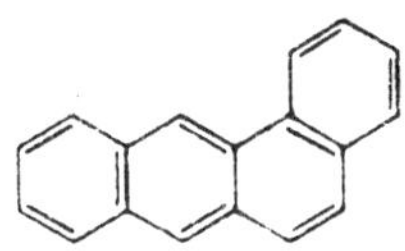

3,4-Benzpyrene

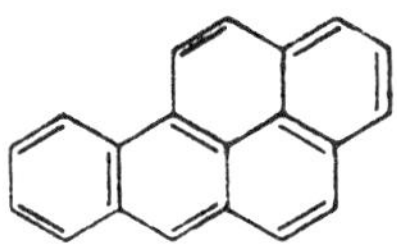

1,2,4,5-Dibenzpyrene

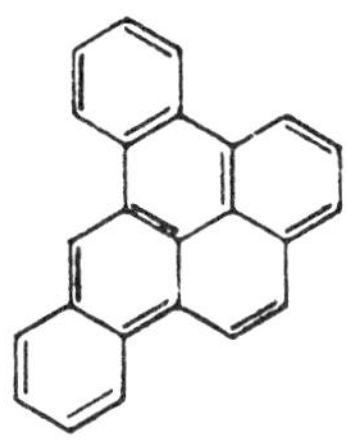

1,2,5,6-Dibenzanthracene

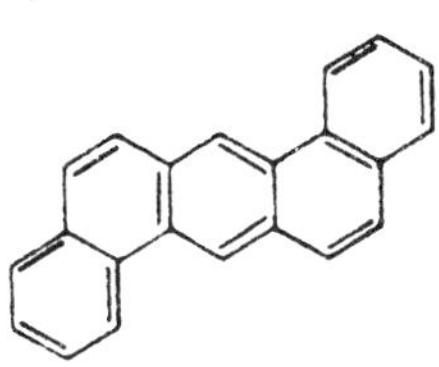

20-Methylcholanthrene

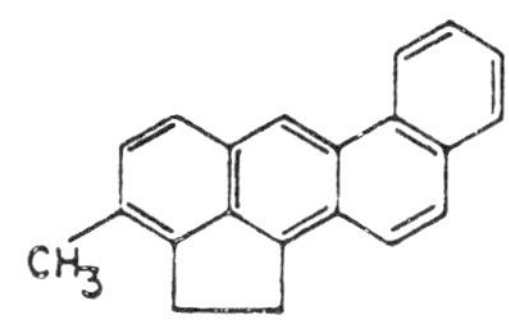

2-Nitrofluorene

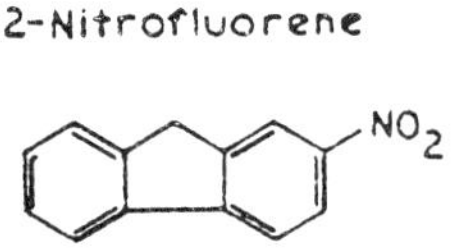

2-Fluoreneamine

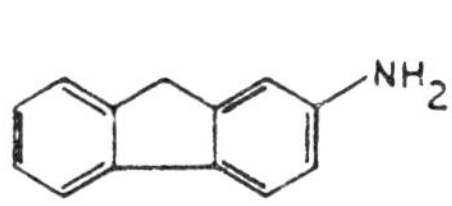

N-2-Fluorenylacetamide

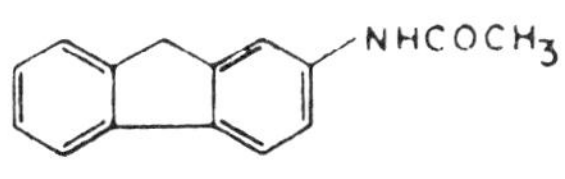

(continued)

Figure 4.2: (continued)

7,9-Dimethylbenz(c)acridine

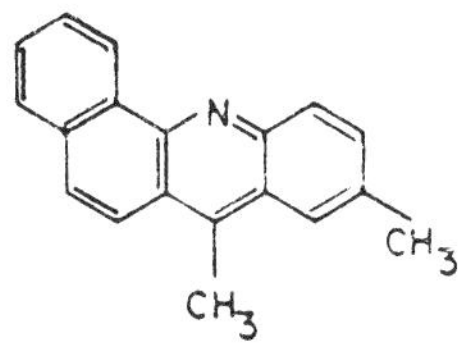

7,10-Dimethylbenz(c)acridine

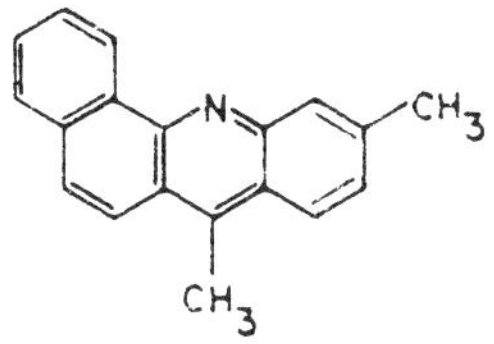

Dibenz(a,h)acridine

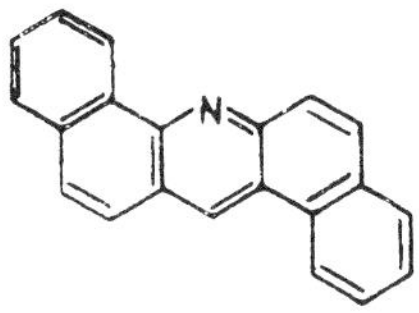

Dibenz(a,j)acridine

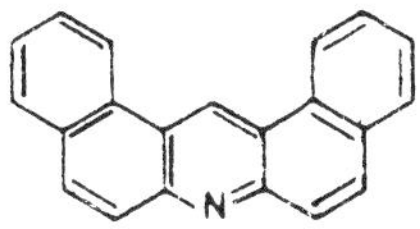

2,3-Butylene oxide

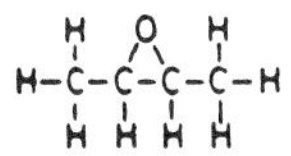

beta-Propiolactone

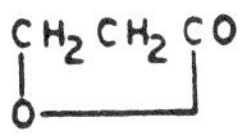

Thiourea

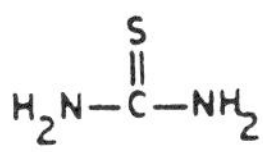

Ethyl carbamate

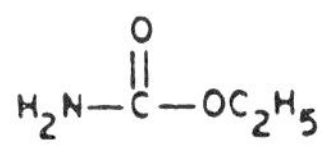

2-Thiouracil

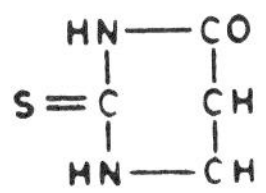

4-Ethoxyphenylurea

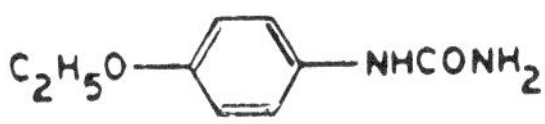

Benzidine

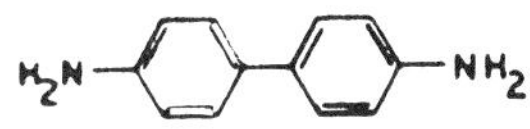

4,4′-Dihydroxy-a,b-diethylstilbene

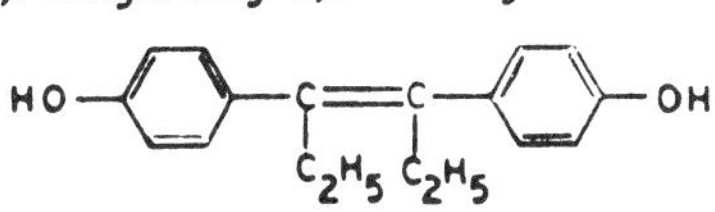

p-Phenylazophenol

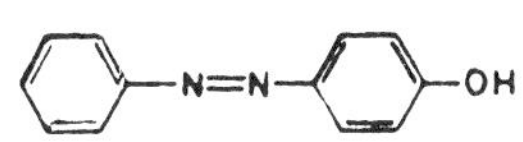

4,4-Bis(dimethylamino)benzophenone

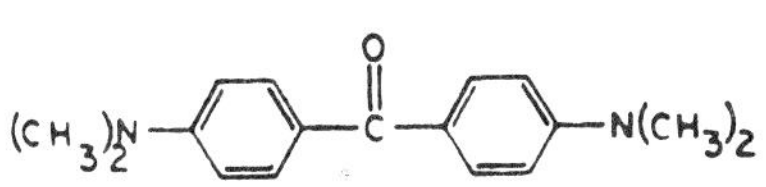

p-Phenylazoaniline

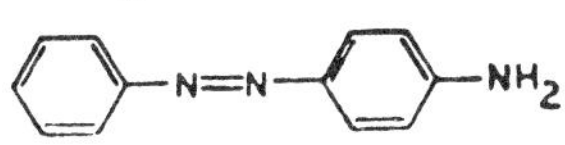

(continued)

Figure 4.2: (continued)

9,10-Dimethylanthracene

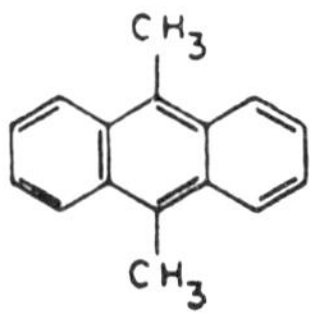

beta-Naphthylamine

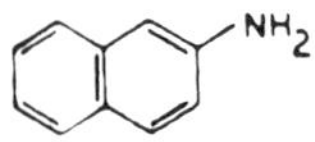

7-Methyl-1,2-benzanthracene

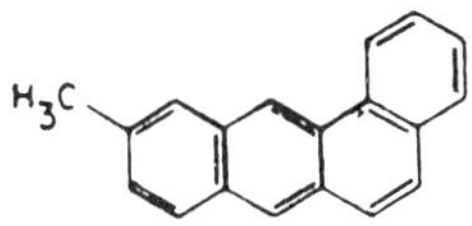

Source: Reference (13) as cited in ORNL/EIS-95

Aeration-tank-mixed liquors were collected on the day of the experiment, blended for 10 seconds, and analyzed for suspended solids concentration (MLSS) by the membrane filter technique. The MLSS concentration was then adjusted to 2500 or 5000 mg/ℓ. 20 ml of the sludge, which was neither washed nor treated with mineral salts, was pipetted into 125-ml Warburg flasks containing the individual organic compounds. Substrate concentration in the Warburg flask reaction compartment was 500 mg/ℓ. Warburg apparatus water bath temperature was 20°C.

Readings were made every 2 hours for the first 24 hours and then every 4 hours for the remaining 120 hours of the experimental run. Results are shown in Tables 4.11, 4.12 and 4.13. Results rapidly indicated that most of the compounds tested were markedly resistant to biological oxidation (13).

Table 4.11: Oxidation of Carcinogens by Ashland City Activated Sludge at 2,500 mg/ℓ*

Compound	Percentage of Theoretical Oxygen Demand**			
	6 Hours	24 Hours	72 Hours	144 Hours
2,3-Butylene oxide	0.3	T	T	2.4
β-Propiolactone	T	T	30.3	55.9
Thiourea	T	T	T	T
Ethyl carbamate	T	T	T	T
2-Thiouracil	T	T	T	T
4-Ethoxyphenylurea	T	T	T	T
Benzidine	T	T	T	T
4,4'-Dihydroxy[ab] diethylstilbene	T	T	T	T
2-Naphthylamine	0.4	T	T	T
4,4'-Bis(dimethylamino)benzophenone	T	T	0.9	0.6
p-Phenylazophenol	T	T	T	T
p-Phenylazoaniline	T	T	T	T
9,10-Dimethylanthracene	0.1	0.2	0.0	0.0

(continued)

Table 4.11: (continued)

Compound	Percentage of Theoretical Oxygen Demand** 6 Hours	24 Hours	72 Hours	144 Hours
1,2-Benzanthracene	0.1	0.0	0.0	0.3
7-Methyl[1,2] benzanthracene	0.1	T	0.0	0.4
9,10-Dimethyl[1,2] benzanthracene	0.1	0.2	0.3	0.5
1,2,5,6-Dibenzanthracene	0.1	T	0.0	T
3,4-Benzpyrene	0.2	1.5	1.6	1.7
1,2,4,5-Dibenzpyrene	0.0	T	T	T
20-Methylcholanthrene	0.1	0.1	0.0	T
2-Nitrofluorene	T	T	T	T
2-Fluoreneamine	T	T	T	T
N-2-Fluorenylacetamide	T	T	T	T
7,9-Dimethylbenz[c] acridine	0.1	0.1	0.7	T
7,10-Dimethylbenz[c] acridine	T	0.0	0.0	T
Dibenz[ah] acridine	0.3	0.3	0.5	0.7
Dibenz[aj] acridine	T	0.1	0.0	T

*Figures in tables obtained using following calculation:

$$\% \text{ TOD} = \frac{O_2 \text{ uptake substrate} - O_2 \text{ uptake control}}{\text{TOD}} \times 100$$

**T is toxic, i.e., O_2 uptake control was greater than O_2 uptake substrate, indicating that the substrate was inhibitory or lethal to the sludge organisms.

Source: Reference (13) as cited in ORNL/EIS-95

Table 4.12: Oxidation of Carcinogens by Nashville Activated Sludge at 2,500 mg/ℓ

Compound	. Percentage of Theoretical Oxygen Demand . 6 Hours	24 Hours	72 Hours	144 Hours
2,3-Butylene oxide	0.5	2.0	4.1	9.6
β-Propiolactone	T	T	T	T
Thiourea	T	T	T	T
Ethyl carbamate	0.1	1.6	T*	T
2-Thiouracil	0.8	2.4	12.2	12.8
4-Ethoxyphenylurea	1.9	6.3	13.0	9.4
Benzidine	1.0	4.1	T	T
4,4′-Dihydroxy[ab] diethylstilbene	T	T	T	T
2-Naphthylamine	0.3	2.8	T*	T
4,4′-Bis(dimethylamino)benzophenone	T	0.2	3.7	4.9
p-Phenylazophenol	0.1	4.7	T*	T
p-Phenylazoaniline	1.5	4.3	T*	T
9,10-Dimethylanthracene	0.5	1.6	3.2	19.5
1,2-Benzanthracene	0.5	1.1	1.6	2.1
7-Methyl[1,2] benzanthracene	T	T	1.6	3.1
9,10-Dimethyl[1,2] benzanthracene	0.5	1.4	9.4	12.7
1,2,5,6-Dibenzanthracene	0.8	1.8	5.5	7.9
3,4-Benzpyrene	0.3	1.4	3.5	2.7
1,2,4,5-Dibenzpyrene	0.3	0.9	1.0	1.8
20-Methylcholanthrene	0.7	1.7	6.1	9.3
2-Nitrofluorene	1.4	2.8	10.3	13.7
2-Fluoreneamine	0.9	4.2	T*	T

(continued)

Table 4.12: (continued)

Compound	. Percentage of Theoretical Oxygen Demand .			
	6 Hours	24 Hours	72 Hours	144 Hours
N-2-Fluorenylacetamide	1.3	3.7	3.2	6.8
7,9-Dimethylbenz[c] acridine	0.3	0.9	1.2	4.1
7,10-Dimethylbenz[c] acridine	0.0	0.1	1.6	4.3
Dibenz[ah] acridine	0.6	0.9	3.1	4.6
Dibenz[aj] acridine	0.0	0.6	3.3	4.7

*Result is consequence of sudden increase in oxygen uptake by sludge alone, perhaps onset of nitrification.

Source: Reference (13) as cited in ORNL/EIS-95

Table 4.13: Oxidation of Carcinogens by Franklin Sludge at 2,500 mg/ℓ

Compound	. Percentage of Theoretical Oxygen Demand .			
	6 Hours	24 Hours	72 Hours	144 Hours
2,3-Butylene oxide	1.8	T	T	T
β-Propiolactone	T	T	T	T
Thiourea	0.5	3.5	1.4	5.7
Ethyl carbamate	T	T	T	T
2-Thiouracil	1.7	2.7	5.5	6.2
4-Ethoxyphenylurea	T	5.8	24.2	20.6
Benzidine	T	T	1.9	1.9
4,4'-Dihydroxy[ab] diethylstilbene	0.4	6.0	T*	5.0
2-Naphthylamine	T	T	T	T
4,4'-Bis(dimethylamino)benzophenone	0.6	2.1	4.1	4.5
p-Phenylazophenol	T	6.4	4.2	T**
p-Phenylazoaniline	T	9.8	4.2	T**
9,10-Dimethylanthracene	0.1	0.2	0.6	1.6
1,2-Benzanthracene	0.5	0.7	1.2	1.6
7-Methyl[1,2] benzanthracene	T	T	T	T
9,10-Dimethyl[1,2] benzanthracene	0.1	0.2	0.5	0.5
1,2,5,6-Dibenzanthracene	0.1	0.1	0.3	0.9
3,4-Benzpyrene	1.2	1.8	3.8	6.1
1,2,4,5-Dibenzpyrene	0.1	0.1	0.0	T
20-Methylcholanthrene	0.8	1.1	2.1	4.4
2-Nitrofluorene	1.2	3.6	7.2	12.4
2-Fluoreneamine	1.1	9.7	8.0	T*
N-2-Fluorenylacetamide	2.6	7.5	10.8	12.3
7,9-Dimethylbenz[c] acridine	0.1	0.4	T	T
7,10-Dimethylbenz[c] acridine	0.1	0.4	0.7	1.1
Dibenz[ah] acridine	0.1	0.2	T	T
Dibenz[aj] acridine	T	T	T	T

*Result is consequence of sudden increase in oxygen uptake by sludge alone, perhaps onset of nitrification.

**Result is consequence of unexplained decrease in oxygen uptake by sludge-substrate mixture.

Source: Reference (13) as cited in ORNL/EIS-95

It could be concluded that the poor oxidation performance of the activated sludge is due to the refractory chemical structure of the carcinogenic compounds. These compounds are condensed nuclear hydrocarbons whose structures are relatively strain-free. The stable structures result from carbon-carbon bonds being in the staggered configuration (the same configuration as methane) and all bond angles being tetrahedral, which produces the least strain (13).

Waste Stream Microorganisms: Saprophytic microorganisms were used in a waste compost to break down organic compounds, including BaP. No degradation of the BaP was observed to at least 99.5% (6).

Fedorenko (14) studied BaP-containing waste streams from a by-product coke plant. The waste streams were subjected to biochemical treatment for purification before being released into a local river. Fedorenko tested the waste stream for BaP content before and after biochemical treatment. Results showed that the treatment, including passage through a biological basin, decreased the BaP content only slightly; the result was contamination of the river at the site of the waste stream discharge.

Gibson and Jerina (15) created a mutant strain of *Beijerinckia* capable of oxidizing BaP. The original *Beijerinckia* species was isolated from a polluted stream by virtue of its ability to grow with biphenyl as its sole source of carbon and energy. Treatment of this organism with N-methyl-N'-nitro-N-nitrosoguanidine led to the isolation of the mutant strain *Beijerinckia* B-836. This organism was then grown on succinate in the presence of biphenyl and, after this procedure, it oxidized BaP and BaA to polar compounds. The compounds were identified as cis-dihydrodiols, which suggests that the intermediates are not formed from arene oxide precursors because trans configuration would then occur. It is possible that eukaryotic microorganisms use a monooxygenase enzyme system for the oxidation of aromatic hydrocarbons (15).

Barnsley (16) studied cultures of four bacteria strains and their metabolism of BaP solutions. The rate of disappearance of BaP increased as a culture entered the stationary phase of growth. The addition of cyanide did not increase the rate of disappearance. The rates observed with the four species during 6 hour incubations in the stationary phase were *Pseudomonas* NCIB 9816, 0.13 μmol/hr; *Pseudomonas* ATCC 17483, 0.05 μmol/hr; *Pseudomonas putida* PpG7, 0.05 μmol/hr; and *Escherichia coli,* not detected.

Synthesis by Microorganisms

Biosynthesis of polybenzene compounds by microorganisms was proposed by Mallet et al (17). Several studies indicate increases in PAH concentrations that could only be attributed to biosynthesis; other investigations using bacterial and plankton extracts showed the synthesis of PAH when fatty acids were the initial substrates.

Brisou (18) showed that at least part of the BaP, a carcinogen, found in nature could have a biological origin. Several common anaerobic soil bacteria isolated from soil samples demonstrated the ability to produce BaP under laboratory conditions. However, a study by Hase and Hites (19) disputed the biosynthesis studies and stated that microorganisms actually act as bioaccumulators.

Detection of PAH in Soils and Deep Sediments: The presence of polybenzene hydrocarbons in plant soils and marine sediments was demonstrated by Mallet and Schneider (17). Samples were taken at different depths below the surface; often samples below the surface had higher BaP concentrations than did surface samples.

Calcareous sediments were studied to determine whether the increased hydrocarbon content was from alteration with air and light. In Gouviex, France, a sample was mechanically cut at a depth of 50 m within an underground quarry with a very homogeneous calcareous base. This sample, which had had no contact with an industrial environment, yielded a BaP content of 1.95 μg/100 g of material, whereas surface soil samples taken in the vicinity of Molineaux, France, yielded only 1.5 μg/100 g of material.(17). The investigations showed that the PAH did not, as might be expected, have an external origin. Mallet and Schneider (17) and Mallet (20) considered the hydrocarbons detected in sediments to have endogenous biosynthetic origins.

Mallet (20) studied both planktonic calcareous sediments, which contain decomposition products of phytoplankton, and boghead coal, which is rich in algae. Transformation of these animal and plant materials into hydrocarbon products, amino acids, fatty acids, and waxes may result in aromatic hydrocarbon production under the influence of bacteria.

Mallet et al (21) proposed that the starting substrates for polycyclic hydrocarbon biosynthesis may be more varied and include sterols, terpenes, and lipids as well as those hydrocarbons previously mentioned. Brisou (22) also proposed that BaP biosynthesis by bacteria has for a starting point lipids, purified fatty acids, and other substances including aliphatic terpenes. Mallet (20) hypothesized that anaerobic microorganisms form polycyclic hydrocarbons at the expense of fatty acids present in sediments. Identification tests showed that *Clostridium putride* was particularly important in the biosynthesis of polycyclic hydrocarbons.

Hase and Hites (19) collected sediment from the Charles River Basin and examined it for PAH and bacterial types. The sediment contained at least 11 different ring structures and abundant alkyl-substituted derivatives of the ring structures. They found indications that at least some of the PAH compounds could originate from natural sources such as biosynthesis by the mixed anaerobic bacteria present in the sediment.

Mixed anaerobic bacteria cultures were prepared from the sediment samples and grown on a chemically defined medium. All equipment, solvents, and glassware were shown to be free of PAH by gas chromatographic mass spectrometry. The media were prepared by mixing with 40 ℓ of tap water and autoclaving at 120°C. Examination (by gas chromatographic mass spectrometry) of extracts from the mixed anaerobic bacteria cultures and blanks showed the presence of several PAH compounds. The PAH in the medium blank could originate from the tap water or the chemicals or from autoclaving the dextrose at 120°C. High-resolution mass spectrometry was used to confirm the presence of PAH and their many alkyl homologs in the bacterial and blank extracts.

Results of the high-resolution mass spectrometry showed that the PAH composition and distribution in the bacteria were about the same as in the media in which they were grown. This fact shows that the bacteria bioaccumulated the PAH from 200 ℓ of total medium to about 400 ml in the total wet weight of the bacteria (19).

Culture Studies: Mallet et al (23), working with planktonic sediments, studied the production of polybenzene hydrocarbons with *Clostridium*-type microorganisms grown in a lipid medium produced from marine plankton. They placed 2 to 4 mg of total lipids extracted from marine plankton in a sterile Erlenmeyer flask, added 10 ml of nutritive broth with seawater, and inoculated the flask with the bacterial culture.

Ten flasks were prepared as described; three were inoculated with strict anaerobic bacteria, to which were added 5 ml of thioglycolate medium with 0.5% agar, whereas the seven remaining flasks were inoculated with aerobic bacteria. Incubation of the groups was carried out by placing the anaerobic cultures in a vacuum incubator at 37°C for 15 days and placing the aerobic cultures in a shaker at 20°C for 15 days. Results of the experiment are shown in Table 4.14.

Table 4.14: Biosynthesis of BaP by Microorganisms on a Plankton Lipid Extract

Plankton	Lipid Extract	BaP for the Whole of the Samples (μg)
Plankton 148		none detected
Plankton 148	Anaerobic 57 *(Clostridium)*	0.02
Plankton 185		traces
Plankton 185	Anaerobic 481 *(Clostridium)*	0.05
Plankton 115		—
Plankton 115	Anaerobic 11 *(Clostridium)*	0.04

Source: Reference (23) as cited in ORNL/EIS-95

Results of the Mallet et al work emphasized the possibility of biosynthesis of BaP by anaerobic bacteria. The anaerobic bacteria used in this work were *Clostridium* types isolated from different coastal plankton. The aerobic bacteria in culture No. 3598 and 3599 were *Flavobacterium*, those in culture No. 3490, 3491, 3547 and 3600 were *Phytobacterium*, and the bacteria in culture No. 3493 was a *Pseudomonas*. The results seem to confirm the biosynthesis of BaP from plankton lipids by anaerobes of the *Clostridium* genera.

Knorr and Schenk (24) measured the BaP-synthesizing capability of various bacteria. Their work concentrated on strains of *Mycobacterium smegmatis*, which were believed to convert cholesterol into one of the structurally related carcinogens. Results, given in Table 4.15, show the BaP produced by strains of *Mycobacterium smegmatis* grown on agar. The same or higher yields of BaP were obtained with bacterial types other than *Mycobacteria* such as *E. coli*, *Proteus, Pseudomonas, Serratia, and Myxobacteria* (Table 4.16).

Brisou (18) used bacteria of the genera *Welchia* and *Clostridium* to identify the substrate that would be most favorable for biosynthesis of BaP by the bacteria. The substrates studied and the amounts of BaP detected are shown in Table 4.17. The bacterial cultures were grown at 37°C either in a reducing medium with thioglycolate or in brain-heart broth maintained in a vacuum oven. The results varied from one bacterial strain to the other for the same substrate.

Table 4.15: BaP Biosynthesis by Strains of *Mycobacterium smegmatis**

Strain	Dry Weight (g)	BaP (μg)	Corrected to 100 g Dry Mass
SN 1	50	0.5	2.8
SN 2	33	0.4	3.4
SN 7	40	0.7	4.8
SN 10	25	0.3	3.4
1675	45 (25)	0.4 (0.3)	2.4 (2.8)
1970	35	0.25	2.0
170	70	0.2	2.0
1619	30 (40)	0.3 (0.5)	2.8 (3.6)

*Values in parentheses are BaP-free control values.

Source: Reference (24) as cited in ORNL/EIS-95

Table 4.16: BaP Formation by Bacterial Types Other than *Mycobacterium*

Strain	Dry Weight (g)	BaP (μg)	Corrected to 100 g Dry Mass
E. coli 91	37	0.5	4.6
E. coli 92	27	0.5	5.0
Proteus	37	0.75	5.6
Pseudomonas fluorescens	45	0.5	3.0
Serratia marcescens	67	0.5	2.0
Myxobacteria Z6	40	0.9	6.0

Source: Reference (24) as cited in ORNL/EIS-95

Table 4.17: Biosynthesis of BaP on Different Substrates

Substrate Added	BaP Detected (μg/100 g dry medium)
Pentadecanoic acid	Traces
Margaric acid	Traces
Leneicosanoic acid	Traces
Nonadecanoic acid	0.0025
Sodium oleate	0.02
Thioglycolate broth	0
Cholesterol	0
Equigyne (mixture of estrogens)	0.06
p-Aminobenzoic acid + succinic anhydride	0

Source: Reference (18) as cited in ORNL/EIS-95

Brisou showed that part of the carcinogenic BaP found in nature might have a biological origin. The production of polybenzenes seems dependent on anaerobic microbes, which are commonly found in soils and sediments. The results obtained with pure cultures are clearly lower than those recorded with mixtures of lipids of plankton, which permits the assumption of the intervention of several substrates, and of complex organic materials (18).

Soil Studies: Mallet and Tissier (25) used *Clostridium putride* isolated from compost soils to show the biosynthesis of BaP in sterilized soils inoculated with the bacteria. The oven-sterilized humus was inoculated with *C. putride* and left for four months at ambient laboratory temperatures. The BaP was measured at the start of the experiment and found to be 80 to 100 μg/100 g of soil; the concentration at the end of the four-month period was 160 μg/100 g of soil.

The large initial amount of polybenzene hydrocarbons lead Brisou (18) to collect some forest soil far from large industrial centers. The BaP measurements at the time of collection were 1 to 2 μg/100 g of soil. The soil was ground, dry-sterilized at 170°C, and set to culture under anaerobic conditions with *C. putride* at laboratory temperatures for six months. At the end of the six-month period, the BaP content was 4 μg/100 g of soil, which is an appreciable increase. Significantly, an experiment using *Escherichia coli,* a facultative aerobic-anaerobic common in the mammalian gastrointestinal tract, produced comparable results.

Effect on Microorganism Growth

Hass and Applegate (26) studied the effect of 3-, 4-, and 5-ring unsubstituted PAH on the growth of *Escherichia coli* (ATCC No. 26922). The PAH studied were BaA, 1,2,5,6-dibenzanthracene (DBA), BaP, tetracene, pyrene, anthracene, phenanthrene, chrysene, 1,2,3,4-DBA, and pentacene (Figure 4.3).

Figure 4.3: Structures of Organic Compounds Studied by Hass and Applegate

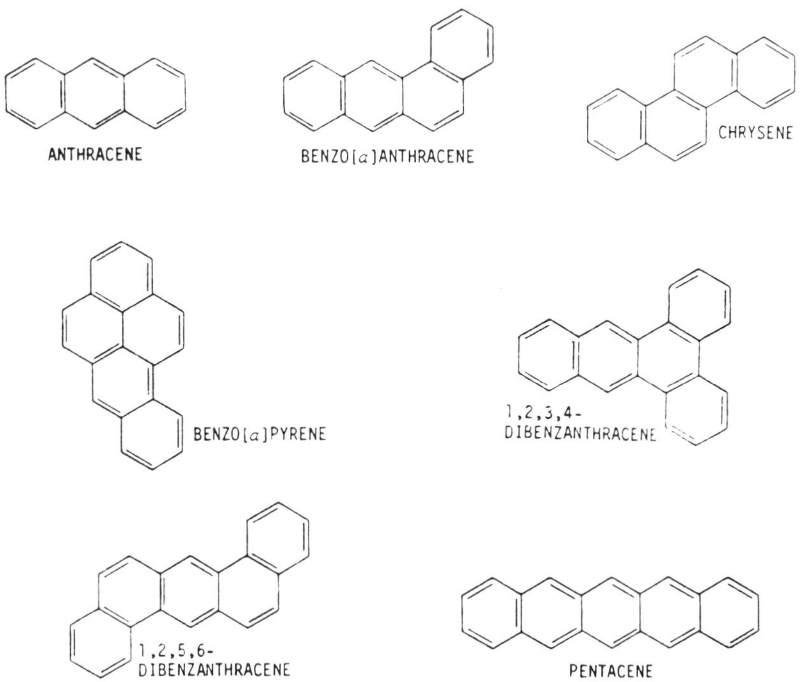

(continued)

Figure 4.3: (continued)

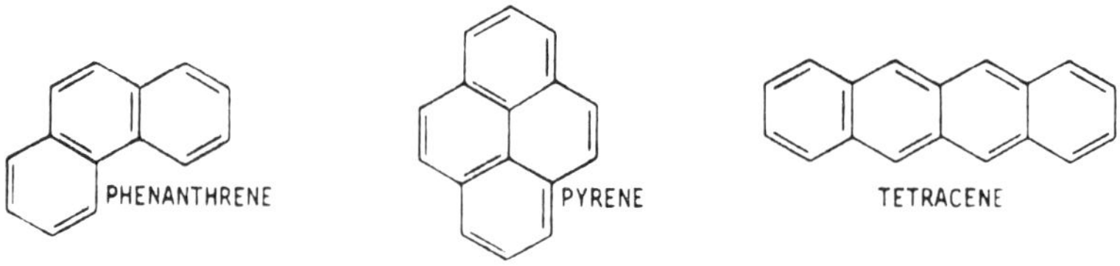

Source: Reference (26) as cited in ORNL/EIS-95

The rate of bacterial growth on nutrient broth medium (with added PAH) was measured by determining the logarithmic phase of the growth curve (k_{PAH}) and comparing it with the rate of growth in flasks without PAH ($k_{control}$). The comparison indicated growth enhancement ($k > 1.0$) or growth retardation ($k < 1.0$). The results were plotted as graphs and are presented in Figures 4.4, 4.5 and 4.6. The angular hydrocarbon configurations of BaA, 1,2,5,6-DBA, and BaP promote growth on *E. coli*; tetracene and pyrene had little or no effect on growth; anthracene, phenanthrene, chrysene, 1,2,3,4-DBA, and pentacene inhibited cell growth at most concentrations. Apparently, on the basis of these results, angular configuration is necessary in the PAH for *E. coli* growth stimulation.

Figure 4.4: Effect of Phenanthrene (•) and Anthracene (○) on Growth of *E. coli*

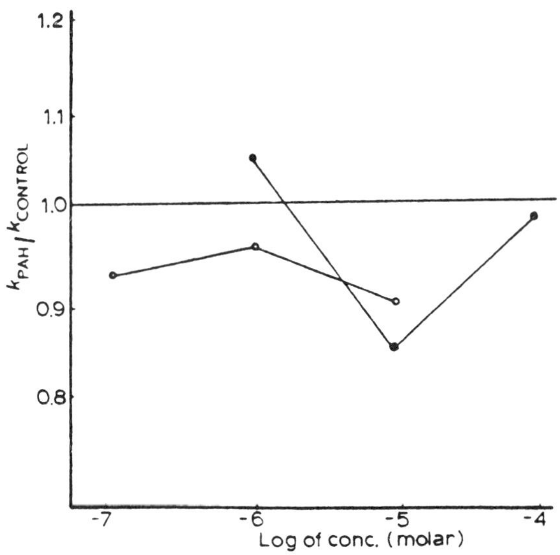

Source: Reference (26) as cited in ORNL/EIS-95

Figure 4.5: Effect of BaA (●), Pyrene (○), Tetracene (■), and Chrysene (□) On Growth of *E. coli*

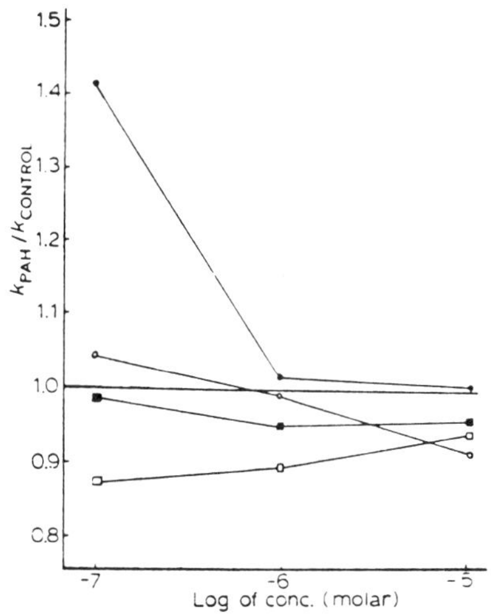

Source: Reference (26) as cited on ORNL/EIS-95

Figure 4.6: Effect of 1,2,5,6-DBA (●), BaP (○), 1,2,3,4-DBA (■), and Pentacene (□) on Growth of *E. coli*

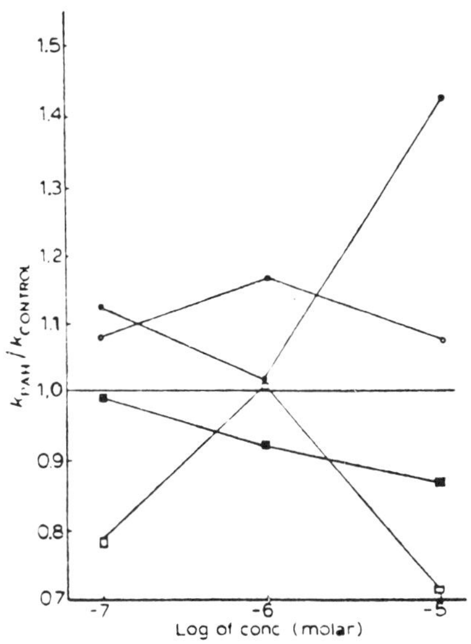

Source: Reference (26) as cited in ORNL/EIS-95

PHOTOCHEMICAL TRANSFORMATION OF PNA

Information about the fate of particulate polynuclear aromatic hydrocarbons (PAH) released to the atmosphere is presently fragmentary and unclear, however, it is widely accepted that POM present in atmospheric aerosols is highly susceptible to photochemical transformation in the presence of sunlight. Certainly, there is ample evidence that most PAH's will undergo photooxidation in solution, as the pure solid, and when adsorbed onto certain solid substrates such as alumina (27) and it has been inferred that similar processes take place when the compounds are adsorbed on airborne particulates (27)-(29). Indeed, it has been suggested that the half-lives of such PAH's in the presence of sunlight may be "only hours or days" (27).

Contrary to this expectation, it is now reported (NTIS COO-4347-1) that the rate and extent of photodecomposition of PAH's may be decreased substantially by adsorption onto coal fly ash particles. Certain PAH's are, however, found to undergo rapid oxidation in the dark when adsorbed onto fly ash.

Studies were conducted using a model system in which individual PAH's were adsorbed onto the surface of fly ash collected from the electrostatic precipitators of several coal fired power plants. Known gas phase concentrations of individual PAH's were generated using a simple diffusion cell (30)(31) and passed through an expanded bed (32)(33) of fly ash. Regulation of the bed temperature and the time of exposure provided a means of controlling the amount of a PAH adsorbed.

All fly ash samples were size fractionated by sieving the 44 to 74 μm physical size range utilized. Exhaustive preextraction with several solvents (cyclohexane, benzene, methanol) showed that none of the fly ashes employed contained more than 50 ng/g of total PAH.

Sample irradiation was achieved using outdoor sunlight and several artificial light sources (150W xenon arc lamp, 275W commercial sunlamp, 5-OT3/C1 General Electric "quartz line" lamps; all unfiltered). The fly ash was tumbled in a quartz container to provide equal exposure of all surfaces and radiant fluxes were estimated to be equal to or greater than that of midday summer sunlight (45° N latitude) over the wavelength range 300 to 800 nm. Portions of each fly ash sample were exhaustively extracted directly following exposure to gaseous PAH and then after irradiation or dark storage.

Analyses for the original PAH and for degradation products were performed by high performance liquid chromatography (UV absorbance detection) and by UV absorption and fluorescence spectrometry. The extent of decomposition was determined both from the disappearance of the original PAH and from the appearance of its reaction products. (Reproducibility of multiple analyses was established to be ±8% RSD.) Individual experiments were conducted to determine the influence of PAH type and concentration, irradiation time, intensity, and wavelength distribution, dark stability, and the nature of the adsorbing substrate on the rate and extent of PAH decomposition.

Rather unexpectedly none of the PAH's which were irradiated following adsorption onto fly ash exhibited appreciable photodegradation (Table 4.18, Class I).

Table 4.18: Decomposition of Polycyclic Aromatic Hydrocarbons Absorbed on Coal Fly Ash

Class I Irradiated PAH				
Compound	Irradiation Time (hr)	Concentrated PAH Adsorbed (μg/g)	Number of Samples	Average Change (%)
Pyrene	5-126	67-291	19	-6.3 (±14.7)
Phenanthrene	2-113	129-186	5	+10 (±12.3)
Fluoranthene	2.5-18	49,115	2	-7.2 (±2.0)
Anthracene	3-24	300	3	-11.0 (±7.0)
Benzo[a]pyrene*	2.5-36.5	44-110	5	-8 (±8)
Benzo[a]pyrene**	21-100	15-44	4	-14 (±7)

Class II Nonirradiated PAH			
Compound	Change	Compound	Change
Acenaphthene	none	Fluorene	decomposed
7,12-Dimethylbenz[a]anthracene	none	Benzo[b]fluorene	decomposed
Carbazole	none	Benzo[a]fluorene	decomposed
Acridine	none	9,10-Dimethylanthracene	decomposed
Phenazine	none	9,10-Dihydroanthracene	decomposed
Fluoranthene	none	4-Azafluorene	decomposed
Benzo[a]pyrene	none	Anthracene	decomposed
Phenanthrene	none		

*Quartzline lamp.
**Xenon arc lamp.

Source: NTIS COO-4347-1

Such degradation as did occur took place very rapidly and did not proceed thereafter with time up to 100 hours (Figure 4.7). Furthermore, no significant dependence on irradiation intensity nor amount of PAH adsorbed was observed.

Under similar illumination conditions, however, extensive photodecomposition was observed for all PAH's irradiated in solution (Figure 4.7) and for both anthracene and benzo[a]pyrene adsorbed onto alumina from methanolic solution. Additional evidence for the apparent influence of adsorbent substrate on photodecomposition was obtained by adsorbing anthracene and benzo[a]pyrene vapors onto both fly ash and alumina which were then coated on thin layer chromatographic plates and exposed to identical irradiation.

In a typical experiment of this type benzo[a]pyrene present on the alumina surface underwent 50% decomposition upon exposure for 80 minutes to an unfiltered 150W xenon arc lamp. By contrast, only 15% decomposition was observed for the benzo[a]pyrene adsorbed on fly ash and irradiated under identical conditions.

While the extent of photodegradation of PAH's adsorbed on fly ash is apparently small, a number of the compounds studied (Table 4.18, Class II) were found to undergo quite extensive oxidation in the absence of light. In particular, different fly ashes were found to give different oxidation rates. In all cases, however, the major oxidation product was the same ketone or quinone as obtained from the photooxidation process.

Figure 4.7: Conversion of Irradiated Benzo[a]pyrene

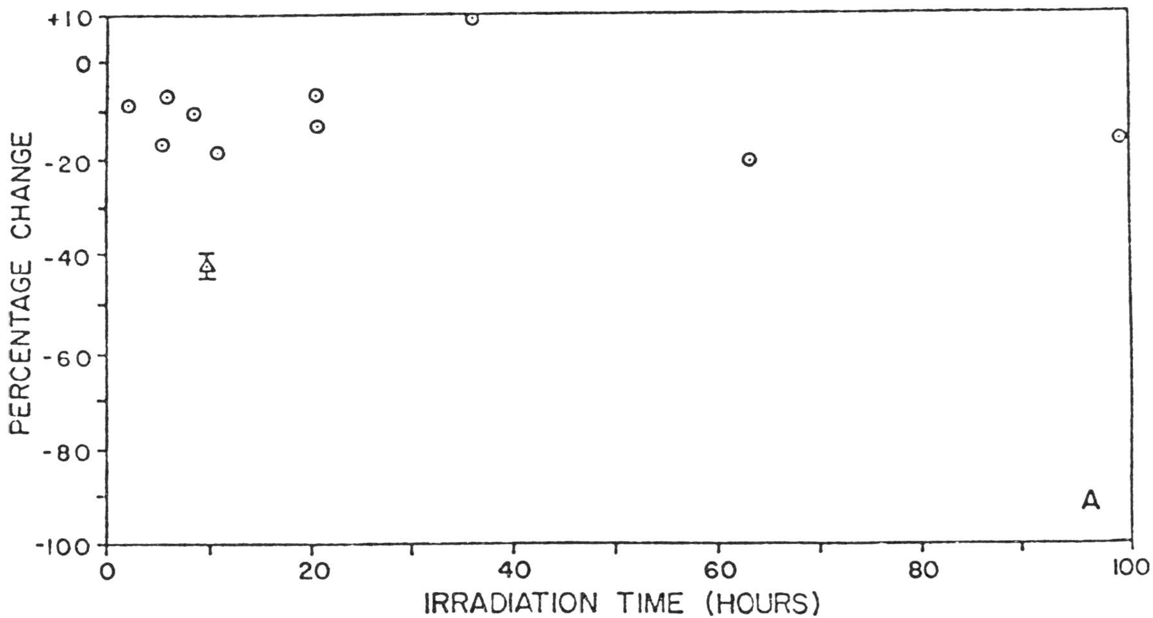

Note: Percent conversion of irradiated benzo[a]pyrene compared to nonirradiated sample as a function of time ⊙ adsorbed onto fly ash △ solution.

Source: NTIS COO-4347-1

Figure 4.8: Conversion of Fluorene Adsorbed on Coal Fly Ash

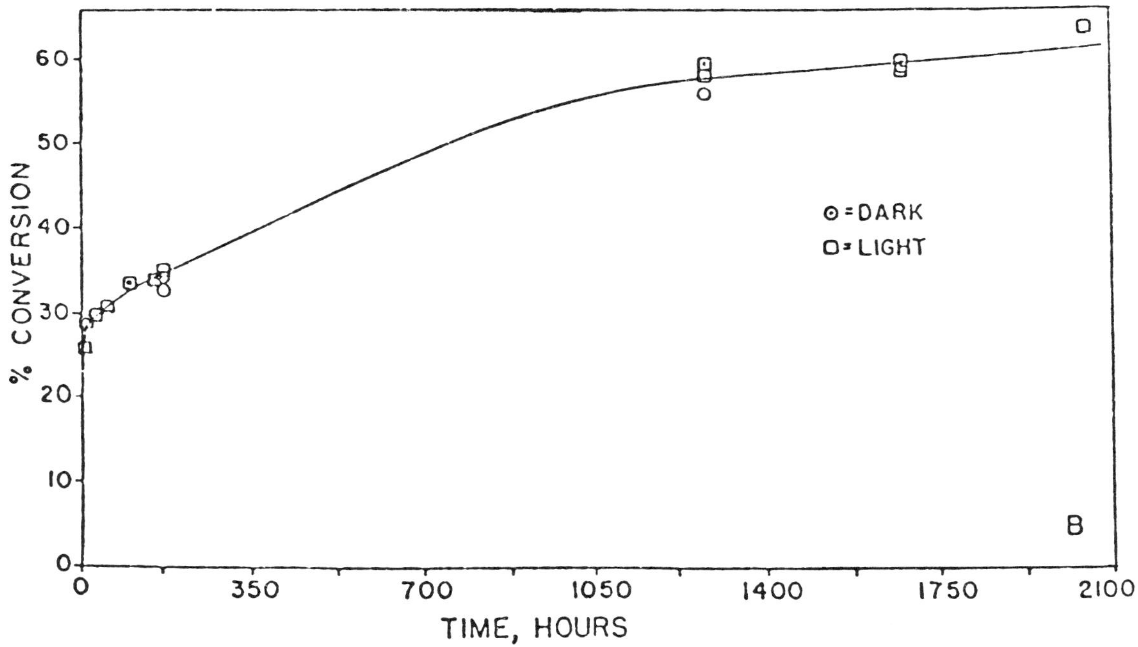

Note: Percent conversion of fluorene adsorbed on coal fly ash as a function of time.

Source: NTIS COO-4347-1

As illustrated in Figure 4.8 for fluorene, which oxidizes to 9-fluorenone, considerable conversion occurs during the actual PAH adsorption step but subsequent irradiation has no effect upon the rate or extent of further reaction. The extent of initial reaction, which occurs at the elevated adsorption temperature, varies from 7 to 90% decomposition depending on the type of fly ash used as the adsorbent substrate and was shown not to be due simply to passage of a PAH through the adsorption apparatus.

It is apparent, therefore, that the nature of the adsorbent is an important factor in determining the rate of this dark reaction. This observation is borne out by the findings that nonphotochemical conversion does not occur for solid fluorene, for fluorene dissolved in methanol or cyclohexane, nor for fluorene adsorbed onto alumina, silica gel, Linde 13X molecular sieve, glass and Ambersorb XE-340 adsorbent resin. However, fluorene adsorbed onto activated carbon and graphite was observed to decompose in a similar manner to that adsorbed on coal fly ash.

Finally, it is noteworthy that all of the compounds which undergo spontaneous oxidation as a result of adsorption onto coal fly ash (Table 4.18, Class II) contain a benzylic carbon atom indicating that this structure is especially susceptible to nonphotochemical oxidative attack. Possible exceptions to this rule are afforded by anthracene and benzo[a]pyrene which exhibit very slow nonphotochemical oxidation when adsorbed onto fly ash.

It is apparent from the foregoing results that PAH adsorbed onto the surface of coal fly ash exhibits quite different chemical behavior from that adsorbed onto other solid substrates, or present either as the pure solid or in solution. The net result is stabilization of PAH against photochemical oxidation on the one hand, but promotion of spontaneous nonphotochemical oxidation for a limited class of compounds on the other.

The reasons for this behavior are not clear. It is conceivable that the limited photochemical decomposition is due either to efficient competitive absorbance of radiation by the fly ash matrix (34), or to the PAH being largely present in pores where light cannot penetrate. A more convincing suggestion, however, is that energetic adsorption, such as apparently occurs on fly ash, activated carbon and graphite surfaces, effectively stabilizes PAH's against photooxidation by molecular orbital rearrangement which either increases the electronic excitation energy or decreases the lifetime of the excited state (34).

The small amount of rapid photooxidation which is observed (Table 4.18, Class I) could then be due to a fraction of the PAH which is not directly bonded to the active surface but rather to another PAH molecule or to a weakly adsorbing mineral impurity. The promotion of nonphotochemical oxidation could, in turn, be due either to extensive molecular orbital rearrangement or simply to the ready availability of active oxygen on strongly adsorbing surfaces (35). Not unexpectedly the points of highest Bronsted acidity (e.g., position 9 in fluorene) are most susceptible to oxidative attack.

REFERENCES

(1) Khesina, A. Ya., Shcherbak, N.P., Shabad, L.M., and Vostrov, I.S., "Benzpyrene Breakdown by Soil Microflora," *Bull. Exper. Biol. Med.* 68(7):1139-41 (1969).

(2) Poglazova, M.N., Fedoseeva, G.E., Khesina, A. Ya., Meisel, M.N., and Shabad, L.M., "On the Possible Modification of Benz(a)pyrene by Soil Microorganisms," *Dokl. Akad. Nauk SSSR* 169(5):1174-77 (1966).

(3) Poglazova, M.N., Fedoseeva, G.E., Khesina, A. Ya., Meisel, M.N., and Shabad, L.M., "Further Investigations of the Decomposition of Benz(a)pyrene by Soil Bacteria," *Dokl. Akad. Nauk SSSR* 176(5):1165-67 (1967).

(4) Poglazova, M.N., Fedoseeva, G.E., Khesina, A. Ya., Meisel, M.N., and Shabad, L.M., "The Oxidation of Benz(a)pyrene by Microorganisms in Relation to Its Concentration in the Medium," *Dokl. Akad. Nauk SSSR* 179(6):1460-62 (1968).

(5) Poglazova, M.N., Fedoseeva, G.E., Khesina, A. Ya., Meisel, M.N., and Shabad, L.M., "Metabolism of Benzo(a)pyrene by Various Soil Microflora and Isolated Microorganisms," *Dokl. Akad. Nauk SSSR* 198(5):1211-13 (1971).

(6) Muller, W.P., and Korte, F., "Microbial Degradation of Benzo-a-pyrene, Monolinuron and Dieldrin in Waste Composting," *Chemosphere* 4(3):195-98 (1975).

(7) Fedoseeva, G.E., Khesina, A. Ya., Poglazova, M.N., Shabad, L.M., and Meisel, M.N., "The Oxidation of Aromatic Polycyclic Hydrocarbons by Microorganisms," *Dokl. Akad. Nauk SSSR* 183(1):208-11 (1964).

(8) Rogoff, M.H., and Wender, I., "The Microbiology of Coal: The Bacterial Oxidation of Phenanthrene," *J. Bacteriol.*, 73:264-68 (1957).

(9) Tausson, W.O., "Die Oxydation des Phenanthrens Durch Bakterien," *Planta* 5:239-73 (1927).

(10) Rogoff, M.H., "Chemistry of Oxidation of Polycyclic Aromatic Hydrocarbons by Soil Pseudomonads," *J. Bacteriol.*, 83:998-1004 (1962).

(11) Sisler, F.D., and ZoBell, C.E., "Microbial Utilization of Carcinogenic Hydrocarbons," *Science* 106:521-22 (1947).

(12) Sikka, H.C., Miyazaki, S., and Lynch, R.S., "Degradation of Carbaryl and 1-Naphthol by Marine Organisms," *Bull. Environ. Contam. Toxicol.*, 13(6):666-72 (1975).

(13) Malaney, G.W., Lutin, P.A., Cibulka, J.J., and Hickerson, L.H., "Resistance of Carcinogenic Organic Compounds to Oxidation by Activated Sludge," *J. Water Pollut. Control Fed.* 39(12):2020-29 (1967).

(14) Fedorenko, Z.P., "The Effect of Biochemical Treatment of Waste Water of a By-Product Coke Plant on the 3,4-Benzpyrene Content," *Hyg. San.* 29(3):19-21 (1964).

(15) Gibson, D.T., and Jerina, D.M., "Oxidation of the Carcinogens Benzo(a)pyrene and Benzo(a)anthracene to Dihydrodiols by a Bacterium," *Science* 189:295-97 (1975).

(16) Barnsley, E.A., "The Bacterial Degradation of Fluoranthene and Benzo[a]pyrene," *Canadian J. Microbiol.*, 21(7):1004-8 (1975).

(17) Mallet, L., and Schneider, C., "Presence of Polybenzene Hydrocarbons of the Benzo-3,4-pyrene Type in Geological and Archeological Levels," *Compt. Rend.*, 259(3):675-76 (1964).

(18) Brisou, J., "Biosynthesis of 3,4-Benzopyrene and Anaerobiosis," *C. R. Soc. Biol.* 163(3):772-74 (1969).

(19) Hase, A., and Hites, R.A., "On the Origin of Polycyclic Aromatic Hydrocarbons in Recent Sediments: Biosynthesis by Anaerobic Bacteria," *Geochem. Cosmochem. Acta* 40(9):1141-43 (1976).

(20) Mallet, L., "Investigation of Polybenzene Hydrocarbons in Ancient Sediments," *C. R. Soc. Biol.* 163(2):319-20 (1969).

(21) Mallet, L., Heros, M., and Brisou, J., "Biosynthesis and Bioregression of the Carcinogenic Polybenzene Hydrocarbons of the Benz-3,4-pyrene Type, Taken as Controls, at the Expense of Lipids," *Pollutions des Milleux Vitraux par les Hydrocarbures Cancerigenes,* pp 177-80, Paris: Librairie Maloine (1972).

(22) Brisou, J., "Lipids, Sterols, Terpenes, and Bacterial Biosynthesis of 3,4-Benzopyrene," *Pollutions des Milleux Vitraux par les Hydrocarbures Cancerigenes,* (editor) L. Mallet, pp 181-83, Paris: Librairie Maloine (1972).

(23) Mallet, L., Zanghi, J., and Brisou, J., "Investigations of the Possibilities of Biosynthesis of Polybenzene Hydrocarbons of the Benzo-3,4-pyrene Type by a *Clostridium putride* in the Presence of Marine Plankton Lipids," *Acad. Sci., Paris Ser. D.,* 264(11):1534-37 (1967).

(24) Knorr, M., and Schenk, D., "The Question of the Synthesis of Polycyclic Aromatics by Bacteria," *Arch. Hyg. Bakteriol.,* 152(3):282-85 (1968).

(25) Mallet, L., and Tissier, M., "Experimental Biosynthesis of Polybenzene Hydrocarbons of the Benzo-3,4-pyrene Type at the Expense of Forest Soils," *C. R. Soc. Biol.* 163(1):63-65 (1969).

(26) Hass, B.S., and Applegate, H.G., "The Effects of Unsubstituted Polycyclic Aromatic Hydrocarbons on the Growth of *Escherichia coli,*" *Chem.-Biol. Interactions* 10:265-68 (1975).

(27) Committee on Biologic Effects of Atmospheric Pollutants: *Particulate Polycyclic Organic Matter,* National Academy of Sciences, Washington, DC (1972).

(28) Falk, H.L., Markul, I. and Kotin, P., "Aromatic Hydrocarbons," *Arch. Ind. Health,* 13 (1956).

(29) Thomas, J.F., Mukai, M., and Tebbins, B.D., "Fate of Airborne Benzo[a]pyrene," *Environ. Sci. Tech.,* 2, 33 (1968).

(30) Natusch, D.F.S., and Miguel, A.H., "A Diffusion Cell for the Preparation of Dilute Vapor Concentrations," *Anal. Chem.,* 47, 1705 (1975).

(31) Miguel, A.H., "Studies of Gas Solid Reactions of Environmental Significance: I. Pyrene Adsorption onto Fly Ash; II. Thiol Oxidation and Adsorption by Activated Charcoal," Ph.D. Thesis, University of Illinois (1976).

(32) Pitts, J.N., Jr., Van Cauwenberghe, K.A., Grosjean, D., Schmid, J.P., Fitz, D.R., Belser, W.L., Jr., Knudson, G.B., and Hynds, P.M., "Atmospheric Reactions of Polycyclic Aromatic Hydrocarbons: Facile Formation of Mutagenic Nitro Derivatives," *Science,* 202, 515 (1978).

(33) Korfmacher, W.A., Miguel, A.H., Mamantov, C., Wehry, E.L. and Natusch, D.F.S., "Apparatus for Vapor-Phase Adsorption of Polycyclic Organic Matter onto Particulate Surfaces," *Environ. Sci. Tech.*

(34) Wehry, E.L., *Analytical Photochemistry and Photochemical Analysis,* J.M. Fitzgerald, (editor), Marcel Dekker, New York, 173 (1971).

(35) Bull, H.I., Hall, M.H., and Garner, W.E., "The Reaction Between Carbon and Oxygen at Low Pressures and Room Temperature," *J. Chem. Soc.,* 837 (1931).

HEALTH EFFECTS

The information in this chapter is based on *Ambient Water Quality Criteria for Polynuclear Aromatic Hydrocarbons,* EPA Report 440/5-80-069, October 1980, prepared by the U.S. Environmental Protection Agency.

Environmental, Health and Control Aspects of Coal Conversion: Volume 2, An Information Overview, ORNL/EIS-95, April 1977, prepared by H.M. Braunstein, E.D. Copenhaver and H.A. Pfuderer of Oak Ridge National Laboratory for Energy Research and Development Administration.

Cell-Mediated Mutagenesis by Chemical Carcinogens, NTIS CONF-780227-4, 1978, prepared by E. Huberman and R. Langenbach of Oak Ridge National Laboratory for National Cancer Institute and the Department of Energy.

Biologically Important Compounds in Synfuels Processes NTIS CONF-800303-13, presented by B.R. Clark, C.-h. Ho, W.H. Griest and M.R. Guerin of Oak Ridge National Laboratory at the ACS Symposium on Environmental Control in Synfuels Processes, March 25, 1980, Houston, Texas.

EXPOSURE

Ingestion from Water

The uptake of polycyclic aromatic hydrocarbons (PAH) in humans from water occurs through the consumption of drinking water. In the United States, the sources of drinking water are groundwaters and surface waters, such as lakes and rivers. Although a small amount of PAH originates from natural or endogenous sources, the predominant sources of PAH in surface waters are man-made. The discharges of raw and industrial sewage, atmospheric fallout and precipitation, road run-off, and leaching from polluted soils, all of which contain substantial PAH concentrations (1), contribute to the PAH contamination in surface waters.

Other than leaching from soils, the only source of PAH in groundwater is of endogenous origin. Borneff (2) estimated that low-level contaminated river and lake waters contain 5 times higher PAH concentration than groundwater, whereas in medium-level polluted river and lake waters this value may be 10 to 20 times higher. The concentration of PAH in groundwater obtained by various authors is given in Table 5.1.

Table 5.1: PAH Concentration in Groundwater

	...... Concentration, μg/ℓ			
Source	BaP	Carcinogenic PAH	Total PAH	Reference
G. Finthen, Germany		0.002		(3)
Mainz, Germany		0.005		(3)
Unspecified German locations	0.0004	0.003	0.04	(4)
Average of 12 German groundwaters*			0.06	(5)
Champaign, IL*	**	0.003	0.007	(6)
Fairborn, OH*	0.0003	0.0008	0.003	(6)
Elkhart, IN*	0.004	0.004	0.002	(6)

*These are results of six specified PAH.
**Not detected.

Source: EPA-440/5-80-069

The PAH level in surface waters was determined by a number of German, English and Russian workers. In all of these methods, the PAH were solvent extracted from the water, subjected to cleanup procedures and analyzed either by TLC-spectrofluorometry or by UV-spectrophotometry. These values are presented in Table 5.2.

Table 5.2: Concentration of PAH in Surface Waters

	.... Concentration, μg/ℓ			
Source	BaP	Carcinogenic PAH	Total PAH	Reference
Rhine River at Mainz	0.08	0.49	1.12	(3)
River Main at Seligenstadt	0.0024	0.155	0.48	(3)
River Danube at Ulm	0.0006	0.067	0.24	(3)
River Gersprenz at Munster	0.0096	0.047	0.14	(3)
River Aach at Stockach	0.017	0.95	2.5	(8)
River Schussen	0.01	0.20	1.0	(8)
River Plyussa:				
at shale-oil effluent discharge site	12			(9)
3,500 m downstream	1			(9)
at Narva water intake	0.1			(9)
A river:				
15 m below coke by-product discharge site	8-12			(10)
500 m downstream	2-3			(10)

(continued)

Table 5.2 (continued):

Source	BaP	Carcinogenic PAH	Total PAH	Reference
	Concentration, μg/ℓ			
Thames River:				
at Kew bridge	0.13	0.18	0.50	(11)
at Albert bridge	0.16	0.27	0.69	(11)
at Tower bridge	0.35	0.56	1.33	(11)

Source: EPA-440/5-80-069

Keegan (12) analyzed the PAH content in three relatively unpolluted U.S. river waters by removing the PAH from water by solvent extraction. The extract was subjected to cleanup and the PAH were analyzed by TLC-spectrofluorometry. Only samples from the Oyster River showed detectable amounts of four PAH. No PAH could be detected in the other two water samples from the Cocheco and Winnepesaukee Rivers.

The PAH levels in surface waters used as raw water sources for drinking water, and the effects of treatments of these waters on PAH levels, are shown in Table 5.3.

According to Borneff (2) in surface waters, one-third of the total PAH is bound to larger suspended particles, a third is bound to finely dispersed particles, and the last third is present in dissolved form. The particle-bound portion of PAH can be removed by sedimentation, flocculation and filtration processes. The remaining one-third dissolved PAH usually requires oxidation for partial removal/transformation. The use of Cl_2, ClO_2, O_3, and UV light for this purpose has been studied.

According to Borneff (2), 50 to 60% of benzo[a]pyrene (BaP) can be removed by chlorination of water. However, the total PAH is reduced to a smaller degree by chlorination. ClO_2 on the other hand, reduces BaP concentration by 90%. At BaP concentrations lower than 10 ppt, ClO_2 no longer functions as an oxidant for the transformation of BaP. The transformation of PAH is faster with O_3, but the use of O_3 requires intensified prepurification to prevent oxidation of other chemicals.

Filtration with activated carbon has been suggested by Borneff as the best method for PAH removal/transformation during water treatment. The reduction of BaP concentration with activated carbon was 99% efficient in actual field tests. With the exception of Appleton, Wis. drinking water, this finding of Borneff has been validated by the work of Basu and Saxena (6)(7), who demonstrated an 88 to 100% reduction of PAH in U.S. drinking waters by the use of activated carbon. In the case of Appleton, Wisc. water, the initial PAH level in raw water was very low. Therefore, it can be concluded that below a certain minimum concentration, activated carbon may not be very effective for PAH removal/transformation.

Table 5.3: Concentrations of PAH in Raw and Treated Surface Water Used as Drinking Water Sources

Source/Treatment	Concentration, μg/ℓ BaP	Carcinogenic PAH	Total PAH	Reference
River Rhine				
Untreated	0.082	0.485	1.11	(4)
Bank and activated carbon filtered	0.0005	0.015	0.13	(4)
Lake Constance				
Untreated	0.0013	0.030	0.065	(4)
Rapid sand filtration and chlorination*	0.0017	0.017	0.053	(4)
English River				
Untreated	0.06**	0.37***	0.73**	(11)
Filtration and chlorination	0.009	0.051***	0.24	(11)
Monongahela River at Pittsburgh				
Untreated	0.04	0.14	0.60	(7)
Treated†	0.0004	0.002	0.003	(7)
Ohio River at Huntington, WV				
Untreated	0.006	0.020	0.058	(7)
Treated†	0.0005	0.002	0.007	(7)
Ohio River at Wheeling, WV				
Untreated	0.21	0.57	1.59	(6)
Treated†	0.002	0.011	0.14	(6)
Delaware River at Philadelphia				
Untreated	0.04	0.16	0.35	(7)
Treated†	0.0003	0.002	0.015	(7)
Lake Winnebago at Appleton, WI				
Untreated	0.0006	0.002	0.007	(6)
Treated†	0.0004	0.002	0.006	(6)

*These are average of five determinations with the exclusion of a sixth high value.

**These values are estimates on the basis of average PAH adsorption in reservoir.

***These values may be a little higher due to the inability of separation of all the carcinogenic from noncarcinogenic PAH.

†The treatment included flocculation, activated carbon addition, filtration, pH control, chlorination and fluoridation.

Source: EPA-440/5-80-069

As some derivatives of BaP and other PAH are formed during the disinfection of water with oxidizing agents and UV radiation, it is of interest to examine briefly the carcinogenicity of such derivatives. With the exception of alkylated derivatives, most BaP derivatives at best have only weak carcinogenic activity (13). However, 10-chloro compounds do cause tumors (1). The quinones, some of which are also formed during chlorination (1) do not produce tumors (13), and may, in fact, inhibit the activity of other carcinogens (14). The possibility of transformation of PAH into other carcinogenic compounds during water treatment processes is an area which remains largely unexplored.

The PAH content in U.S. drinking waters was analyzed by Basu and Saxena (6) (7). Six representative PAH recommended by the World Health Organization (15) as the measure of PAH contamination in drinking water were monitored in this study [benzo[b]fluoranthene (BbFL) was replaced by benzo[j]fluoranthene (BjFL)] and the average concentration of PAH was found to be 13.5 ng/ℓ.

The U.S. EPA also conducted the National Organic Monitoring Survey (NOMS, 1977) to determine the frequency of occurrence and the levels of PAH in U.S. drinking water supplies. Of the 110 water samples analyzed, none showed any PAH other than fluoranthene. 17 out of 110 samples analyzed showed positive fluoranthene values with an average of 20 ng/ℓ concentration. It should be mentioned that the detection limit of PAH in this study was as high as 50 ng/ℓ. The PAH levels in various drinking waters are shown in Table 5.4.

Table 5.4: PAH Levels in a Few Drinking Waters

	...........Concentration, ng/l			
Source	**BaP**	**Carcinogenic PAH**	**Total PAH**	**Reference**
Mixed tap water at:				
Mainz, Germany			7.0	(3)
Water at:*				
Syracuse, NY	0.3	0.3	1.1	(7)
Buffalo, NY	0.2	0.2	0.9	(7)
New York, NY	0.5	3.9	6.4	(7)
Lake George, NY	0.3	1.5	4.2	(7)
Endicott, NY	0.2	1.1	8.3	(7)
Hammondsport, NY	0.3	1.5	3.5	(7)
Pittsburgh, PA	0.4	1.9	2.8	(7)
Philadelphia, PA	0.3	2.0	14.9	(7)
Huntington, WV	0.5	2.0	7.1	(7)
Wheeling, WV	2.1	11.3	138.5	(6)
New Orleans, LA	1.6	1.6	2.2	(7)
Appleton, WI	0.4	2.4	6.1	(6)
Champaign, IL	ND	1.2	2.8	(6)
Fairborn, OH	0.1	0.8	2.9	(6)
Elkhart, IN	ND	0.3	0.3	(6)

Note: ND: Not detected.

*Only the six WHO-recommended PAH (15) were analyzed, with the exception BbFL replaced BjFL. PAH were concentrated by passing 60 ℓ of drinking water through polyurethane foams. The eluate from the foams was subjected to cleanup and analyzed for PAH by TLC-spectrofluorometry.

Source: EPA-440/5-80-069

Finished waters from various treatment sites are transported to the consumers through a variety of pipelines. Borneff (2) reported a tenfold increase in PAH concentration from beginning to end of a water supply pipe that resulted from the paint used on the water pipes. Leaching of PAH from the coating materials used on the pipes could possibly cause an increase in their concentration in the water reaching consumers. In other instances, PAH could be adsorbed from the water onto the surface of the pipes causing a decrease in their concentration.

In the United States, two kinds of pipes are commonly used as distribution lines for transporting treated waters. These are cast/ductile iron, asbestos/cement pipes and a combination of these. The effect of contact with these pipes on the quality of drinking water in terms of PAH concentration was studied by Basu and Saxena (6).

Because of the intermixing of the pipes, it is difficult to draw definite conclusions from their results. However, it seems likely that in instances where an enhancement of PAH concentration was observed, the tar/asphalt coating of the pipes was responsible for the increase. Cement-coated pipes, on the other hand, produced lower PAH concentrations, possibly due to adsorption of PAH from the water.

There are very few epidemiological studies concerning the correlation between cancer and drinking water. It was, nevertheless, noted that four London boroughs, supplied largely by well water, had lower cancer mortalities than most of the other boroughs, which were supplied with surface water (16). Another study concluded that the highest cancer death rates occurred in communities supplied by river water, followed by communities supplied by well water, and health water (17)(18). However, none of these studies attempted to correlate cancer morbidity with concentrations of PAH. Finally, it should be noted that one epidemiological study of the incidence of gastric cancer concluded that social factors and the kinds of soils present reduced the correlations otherwise obtained with the type of domestic water supply (19)(20).

Although the levels of PAH detected in U.S. drinking waters are well below the WHO (15) recommended limit of 200 ppt, the health hazards associated with repeated exposure (more effective than an equivalent single dose) (21) of carcinogens through drinking water should not be underestimated.

Shabad and Il'nitskii (22) stated that the amount of carcinogenic PAH consumed by man from water is typically only 0.1% of the amount he consumes from foods. If the total PAH uptake from food is taken as 4.15 mg/yr (2), the human uptake of PAH from drinking water should not exceed 4 μg/yr. Assuming the PAH concentration value of 13.5 ng/ℓ in U.S. drinking water (6)(7), and a daily consumption of 2.5 liters of drinking water, the yearly intake of PAH from U.S. drinking water would be 12.3 μg/yr or 0.3% of the total food intake.

Nevertheless, the accumulation of PAH in edible aquatic organisms through polluted surface waters can greatly increase their amount in foods, including fish, some mollusks, and edible algae (23). The use of contaminated water for irrigation can also spread PAH into other vegetable foodstuffs (22). Therefore, it is important to monitor the PAH levels in surface waters not used as drinking water sources as well as drinking waters, in order to estimate accurately the human intake of PAH.

Ingestion from Foods

PAH formed through both natural and man-made sources can enter the food chain of man. There is considerable disagreement, however, concerning the contribution of each of these sources to the total PAH contamination in foods.

From their work with marine algae and fishes obtained from polluted and unpolluted sources, Harrison, et al (11) concluded that endogenous synthesis may be the important factor for PAH contamination in these species. Others, however, believe that the endogenous formation of PAH occurs to such a limited extent that it is completely masked by the accumulation of PAH from the environment (24). The latter conclusion was verified by Shabad and Smirnov (25). It has been demonstrated by these authors that plants near an airport contained 10 to 20 times more BaP than areas remote from the runway.

The results of Dunn and Stich (26) indicated a correlation between the PAH level in mussels with industrial, urban, and recreational activity. The highest occurrence of BaP in marine organisms in the areas adjacent to the sea lanes tends to support the view that exogenous sources are the predominant factor for PAH contamination in foods.

The primary routes of entry for PAH in foods are surface adsorption and biological accumulation from the environment (27). The adsorption of PAH from the soil by various plant roots and translocation to the shoots is well documented (28). Similarly, the absorption of PAH by other marine organisms has been demonstrated by Lee, et al (29). Oysters and clams collected from moderately polluted waters also concentrate PAH via absorption (30)(31). The waxy surface of some plant leaves and fruits can concentrate PAH through surface adsorption (32)(33).

Kolar, et al (33) have shown that the concentration of BaP in vegetation is proportional to the exposure time during the growing season (bioaccumulation through adsorption) and the structure of the surface of the plant (surface adsorption). The above-ground parts of the vegetables contain more BaP than underground parts. Only about 10% of the externally deposited BaP in lettuce, kale, spinach, leeks, and tomatoes can be removed by cold water washing (33).

Food additives and food packaging materials such as paraffin waxes containing PAH, contribute to the enhancement of PAH levels in processed foods. For example, Swallow (34) found that paraffin wax wrapping for food contained PaA, CH, BeP, and BaP at levels of 29 ppb, 2 ppb, 0-48 ppb, and 2 ppb, respectively. Certainly, some of these PAH in the packing material can diffuse into the food. Hexane, a solvent used to extract edible vegetable oils, is also a source of PAH contamination. PAH present in food-grade carbon blacks used for food processing can be transported to the food products. Curing smoke and other pyrolysis products used during cooking add to the level of PAH in food. However, in raw foods which require cooking, the largest source of PAH contamination originates from the cooking process itself.

To summarize the available data on PAH levels, various foods have been categorized following the pattern of USDA-FDA for total diet samples (35). These are shown in Tables 5.5 through 5.12. In these tables the following abbreviations are used: A, anthracene; ANT, anthranthrene; BaA, benz[a]anthracene; BaP, benzo[a]pyrene; BbFL, benzo[b]fluoranthene; BeP, benzo[e]pyrene; BjFL, benzo[j] fluoranthene;

BkFL, benzo[k] fluoranthene; BPR, benzo[g,h,i] perylene; CH, chrysene; CR, coronene; DBA, dibenz[a,h] anthracene; DMBA, 7,12-dimethylbenz[a] anthracene; F, fluorene; FL, fluoranthene; IP, indeno[1,2,3-cd] pyrene; NA, naphthalene; P, pyrene; PA, phenanthrene, and PR, perylene. It should be recognized that the data in the tables are neither exhaustive nor absolute. Only the most frequently detected PAH are listed.

The concentration values given in these tables are subject to considerable variation. The PAH concentrations in uncooked foods largely depend on the source of food. For example, vegetables, fruits, and fishes obtained from a polluted environment can be expected to contain higher concentrations of PAH. Therefore, the PAH content is subject to regional variation. In the case of raw foods which require cooking, the method of cooking is largely responsible for the PAH content in the food and is subject to regional or even personal variation. Therefore, the frequency of occurrence of PAH in a particular food is dependent on a number of factors. The results presented in Tables 5.5 and 5.6 represent only the values where the sample showed detectable levels of PAH.

It has been claimed by Zitko (43) that PAH are not bioaccumulated along the food chain. However, Bjørseth (44) demonstrated that both common and horse mussels bioaccumulated PAH, although not to the same degree. Dunn and Stitch (26) have shown that mussels cannot metabolize BaP upon their removal from water. In water, mussels released 79% of naphthalene in 3 days, with a half-life of 1.3 days. The BaP released from both clams and mussels in water takes place with a half-life of 2 to 5 weeks (26). The human intake of PAH through the digestive system has been estimated by Borneff (2). According to this estimate the human intake of PAH per year is about 3 to 4 mg from fruits, vegetables, and bread, 0.1 mg from vegetable fats and oils, and about 0.05 mg from smoked meat or fish and drinking water.

Vegetable Fats, Oils, and Shortening: Several PAH have been found in edible oils by European workers (45). The PAH levels in a few vegetable oils and margarine are presented in Table 5.5. PAH other than those shown in Table 5.5 have been reported in these oils (34). Since the concentration of PAH in vegetable oils depends on the nature of refinement of the crude oil (46), one can expect variations in their concentrations. Heating of the oils also leads to a slight increase in PAH concentrations. For example, Lijinsky and Shubik (47) did not detect any PAH in uncooked Wesson and Crisco oil. However, oil used previously for deep-frying of food showed 1.4 ppb BaP, 12 ppb FL, and 6 ppb pyrenes (41) (48).

Swallow (34) determined the level of PAH in butter and found the concentration of BaA + CH, BaP, IP + DBA, and BPR to be 1 ppb. In a total diet study with a composite sample containing the fats, oils, and shortening, Howard, et al (49) found less than 0.5 ppb of seven PAH. However, Borneff (2) estimated that the human intake of PAH from vegetable fats and oils amounted to 0.1 mg/yr.

Fish and Other Marine Foods: Raw fish from unpolluted waters usually do not contain detectable amounts of PAH, but smoked or cooked fish contain varying levels of PAH. In addition to the origin of the fish, (polluted or unpolluted water), the amount of PAH in smoked fish depends on various parameters, such as type of smoke, temperature of combustion, and degree of smoking (50).

Table 5.5: PAH Concentrations (ppb) in a Few Vegetable Oils and Margarine

	A	PA	FL	P	BaA	BeP	BaP	PR	BPR	CH
Corn*	–	–	–	3.1	0.8	0.7	0.7	–	0.6	–
Coconut**	36	51	18	15	–	2	2	–	–	12
Margarine***	–	–	–	–	1.4-29.5	0.5-1.2	0.2-6.8	–	–	–
Sunflower***	–	–	–	–	13†	4	8	–	4	–
Soybean*	–	–	1.3	1.6	0.9	1.6	1.4	–	1.0	–
Olive*	–	–	3.2	2.6	1.0	0.4	0.5	–	0.9	–
Peanut*	–	–	3.3	2.9	1.1	–	0.6	–	0.9	–

*(36)
**(37
***(34)
†This value represents concentration of BaA and CH.

Table 5.6: PAH Concentrations (ppb) in Smoked and Nonsmoked Fish

	F	A	PA	FL	P	BaA	BeP	BaP	PR	BPR
Smoked eel*	9	4	37	4	6	–	Trace**	1.0	–	–
Smoked lumpfish*	5	Trace	10	2	1	Trace	Trace	0	–	–
Smoked trout*	67	26	52	12	5	–	Trace	Trace	–	–
Smoked herring**	–	–	–	3	2.2	–	–	–	–	–
Smoked herring (dried)**	–	–	–	1.8	1.8	1.7	1.2	1.0	–	1.0
Smoked salmon**	–	–	–	3.2	2.0	0.5	0.4	–	–	–
Smoked sturgeon**	–	–	–	2.4	4.4	–	–	0.8	–	–
Smoked whitefish**	–	–	–	4.6	4.0	–	–	4.3	–	–
Smoked whiting***	–	–	–	–	<0.5	–	–	6.6	0.7	2.4
Smoked redfish†	–	1.5	4.1	4.0	3.0	–	0.3	0.3	–	–
Smoked cod†	–	–	–	–	0.6	–	–	4.0	0.4	2.2
Electric smoked mackerel†	2.6	1.9	9	5.2	3.6	1.2	0.5	0.2	Trace	0.2
Gas smoked mackerel†	8.2	2.3	11	2.6	4.0	0.6	0.2	0.3	Trace	0.3
Nonsmoked haddock**	–	–	–	1.6	0.8	–	–	–	–	–
Nonsmoked herring (salted)**	–	–	–	0.8	1.0	–	–	–	–	–
Nonsmoked salmon	–	–	–	1.8	1.4	–	–	–	–	–

*(38)(39)
**(40)
***(41)
†(42)

Source: EPA-440/5-80-069

The skin of fish apparently serves as a barrier to the migration of PAH into the body tissues. This was postulated by Malanoski, et al (41) from their observations that the BaP level in the skin was much higher than in the interior of cooked fish.

The PAH levels in various smoked and unsmoked fish are shown in Table 5.6. In addition to the fishes presented in this table, various other marine organisms had been tested for PAH content. For example, cooked squid and prawns had BaP concentrations of 1.04 ppb and 0.08 ppb, respectively (51). Various other edible marine organisms were investigated and found to contain PAH.

Swallow (34) analyzed smoked oysters and determined the levels of BaA + CH, BbFL + BkFL + BjFL, IP + CBA, BaP, and BPR. Cooked scallops were found to contain 9.9 ppb BaP (51). Shiraishi, et al (51) detected 0 to 31.3 ppb BaP in various Japanese seaweeds. However, no BaP was detected in crab. The absence of BaP in crab is corroborated by the work of Lee, et al (52), who found no evidence of PAH storage by any of the crab tissues.

A bioconcentration factor (BCF) relates the concentration of a chemical in aquatic animals to the concentration in the water in which they live. The steady-state BCFs for a lipid-soluble compound in the tissues of various aquatic animals seem to be proportional to the percent lipid in the tissue. Thus the per capita ingestion of a lipid-soluble chemical can be estimated from the per capita consumption of fish and shellfish, the weighted average percent lipids of consumed fish and shellfish, and a steady-state BCF for the chemical.

Data from a recent survey on fish and shellfish consumption in the United States were analyzed by SRI International in their 1980 Report for EPA, *Seafood Consumption Data Analysis.* These data, according to a 1980 memorandum from C.E. Stephan to EPA, were used to estimate that the per capita consumption of freshwater and estuarine fish and shellfish in the United States is 6.5 g/day. In addition, these data were used with data on the fat content of the edible portion of the same species to estimate that the weighted average percent lipids for consumed freshwater and estuarine fish and shellfish is 3.0 percent.

No measured steady-state BCF is available for any of the following compounds, but the equation "Log BCF = (0.85 Log P) - 0.70" can be used (53) to estimate the steady-state BCF for aquatic organisms that contain about 7.6 percent lipids (G.D. Veith in a 1980 memorandum to EPA) from the octanol/water partition coefficient (P). The log P values were obtained from Hansch and Leo (54) or were calculated by the method described therein.

The adjustment factor of 3.0/7.6 = 0.395 is used to adjust the estimated BCF from the 7.6 percent lipids on which the equation is based to the 3.0 percent lipids that is the weighted average for consumed fish and shellfish in order to obtain the weighted average bioconcentration factor for the edible portion of all freshwater and estuarine aquatic organisms consumed by Americans (Table 5.7). Caution must be exercised in application of common practice in obtaining the BCF as described here, because the ecological impact of PAH is not well understood at this time. Numerous studies show that despite their high lipid solubility, PAH show little tendency for bioaccumulation in the fatty tissues of animals or man (29)(55)(56). This observation is not unexpected, in light of convincing evidence to show that PAH are rapidly and extensively metabolized.

Since only low levels of PAH are detected in plants and lower organisms (57), transfer of PAH through the food chain does not seem likely. The direct impact of PAH on plants, animals, or the ecological balance of nature is difficult to evaluate, since few data are available which suggest that adverse effects may occur. Thus it is inappropriate in the case of PAH to use the octanol/water partition coefficient for estimation of the BCF. Instead a more realistic value of 30, based on the work of Lu, et al (58), in fish, is recommended for criteria derivation.

Table 5.7: Calculated Bioconcentration Factors of PAH Based upon the Octanol/Water Partition Coefficient

Chemical	Log P	Estimated Steady State BCF	Weighted Average BCF
Acenaphthalene	3.74	301	119
Anthracene	4.45	1,210	478
Benzo[a] pyrene	6.06	28,200	11,100
Phenanthrene	4.46	1,230	486
3-Methylcholanthrene	6.97	168,000	66,400
1-Methylphenanthrene	5.00	3,550	1,400
Dibenzofuran	4.12	634	250
Fluoranthene	4.90	2,920	1,150
Fluorene	4.18	713	282
Benz[a] anthracene	5.61	11,700	4,620
1,12-Benzoperylene	6.51	68,200	26,900
Benzo[k] fluoranthene	6.06	28,200	11,100
Benzo[b] fluoranthene	6.06	28,200	11,100
Chrysene	5.61	11,700	4,620
Dibenz[a,h] anthracene	6.77	113,000	4,460
2,3-Phenylene pyrene	6.51	68,200	26,900
Pyrene	4.88	2,800	1,110
Dibenz[a,h] acridine	5.73	14,800	5,850

Source: EPA-440/5-80-069

Meat and Meat Products: Raw meat does not normally contain PAH, but smoked or cooked meat may contain varying amounts of PAH (28). Table 5.8 shows the concentration of PAH detected in a few meats and meat products.

The higher concentration of PAH in charcoal broiled ribs (containing more fat) than in charcoal broiled steaks tends to support the idea that the most likely source of PAH is the melted fat. These fats drip on the heat source and are pyrolyzed. The PAH compounds in the smoke are then deposited on the meat as the smoke rises (59).

Many factors, such as degree of smoking, and the temperature of combustion affect the composition and concentration of PAH in cooked meat (40). In addition to the pyrolysis of fats, incomplete combustion of charcoal can also contribute to the PAH content in broiled meat. Thus, the source of heat used for cooking is responsible for the PAH concentration in cooked meats. These effects are indicated in Table 5.9.

Table 5.8: PAH Concentrations in a Few Smoked Meat and Meat Products

Meat	A	PA	FL	P	BaA	BeP	BaP	PR	BPR	CH	Reference
	(ppb)										
Charcoal broiled steaks	–	21.0	43.0	35.0	1.4	5.5	5.8	0.9	6.7	0.6	(59)
Smoked beef (chipped)	–	–	0.6	0.5	0.4	–	–	–	–	–	(40)(60)
Smoked ham	–	–	14.0	11.2	2.8	1.2	3.2	–	1.4	–	(40)(60)
Smoked pork (roll)	–	–	3.1	2.5	–	–	–	–	–	–	(40)(60)
Smoked frank-furters	–	–	6.4	3.8	1.5	–	2.0	–	–	–	(40)(60)
Barbecued beef	–	–	2.0	3.2	13.2	1.7	3.5	–	4.3	9.6	(60)(61)
Smoked hot sausages	–	–	–	1.5	0.5	–	0.4	–	–	1.0	(41)
Smoked mutton	13.0	104.0	18.0	8.0	2.0	5.0	*	–	–	–	
Smoked mutton sausages	2.0	17.0	6.0	2.0	0.5	*	*	–	–	–	(38)
Smoked bologna	–	–	–	–	0.04-	5.0	0.04-	0.04-	0.04-	0.15-	(60)(61)
	–	–	–	–	0.55	–	0.08	0.07	0.20	1.20**	
Barbecued ribs	7.1	58.0	49.0	42.0	3.6	7.5	10.5	1.5	4.7	2.2	(59)
Smoked salami	0.7	***	5.6	5.2	0.6	0.2	0.8	3.2	***	1.2	(28)
Smoked morta-della	2.6	***	22.0	15.0	2.8	1.8	0.7	0.1	0.4	3.4	(28)
Heavily smoked bacon	20.0	***	35.0	27.0	29.0	***	3.6	0.9	3.0	***	(28)

*Trace.
**Compound unseparated.
***Detected.

Source: EPA-440/5-80-069

Table 5.9: Effect of Different Cooking Variables on the Concentration of PAH (ppb) in Cooked Meat

Meat	Variable	FL	P	BaA	BeP	BaP	BPR	CH	CR
Charcoal broiled hamburger*	Fat content								
Fat, hot		13.3	7.7	2.7	–	2.6	14.9	1.7	1.0
Lean, hot		0.3	1.6	–	–	–	0.9	–	–
No-drip pan		0.2	0.1	–	–	–	Trace	Trace	–
Charcoal broiled hamburger*	Heating temperature								
Lean, hot		0.3	1.6	–	–	–	0.9	0.3	–
Lean, cool		1.3	0.6	–	–	–	–	–	–
Broiled T-bone steak*	Heat source								
Charcoal, hot		19.8	19.1	31.0	17.6	50.4	12.4	25.4	8.0
Flame, hot		19.0	20.0	3.9	5.7	4.4	6.2	2.0	9.0
Smoked ham**	Degree of smoke								
Light		4.0–14.0	2.0–11.0	0.5–3.0	0–2.0	3.0–4.0	0–1.4	0–3.0	–
Heavy		48.0–156.0	35.0–161.0	6.0–33.0	4.0–26.0	3.8–55.0	2.5–25.0	12.0–66.0	–

Note: Fat: 21% fat.
Hot: 7 cm from heat source.
Lean: 7% fat.
Cool: 25 cm from heat source.

*(48)
**(64)(65)

Source: EPA-440/5-80-069

In North America, except for smoked ham, most smoked meats contain much less carcinogenic PAH than European samples (40)(62). The high incidence of stomach carcinoma in Iceland has been explained by the high concentration of BaP in smoked trout and mutton which are consumed in large quantities in the area (63). On the other hand, very low concentrations of PAH in Norwegian bologna sausages (see Table 5.8) are probably indicative of the tradition of light smoking of food in Norway (61).

About 60 to 75% of the BaP in smoked food has been found to be in the superficial layer of meat (38). This low penetration has also been noted by Rhee and Bratzier (66), who observed that in smoked bologna sausages, the BaP is located within 1.5 mm from the surface. Cellulose casings can be used as a more effective barrier to BaP permeation during smoking of frankfurters than animal casing (67).

In addition to meat and meat products, liquid smoke flavorings used during the cooking of meat have been found to contain a variety of PAH. Lijinsky and Shubik (47) have detected BaP, FL, P, BPR, BaA, and CH in liquid smoke at concentrations of 1 ppb, 16 ppb, 7 ppb, 1 ppb, 12 ppb, and 6 ppb, respectively. Youngblood and Blumer (68) have studied the concentration of PAH in hickory liquid smoke flavoring. The high level of PAH present in the resinous condensate in liquid smoke flavoring indicates the importance of its efficient removal from the aqueous flavoring prior to its use in foodstuffs (69).

Vegetables, Fruits, Grains and Cereal Products, Sugar and Adjuncts, and Beverages: Various European and Japanese workers have reported the presence of BaP and other PAH in these products; their results are summarized in Tables 5.10 through 5.12. Studies in this field in North America are lacking.

Test results indicate that surface adsorption and root uptake are the principal modes of PAH accumulation in vegetables (27). The frizzy leaf of kale, for example, has a large surface area and holds dust particularly well. PAH are adsorbed by the wax layer and protected against solar reactions (32).

In kale, Hetteche (32) found the concentration of PAH to be the following: PA, 70 to 586 ppb; A, 2.4 to 97.5 ppb; P, 36.2 to 510 ppb; FL, 53.6 to 1,196 ppb; BaA, 11.2 to 230 ppb; CH, 28.6 to 395 ppb; BeP, 3.8 to 67.2 ppb; BaP, 0.9 to 48.6 ppb; PR, N.D. to 7 ppb; BPR, 1.2 to 46.4 ppb; and CR, 0.1 to 7.2 ppb.

The concentration of BaP in vegetables is directly proportional to exposure time during the growing season and structure of the surface of the plant. The aboveground parts contain more BaP than underground parts. Washings with cold water do not remove more than 10% of the BaP (33).

Fruits grown in polluted environments show a high degree of PAH contamination mainly through adsorption on the waxy surface.

In smoked Gouda cheese, Panalaks (60) found 0.5 ppb BaP and Howard, et al (40) found 2.8 ppb FL and 2.6 ppb P. The unsmoked cheese contained lower levels of PAH (40). Grimmer (81) analyzed bakers' yeasts and determined the level of PAH. The values are shown in Table 5.13.

Table 5.10: BaP Content in Fruits and Other Foods

Product	Concentration (ppb)	Comments	References
Fruits			
Apple	0.02		(51)
Apple	8.3	*	(33)
Banana	0.02		(51)
Banana peel	0.03		(51)
Grape	0.2	*	(33)
Grape	0.02		(51)
Japanese pear	0.05		(51)
Pear	1.9	*	(33)
Persimmon	0.02		(51)
Pineapple	0.02		(51)
Plums	0.04		(51)
Plums	29.7	*	(33)
Dried prunes	0.2-1.5		(72)
Mandarin orange	0.03		(51)
Orange peel	0.15		(51)
Strawberry	N.D.		(51)
Pumpkin	N.D.-trace		(70)
Grains and Cereal Products			
Wheat grain	0.1	*	(33)
Wheat sprouts	60.0		(71)
Cereals	0.2-4.1		(72)
Barley	0.3	*	(33)
Oats	0.2	*	(33)
Polished rice	N.D.		(73)
Rye seedling	10.0-20.0	**	(74)
Lentil seedlings	10.0-20.0	**	(74)
Sesame seeds	N.D.		(73)
Sugar and Adjuncts			
Charred biscuits	11.0-72.0		(75)
Caramel	N.D.		(73)
Chocolate	0.2-1.7	***	(76)

Note: N.D. is not detected.
*Polluted environment.
**8 other PAH identified.
***4 other PAH quantified.

Source: EPA-440/5-80-069

Table 5.11: Concentration (ppb) of a BaP in a Few Vegetables

Vegetable	Concentration	Comments	References
Parsley leaf and stem	24.3	*	(33)
Red clover	7.5	*	(33)
Mushroom	7.0	*	(33)
Lettuce	8.6	*	(33)
Lettuce	N.D.		(70)
Spinach	6.2	*	(33)
Spinach	1.3		(73)
Spinach	7.4		(72)
Radish leaves	5.3	*	(33)
Radish roots	1.2	*	(33)

(continued)

Table 5.11 (continued):

Vegetable	Concentration	Comments	References
Radish roots	N.D.		(70)
Tomatoes	0.1	*	(33)
Tomatoes	0.2		(72)
Cabbage	12.3-20.9	*	(33)
Cabbage	N.D.		(70)
Chinese cabbage	0.05		(70)
Potatoes	N.D.-0.01		(70)
Potatoes	0.2	*	(33)
Sweet potatoes	N.D.		(70)
Sweet pepper	N.D.		(70)
Cauliflower	5.1	*	(33)
Bean paste	N.D.		(73)
Kidney bean	N.D.		(73)
Carrot	N.D.-0.02		(73)
Cucumber	N.D.		(73)
Eggplant	N.D.		(73)
Onion bulb	N.D.-0.01		(70)
Onion greens	0.01		(70)

Note: N.D. is not detected.
*Polluted environment.

Source: EPA-440/5-80-069

Table 5.12: BaP Concentrations (ppb) in Beverages

Beverage	Concentration	Comments	References
Dark rum	1.0		(34)
Whiskey	0.04	3*	(72)(77)
Tea leaves	3.9-21.3		(72)
Black tea aroma**		7*	(78)
Roasted coffee (moderate dark)	N.D.		(79)
Roasted coffee (darkest)	N.D.-4.0		(79)(80)
Coffee soots***	200.0-440.0		(80)

*Quinolines detected.
**This is the volatile component of black tea.
***The soots generated during direct and indirect roasting of coffee beans.

Source: EPA-440/5-80-069

Table 5.13: PAH Concentrations (ppb) in a Variety of Bakers' Yeasts*

PAH	French	German	Scottish	Russian
PA	17.8-34.60	67.0	1,620	7.2
A	2.6-13.6	4.8-10.2	567	4.7
P	11.6-19.6	11.5-35.0	327	16.9
FL	18.5-21.2	17.2-66.8	93	32.1
BaA	9.8-23.3	2.5-15.8	203	10.8
CH	8.1-13.4	4.2-14.0	50	11.1
BeP	8.0-10.6	3.1-14.3	40.4	8.7

(continued)

Table 5.13 (continued):

PAH	French	German	Scottish	Russian
BaP	8.0-12.2	1.8-13.2	6.2	0.5
PR	0.9-1.2	N.D.-0.5	16.7	6.0

*This is bakers' yeast as opposed to dietary or brewers' yeast.

Source: Reference (81) as cited in EPA-440/5-80-069

Inhalation

A variety of PAH have been detected in ambient air in the United States and elsewhere in the world. Because of its carcinogenic properties, BaP has been most extensively monitored and has frequently been used as an indicator of ambient PAH.

The presumed correlation between the concentration of BaP and other PAH, however, does not always exist. For example, a study by Kertesz-Saringer and Morlin (82) found little or no relationship between BaP and other PAH in Budapest air. Gordon (83) and Gordon and Bryan (84) came to a similar conclusion from their work with ambient Los Angeles air.

The concentration and the nature of PAH in ambient air are dependent on a number of factors. In general, the PAH concentration is lowest during the summer months and highest during the winter (85), probably due to commercial and residential heating during winter (86). However, there are some exceptions. Cleveland, for instance, does not follow the high winter-low summer pattern. It has been suggested that this may be due to significant industrial emissions that are uniform throughout the year (86).

The nature and relative amounts of individual PAH in ambient air are also dependent on the source of these compounds. Thus, the content of PAH sampled in an industrial area is a composite of the emissions from various industrial and transportation sources within the area. For example, Gordon (83), from his study of the relative PAH concentration pattern for different areas in Los Angeles, found a correlation between coronene concentration and automobile emissions.

Similarly, Greinke and Lewis (87) had demonstrated that emissions from coke ovens contain lower amounts of certain methyl-substituted PAH than emissions from petroleum pitch volatiles. Bartle, et al (88) also used a PAH profiling technique for the identification of air pollution sources, such as coal burning, vehicular emissions, and oil and gas burning.

Meteorological factors have a dominant effect on PAH concentrations. For example, Lunde and Bjørseth (89) demonstrated that under favorable wind conditions PAH from downtown London could be transported to Norway. The tendency of atmospheric inversion to increase the PAH levels in urban areas has also been shown (90).

The annual average ambient BaP concentrations for different U.S. urban and rural locations during the period 1966-70 have been compiled by a U.S. EPA report (91). The average BaP concentrations in U.S. urban and rural areas obtained from this U.S. EPA study are shown in Table 5.14.

Table 5.14: Average BaP Concentrations (ng/m^3) in U.S. Urban and Rural Areas During 1966-76

	Period		
	1966	1970	1976
Urban	3.2	2.1	0.5
Rural	0.4	0.2	0.1*

*This value is the average of two rural locations.

Source: Reference (91) as cited in EPA-440/5-80-069

An interesting trend has developed from the National Air Surveillance Network (NASN) which monitored BaP values listed in Table 5.14. As can be seen, the average BaP concentrations in urban areas decreased from 3.2 ng/m^3 in 1966 to 2.1 ng/m^3 in 1970, approximately a 30% decrease. The decrease is more dramatic (i.e., >80%) between the period 1966 to 1976. Even the concentrations in rural areas indicate a downward trend.

This decline in BaP concentration is believed to be due primarily to decreases in coal consumption for commercial and residential heating, improved disposal of solid wastes, and restrictions on open burning (92). A further observation that can be made from Table 5.14 is the five- to tenfold difference in BaP concentration between urban and rural locations.

The NASN study did not include the determination of concentrations of other PAH. The summer and winter averages of ambient PAH concentrations for seven urban locations were determined by Sawicki, et al (85). The averages of summer and winter data from this work are presented in Table 5.15.

Table 5.15: Summer-Winter Average of Ambient PAH Concentrations (ng/m^3) in the Air of Selected Cities

City	BPR	BaP	BeP	BkFL	P	CR	PR	A	Total
Atlanta	7.0	4.5	3.1	3.7	3.4	3.4	0.8	0.4	26.3
Birmingham	13.2	15.7	8.0	8.8	9.6	3.0	3.8	1.3	63.4
Detroit	21.3	18.5	14.2	12.5	19.4	4.1	3.9	1.2	95.1
Los Angeles	10.2	2.9	4.4	3.1	3.2	7.1	0.8	0.1	31.8
Nashville	10.2	13.2	7.6	8.0	15.3	3.0	2.3	1.0	60.6
New Orleans	6.0	3.1	4.8	2.9	1.3	14.8	0.6	0.1	33.6
San Francisco	5.1	1.3	1.7	1.0	1.0	3.3	0.2	0.1	13.7

Source: Reference (85) as cited in EPA-440/5-80-069

The average of total PAH concentrations for all cities listed in Table 5.15 is 46.4 ng/m^3. However, these values were obtained from ambient air sampled in 1958-59 and probably have decreased during subsequent years. If an 80% decrease of total PAH concentration is assumed (as in the case of BaP), the ambient PAH concentration in the U.S. urban areas can be extrapolated as 9.3 ng/m^3.

Although the concentration of BaP and some other PAH might have decreased in past decades, the concentration of coronene and some other PAH may not have maintained the same trend. This could be due to the higher number of automobiles in use. Therefore, this 80% decrease figure may or may not be valid for all PAH.

The concentrations of PAH in recent years in individual U.S. cities have been determined by a number of authors. The lowest and highest values of these determinations published during the period 1971-77 are shown in Table 5.16.

Table 5.16: PAH Concentration Range in U.S. Cities Determined by Various Authors in Recent Years

Compound	Concentration Range, ng/m^3	Reference
NA	0.052-0.350	(93)
A	0.068-0.278*	(89)
BaA	0.18-4.6	(83)(94)
PA	0.011-0.340	(93)
FL	0.10-4.1	(90)(94)
BbFL	0.1-1.6	(84)
BjFL	0.01-0.8	(84)
BkFL	0.03-1.3	(84)
P	0.18-5.2	(84)(94)
BaP	0.13-3.2	(94)(95)
BeP	0.9-4.6	(83)(94)
IP	0.03-1.34	(83)(84)
CH	0.6-4.8	(83)(94)
PR	0.01-1.2	(84)
BPR	0.2-9.2	(84)
CR	0.2-6.4	(84)

*This Norwegian value is included because no recent U.S. data are available.

Source: EPA-440/5-80-069

The exact amount of human PAH intake from all modes is difficult to determine because of the different modes of inhalation due to smoking, occupational exposure, or exposure to ambient air. Considering only exposure to ambient air, one needs an average PAH concentration in air in order to determine the PAH intake through inhalation. In the absence of national average data for PAH equivalent to NASN data on national average BaP levels, the yearly average data for Los Angeles are used for the derivation of PAH intake due to inhalation. These values are given in Table 5.17.

Table 5.17: Average Ambient PAH Concentration in U.S. and Daily Intake of PAH Through Inhalation

	BaP	Carcinogenic PAH*	Total PAH
Ambient concentration, ng/m^3	0.5	2.7	10.9
Inhalation intake/day, ng**	5.0	27.0	109.0

*(96)

**These values are based on 10 m^3 inhalation of air/day.

Source: Reference (83) as cited in EPA-440/5-80-069

It can be seen from Table 5.17 that the yearly intake of total PAH, carcinogenic PAH, and BaP through inhalation is 39.8 μg, 9.9 μg, and 1.9 μg, respectively. It should be recognized that these data are based on the average ambient air concentration of one city and probably will not reflect the true U.S. average. It is noteworthy, however, that the total ambient PAH concentration of 10.9 ng/m^3 derived from this work is very close to the earlier extrapolated value of 9.3 ng/m^3.

Dermal

No direct information is available on the importance of dermal absorption in total human exposure to PAH. PAH can be absorbed across the skin by animals. For those humans exposed to only ambient levels of PAH, dermal absorption is not likely to be a significant route of entry.

TOXICITY

There is little published information on the toxicity of PAH in experimental animals other than the mouse or rat (97). The systemic acute dose LD_{50} (the lethal dose for 50% of animals tested) is usually extremely high as compared with the dose necessary to produce carcinogenic effects.

Acute toxicity is difficult to produce orally because PAH are often poorly absorbed from the gastrointestinal tract. Most skin-painting experiments use concentrations of PAH so low (around 1%) that toxic effects are not produced in a single dose.

Systemic Effects

Shubik and Della Porta (98) describe acute effects in mice treated with large doses of several PAH. Groups of twelve 2- to 2½-month-old male Swiss mice received repeated skin paintings of concentrated acetone solutions of benzo[a]-pyrene (BaP) or 7,12-dimethylbenz[a]anthracene (DMBA) or a benzene solution of 3-methylcholanthrene six days a week for periods of up to nineteen weeks. The dose varied with the solubility of the various hydrocarbons; each dose consisted of about 0.08 ml of a 1.5% solution of BaP or 2% of 3-methylcholanthrene or DMBA.

An additional group of mice received a 4% solution of 3-methylcholanthrene twice each day. Five groups of five mice each were given injections of a single dose of 1 g/kg body weight (25 mg) of DMBA, BaP, 3-methylcholanthrene, fluorene, or anthracene as either 3.3% solutions or suspensions in olive oil.

DMBA produced an acute lethal effect when applied topically to the skin and when injected intraperitoneally in mice. BaP and 3-methylcholanthrene produced a similar effect only after intraperitoneal injection of the single large dose. Intraperitoneally administered anthracene and fluorene failed to produce any toxic effects. In the skin-painting studies, all twelve mice receiving BaP developed multiple tumors. Although each animal bore at least one squamous cell carcinoma, the majority of the 71 tumors were papillomas.

In the 3-methylcholanthrene group, two mice without tumors died at the second and fifth weeks; the remaining ten mice developed at least one squamous cell carcinoma, although again most of the tumors were papillomas. All DMBA-painted mice died or were moribund by the sixteenth day, before any tumors had arisen.

In another experiment, a 2% solution of DMBA (400 mg/kg body weight) in acetone was applied to a large shaved area of skin. Two animals died and two were moribund on the seventh day after dosing; five other mice had died or were moribund by the eleventh day. The pathological pictures of tissues of the dying animals after treatment were essentially morphologically similar, but they showed quantitative variations. Myelo-, erythro-, and lymphopoiesis were markedly affected, involving the peripheral blood, spleen, lymph nodes, and bone marrow. Pathological changes were absent in other organs, except for the testes and intestinal mucosa.

In a similar, earlier study, Stowell and Maas (99) painted the backs of Swiss mice with solutions of 1% 3-methylcholanthrene, three, six, and nine times a week. Controls were unpainted or painted with benzene. Malignant tumors developed rapidly. The mice showed systemic effects–inflammation in the liver, kidney, and lungs; hyperplasia of the bone marrow, spleen, and lymph nodes with extramedullary myelo-, lympho-, and erythropoiesis; and leukemia. Some thyroid glands had atypical acini, and some parathyroids were hypertrophic and hyperplastic.

Toxicity of Metabolites

Huggins and Morii (100) observed that a single dose of DMBA (20 to 30 mg by ingestion or 5 mg intravenously) caused selective necrosis of the adrenal cortex in the mature female rat. Massive necrosis occurred in the two inner zones of the cortex, whereas other regions of the adrenal glands were uninjured. The main products of the metabolism of DMBA by rat liver homogenates are the isomeric monohydroxymethyl derivatives, 7-hydroxymethyl-12-methylbenz[a]-anthracene and 12-hydroxymethyl-7-methylbenz[a]anthracene (101). 7-Hydroxymethyl-12-methylbenz[a]anthracene is a more potent adrenocorticolytic agent than DMBA; 12-hydroxymethyl-7-methylbenz[a]anthracene has no adrenocorticolytic effect (102). DMBA is the only PAH known to produce an adrenal necrotic effect.

Fetal Toxicity

Currie et al (103) investigated the effects of DMBA and its monohydroxymethyl derivatives on the pregnant rat and rat fetus. Pregnant Sprague-Dawley rats received a single intravenous injection of DMBA, 7-hydroxymethyl-12-methylbenz-[a]anthracene, 12-hydroxymethyl-7-methylbenz[a]anthracene, or 7-hydroxymethylbenz[a]anthracene at a dose of 2.5 mg/100 g body weight, from day two to day eighteen of pregnancy.

The adrenal glands of every rat treated with 7-hydroxymethyl-12-methylbenz[a]-anthracene were necrotic, and the litters of the rats treated on day seventeen showed extensive necrosis of the central parts of the glands. This dose of DMBA proved highly toxic to the mothers: four of nine rats treated on day eight and seven of eleven rats treated on day thirteen died by day twenty. Varying the dose of 7-hydroxymethyl-12-methylbenz[a]anthracene showed a fairly close correlation between dose and the degree of damage to the maternal adrenal cortex.

Cell Toxicity

Grover, et al (104) and Huberman, et al (105) studied the cytotoxicity of dibenz[a,h]anthracene, benz[a]anthracene (BaA), and several of their derivatives at concentrations of 2.5, 5.0, and 10 and 1.0, 2.5, and 5.0 μg/ml of medium, respectively. Cytotoxicity was expressed as the percentage of colonies of plated rodent cells in PAH-treated dishes as compared with those in acetone-treated controls (cloning efficiency). Neither parent compound was toxic at the concentrations used.

In contrast, the PAH derivatives, the epoxides, and cis-dihydrodiols caused increased transformations of the cells, and the K-region phenols of BaA and dibenz[a,h]anthracene were the most toxic derivatives tested (Tables 5.18 and 5.19). The results confirm the view that metabolism is necessary for hydrocarbons to produce toxicity. An increase in the concentration of the compounds was generally accompanied by an increase in toxicity to the cells (104).

Table 5.18: Transformation and Cloning Efficiencies of Hamster Embryo Cells Treated with K-Region Derivatives of BaA

Compound	Concentration (μg/ml of medium)	Total Colonies Counted	Cloning Efficiency (%)	Transformed Colonies Number	Transformed Colonies Percent
Control (acetone)		818	18.2	0	0
Hydrocarbon	1.0	893	16.2	2	0.2
	2.5	1,039	19.0	2	0.2
	5.0	1,140	22.8	0	0
Epoxide	1.0	1,082	19.5	3	0.3
	2.5	565	10.2	37	6.9
	5.0	0			
Diol	1.0	1,320	22.0	3	0.2
	2.5	867	15.6	10	1.2
	5.0	596	12.0	39	6.5
Phenol	1.0	1,161	21.2	3	0.3
	2.5	272	5.7	3	1.1
	5.0	0			

Source: Reference (104) as cited in ORNL/EIS-95

Table 5.19: Transformation and Cloning Efficiencies of Hamster Embryo Cells Treated with K-Region Derivatives of Dibenz[a,h]anthracene

Compound	Concentration (μg/ml of medium)	Total Colonies Counted	Cloning Efficiency (%)	Transformed Colonies Number	Transformed Colonies Percent
Control (acetone)		1,291	12.1	5	0.4
Hydrocarbon	2.5	1,341	13.4	3	0.2
	5.0	1,363	14.0	11	0.8
	10	1,365	14.5	7	0.5
Epoxide	2.5	895	10.1	7	0.8
	5.0	866	9.3	20	2.3
	7.5	817	9.3	22	2.7
	10	707	7.7	30	4.2
Diol	2.5	739	8.5	3	0.4
	5.0	321	3.7	4	1.2

(continued)

Table 5.19 (continued)

Compound	Concentration (μg/ml of medium)	Total Colonies Counted	Cloning Efficiency (%)	Transformed Colonies Number	Transformed Colonies Percent
	10	2			
Phenol	2.5	729	8.0	0	0
	5.0	2			
	10	0			

Source: Reference (104) as cited in ORNL/EIS-95

In another study, the K-region epoxides of 3-methylcholanthrene, chrysene, and phenanthrene were more active than the parent hydrocarbons in producing cytotoxicity (105).

Using a microfluorometric technique, Lubet, Brown and Kouri (106) found that the intracellular uptake of BaP alone could not account for the toxic effects produced on hamster embryo cells in culture. By either intracellular production or cellular uptake, similar amounts of 3-hydroxy-BaP resulted in similar levels of cell toxicity. Intracellular 3-hydroxy-BaP concentrations of 50 to 100 times less than those of BaP resulted in the same amount of toxicity to the cells.

Toxicity probably was not due to a subsequent metabolite of 3-hydroxy-BaP since established cell lines, which have decreased abilities to metabolize these hydrocarbons, showed an exponential toxicity when exposed to pure 3-hydroxy-BaP (107).

METABOLISM

Knowledge of the metabolism of PAH is crucial to an evaluation of their hazard. From the preceding data, it is clear that the metabolites are more toxic than the pure compounds. Discussions of PAH metabolism can be found in Gerarde (97) and Williams (108). Generally, hydrocarbons of two to five rings undergo reactions, forming (1) phenols, (2) partially saturated rings, (3) quinones, (4) mercapturic acid, and (5) glucuronide and ethereal sulfate (108). Metabolism is carried out primarily by the drug-metabolizing enzyme system of the microsomes.

Rat Liver Microsomes

An example of what is known of the metabolism of a PAH is shown in Figure 5.1. BaP added to rat liver preparations gave the following metabolites: the monophenols (3- and 6-hydroxy), the diphenols (1,6- and 3,6-dihydroxy), the corresponding quinones (3,6- and 1,6-dione), and the two dihydrodiols (1,2 and 9,10). A number of unidentified conjugated products were also present.

The K-regions (4,5 and 11,12 bonds), postulated to be active sites, were not metabolized. None of the identified metabolites are known carcinogens. Wiebel and Gelboin (109) listed the metabolites of BaP formed by rat liver microsomes as three dihydrodiols, two phenols, and two quinones (Table 5.20).

Figure 5.1: Metabolism of BaP

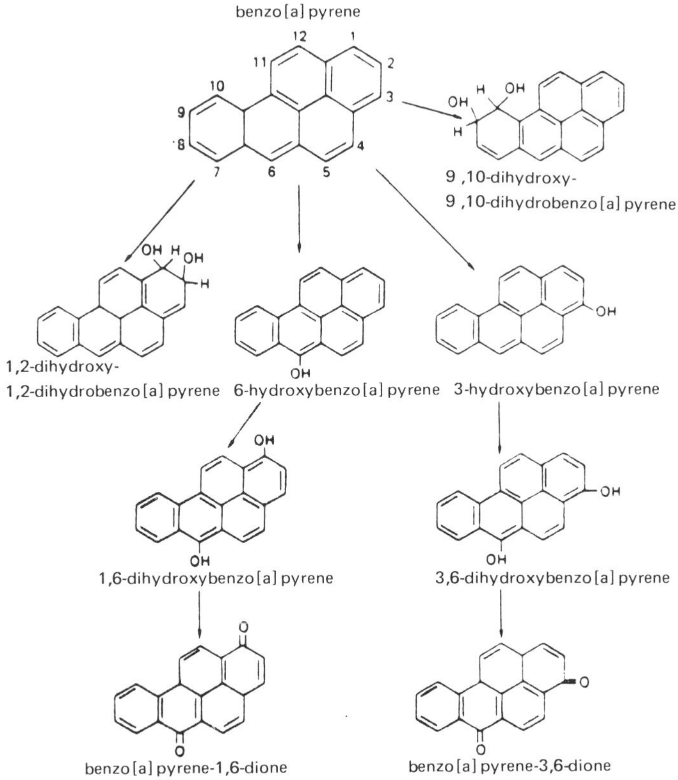

Source: Reference (96) as cited in ORNL/EIS-95

Table 5.20: Metabolites of BaP

7,8-dihydro-7,8-diol
9,10-dihydro-9,10-diol
4,5-dihydro-4,5-diol
3-hydroxy
9-hydroxy
1,6-quinone
3,6-quinone

Source: Reference (109) as cited in ORNL/EIS-95

Wiebel and Gelboin (109) state that no epoxides of BaP have been isolated as products of cell metabolism. An epoxide of BaP may be the carcinogenic metabolite. Waterfall and Sims (110) synthesized epoxides of BaP in the laboratory which appeared to be BaP-7,8- and -9,10-oxide. These oxides (epoxides) were unstable and readily rearranged to phenols. They were converted by rat liver homogenates and microsomal preparations into phenols and dihydrodiols, but not glutathione conjugates. The dihydrodiols thus formed were identical in their chromatographic and spectrographic properties to the dihydrodiols formed when the parent BaP compound was metabolized by rat liver homogenates.

Boyland and Sims (101) and Sims (111)(112) characterize the metabolites of PAH (including BaP and benz[e]pyrene and two methylated PAH—DMBA and 7-methylbenz[a]anthracene) in rat liver homogenates and mouse embryo cell cultures. Metabolism of PAH containing methyl groups such as DMBA and 7-methylbenz[a]anthracene involved either ring hydroxylation or hydroxylation of the methyl groups, the latter being followed by further oxidation of the hydroxymethyl compounds either to carboxylic acids or to ring-hydroxylated products.

Ring hydroxylation occurred mainly in the 8,9 position, with both the parent compounds and the hydroxymethyl derivatives yielding dihydrodiols (112). Epoxides have also been isolated as intermediate metabolites of BaA and dibenz[a,h]anthracene in cell cultures (113)(114). Epoxides are the metabolic precursors of dihydrodiols and phenols (115)-(117).

Human Lymphocytes

Booth, et al (118)(119) demonstrated that cultured lymphocytes metabolized PAH by pathways similar to those described by Sims (120) for rat liver preparation. Thus, dihydrodiols were isolated as metabolites of BaA, 7-methylbenz[a]-anthracene, and BaP. The epoxide BaA-5,6-oxide was converted into the corresponding dihydrodiol and glutathione conjugate by these cells. Grover, Hewer and Sims (115) were first to identify an epoxide formed as a metabolite of BaA by human lung preparations as the K-region derivative BaA-5,6-oxide.

Human Placental Microsomes

The major metabolites of BaP produced by human placental microsomes from smokers were identified as 3-hydroxy-BaP and 7,8-dihydro-dihydroxy-BaP (7,8-diol) (121). Other metabolites included quinones, the 4,5- and 9,10-diols, and some unidentified polar metabolites.

Human Fibroblast and Epithelial Cells

Huberman and Sachs (122) measured the metabolism of BaP to form water- and alkali-soluble metabolites in cultured fibroblast and epithelial cells from human embryos.

Fibroblasts from different organs of the same embryo metabolized similar amounts of BaP into water-soluble products. Fibroblasts derived from different embryos metabolized different amounts of BaP.

The fibroblasts were divided into three groups, which metabolized an average of 350, 850 and 3,400 $\mu\mu$mols of water-soluble products per 1 million cells in three days. Cultures that contained more than 20% epithelial cells metabolized 3 to 25 times more BaP than did fibroblast cultures from the same embryo.

Epithelial cells from different embryos also varied in the degree of BaP metabolism, and low epithelial activity was not necessarily associated with low fibroblast activity in the same embryo.

Huberman and Sachs suggest that there is a genetic heterogeneity in BaP metabolism in fibroblast and epithelial cells and that the higher activity of epithelial cells might be related to the higher incidence of carcinomas rather than sarcomas in humans.

Human, Rat and Hamster Respiratory Tract Tissues

Wistar rat and Syrian hamster trachea metabolize PAH to the same products formed by human bronchial tissue (123). These tissues, maintained in short-term culture, metabolized BaP, BaA, and 7-methyl-BaA in a qualitatively similar manner.

Preliminary results indicate that the amounts of each of the metabolites formed were also of the same order. Thus, the radioactive products formed from BaA were inseparable on thin-layer chromatograms from 5,6-dihydro-5,6-dihydroxy-BaA, 8,9-dihydro-8,9-dihydroxy-BaA, and 9-hydroxy-BaA.

From 7-methyl-BaA, radioactive products having the chromatographic characteristics of the related 5,6- and 8,9-dihydrodiols were formed, together with 7-hydroxymethyl-BaA. BaP yielded products having the chromatographic characteristics of 4,5-dihydro-4,5-dihydroxy-BaP, 7,8-dihydro-7,8-dihydroxy-BaP, and 9,10-dihydro-9,10-dihydroxy-BaP; 3-hydroxy-BaP was also detected.

Mechanism of Metabolism

Keysell, Booth and Sims (124) state that PAH are initially metabolized by microsomal oxidases into epoxides, which are then converted into dihydrodiols, phenols, and glutathione conjugates.

The dihydrodiols are further metabolized by the oxidases to intermediates, which yield water-soluble products with glutathione. Their studies on the metabolism of BaA dihydrodiols showed that the 5,6-dihydrodiol is metabolized at the 8,9 bond and the 8,9-dihydrodiol at the 5,6 and 10,11 bonds. Formation of the 8,9-dihydrodiol-10,11-oxide (epoxide) of BaA by rat liver microsomal fractions was demonstrated.

The role of the liver in the metabolism of BaP and its relation to local carcinogenesis was suggested in experiments by Kotin, Falk and Miller (125). An enhanced subcutaneous sarcoma response occurred in mice in which liver injury had been produced with carbon tetrachloride in comparison with mice receiving only BaP subcutaneously.

Epoxides are highly reactive. Figure 5.2 illustrates the possible pathways for an epoxide produced by mixed-function oxidative reactions. The epoxide may (1) spontaneously rearrange to a phenol, (2) be converted to a trans-dihydrodiol by the action of microsomal epoxide hydrases, (3) conjugate with glutathione, or (4) combine covalently with cellular nucleic acids, histones, and protein.

Figure 5.2: Possible Pathways In Vivo for Reactive Epoxide Intermediates of Xenobiotics or Endogenous Substrates

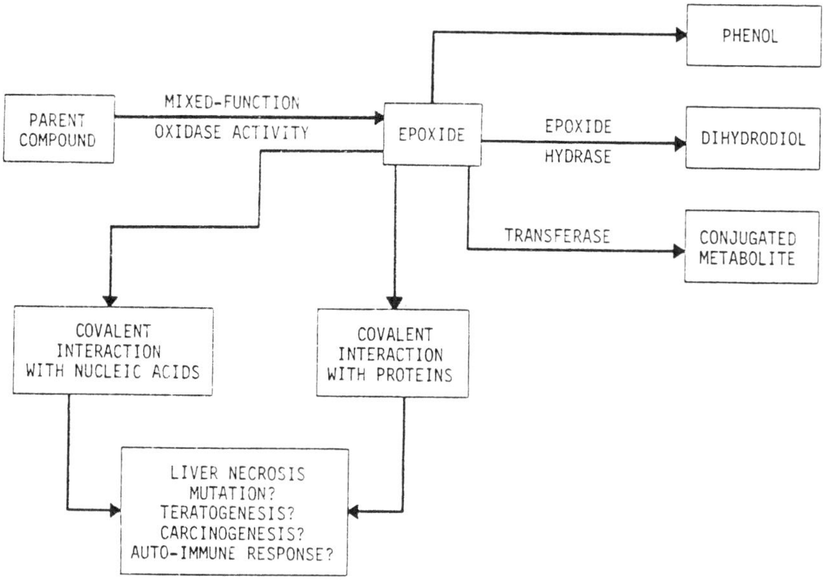

Source: Reference (126) as cited in ORNL/EIS-95

ELIMINATION

The hepatobiliary system and the gastrointestinal tract are the main routes of elimination of PAH and their metabolites in different species, regardless of the route of administration. Because BaP and some of its metabolites are highly fluorescent in ultraviolet light, it was possible, in early studies, to follow the steps in the metabolism of intravenously injected BaP by observing color changes in the fluorescence of experimental animal tissues (127). Later studies used radioactive compounds.

Biliary Excretion

Many observations have been made on the elimination of BaP and its metabolites following intravenous, subcutaneous, and intraperitoneal injections in experimental animals. Kotin, Falk and Busser (128), using radioactive BaP, and Heidelberger and Jones (129), using dibenz[a,h]anthracene, confirmed that, following these different modes of injection, the hydrocarbons are metabolized in the mouse and rat liver and excreted via the bile and intestinal tract into the feces. Radioactivity was completely cleared from the blood within 10 minutes.

Heidelberger and Jones (129) observed the distribution and elimination of dibenz[a,h]anthracene labeled in the 9 and 10 atoms with ^{14}C following three different modes of administration to mice. Following intravenous injection of an aqueous colloid, an essentially quantitative recovery of the dose was found.

There was an insignificant amount of radioactivity in the respiratory carbon dioxide; most of the radioactivity was found in the feces or the gastrointestinal tract. The urinary secretion of radioactivity was small compared with the amount found in the feces. Following administration by stomach tube (less than 1 mg in tricaprylin or in colloid), the radioactive dibenz[a,h]anthracene substances were eliminated almost exclusively in the feces.

There was no detectable activity in the intestines or intestinal contents 48 hours after administration. There was little difference in the fate of dibenz[a,h]anthracene, whether administered as colloid or in tricaprylin, and there was no evidence that any appreciable quantity of material was absorbed in the body.

Following intraperitoneal injection (1 mg dibenz[a,h]anthracene in 0.30 cm^3 tricaprylin) and subcutaneous injection, fecal elimination was again the primary route of removal of dibenz[a,h]anthracene. Dibenz[a,h]anthracene was retained for fairly long periods at the site of subcutaneous injection in fatty solvents. The distribution and rate of elimination depended on the mode of administration. Little material was absorbed into the body.

In another study, the results were the same: Radioactivity following intravenous injection of an aqueous colloid of 0.45 mg BaP-5-^{14}C into mice was recovered quantitatively, and no radioactivity appeared in the respiratory carbon dioxide. The compound and its metabolites were rapidly excreted into the bile, gastrointestinal tract, and feces (130).

Half-Life: The disappearance rates of radioactivity (half-lives) of three labeled hydrocarbons from the sites of subcutaneous injection in mice were 12 weeks for dibenz[a,h]anthracene; 3½ weeks for 3-methylcholanthrene; and 1¾ weeks for BaP (130). The relative carcinogenicity of each compound was directly proportional to the time of retention at the injection site.

Heidelberger and Weiss (130) also administered a single application of radioactive BaP or dibenz[a,h]anthracene in benzene to the shaved sacral region of the skin of mice. BaP or its metabolites, which had two half-lives, were eliminated much more rapidly than dibenz[a,h]anthracene ($t_{1/2}$ = 1⅔ and 4⅓ d vs $t_{1/2}$ = 85 d).

Metabolite Excretion: After intravenous injection into female Wistar rats, ^{3}H-7,12-dimethyl benz[a]anthracene (^{3}H-DMBA) was rapidly taken up by the liver, bound to particulate fractions, and metabolized to polar derivatives (131). The metabolites passed into the cytosol and were rapidly excreted in the bile. Radioactivity appeared in the bile within a few minutes.

Agents that induce or inhibit microsomal drug-metabolizing enzymes increased or decreased, respectively, the biliary excretion of metabolites. Injected metabolites of ^{3}H-DMBA appeared to be taken up directly into the cytosol, and the rate of their excretion was more rapid than when the hydrocarbon itself was injected. Inducing and inhibiting agents did not affect the rate of biliary excretion of the injected metabolites.

Thus, it appeared that transfer into the cytosol is the final process before biliary excretion and that this step is dependent upon metabolism, which, under the conditions of the experiment, was probably the rate-limiting step in biliary excretion. Practically all of the injected ^{3}H-DMBA (96 to 97%) appeared in the bile, almost exclusively in the form of polar metabolites.

Urinary Excretion

Generally, the noncarcinogenic PAH such as naphthalene, anthracene, and phenanthrene are metabolized to a variety of soluble metabolites, which are eliminated in the urine. The carcinogenic hydrocarbons, as represented by BaA and chrysene (weak carcinogens) and dibenz[a,h]anthracene, BaP, and DMBA, are metabolized to a variety of excretory products, which are found principally in the feces. In addition, the carcinogenic hydrocarbons are excreted principally as unconjugated phenols and quinones (132).

Pulmonary Excretion

Although pulmonary exhalation plays an important role in the elimination of volatile hydrocarbons, exhalation of PAH is negligible because their vapor pressure is low (97).

CARCINOGENICITY

Formation of PAH-induced cancers in laboratory animals is well documented (72)(133)-(138). Experimental animal studies in which the carcinogenicity of a PAH compound is clearly demonstrated, irrespective of administration route, have been reported (72).

The studies show that BaP, for example, has produced tumors in mice, rats, hamsters, guinea pigs, rabbits, ducks and monkeys after administration by oral, skin and intratracheal routes. BaP has both a local and systemic carcinogenic effect.

In monkeys, there is some evidence of the ability of BaP to produce local sarcomas following repeated subcutaneous injections and to produce lung carcinomas following intratracheal instillation. BaP is also an initiator of skin carcinogenesis in mice as shown by numerous studies (133)-(138), and it is carcinogenic in single-dose experiments and following prenatal exposure.

The amounts of different PAH necessary to produce cancer in 50% of treated animals (ED_{50}) vary greatly. In mouse skin-painting studies by Boyland (139), the ED_{50} values for dibenz[a,h]anthracene, 3-methylcholanthrene, and BaP were 20, 20 and 80 μg, respectively.

Various polycyclic hydrocarbons that produce malignant tumors in rodents have failed to produce tumors under similar experimental conditions in rhesus, cynomolgus and squirrel monkeys (140). Fibrosarcomas were produced by 3-methylcholanthrene and BaP in tree shrews and galagos after a single subcutaneous injection of 10 mg.

It was postulated (140) that the prosimian primates, which include tree shrews, lemurs, lorises and tarsiers, may be more closely related to the rodent than to the higher primates in their reactions to PAH carcinogens. Primates of the suborder Prosimii appear to be more susceptible to carcinogenesis by PAH and have shorter latent periods than primates of the suborder Anthropoidea, which includes monkeys, apes and man (96).

Only a few reports concerning the effects of a single PAH on man are available, and those describe only BaP. Cottini and Mazzone (141) studied the effects of a 1% solution of BaP in benzene applied daily to normal human skin and to various cutaneous lesions for periods of up to four months. This treatment resulted in definite manifestations localized to the treated areas. Erythema, pigmentation, desquamation, formation of verrucae (warts), and infiltration developed in chronological order. Pigmentary and wartlike lesions were the most commonly observed changes.

After a maximum of 120 applications or when evidence of infiltration appeared, treatment was stopped and the manifestations regressed completely within two to three months. Although reversible and apparently benign, the changes were believed to represent early stages of a process that would have ultimately resulted in neoplastic proliferation.

Changes were more pronounced on unprotected skin surfaces and in older individuals. Treatment of squamous cell skin cancer with the BaP solution resulted in temporary amelioration. Similar changes and a squamous epithelioma were mentioned in two reports concerning accidential exposure to BaP (72).

No case reports or epidemiological studies on the significance of man's exposure to other single PAH are available (72). However, coal tar and other materials believed to be carcinogenic to man contain these hydrocarbons, and many PAH have been detected in other environmental situations.

Dose-Response Relationship

A dose-response relationship for atmospheric pollutants is difficult to show. Doses to experimental animals are many times greater than those received by man, either by skin contact with or inhalation of a polluted atmosphere. A rough estimation is that it would take a man living in a large urban area at least 272 years to retain 30 mg BaP in his lungs (96).

It is also difficult to evaluate factors such as (1) the effects of particles and particle size in polluted atmospheres, (2) individual breathing parameters and clearance mechanisms, (3) cocarcinogens and cigarette smoke, and (4) other less well defined exposures.

Skin Tumors: In an attempt to determine a dose-response relationship and to estimate a threshold dose of BaP on the skin of mice, Schmidt, Schmal and Misfeld (142) painted the shaved skin of the backs of NMRI and Swiss strain mice twice weekly with BaP doses of 0.05, 0.2, 0.8 and 2.0 μg. The dose was delivered in one drop of liquid. Tumors occurred in the two groups receiving the highest doses, whereas no tumors at the site of treatment were observed in the lower-dose and control groups.

Two NMRI mice receiving a dose of 0.8 μg (median total dose of 120.8 μg in 53 weeks) developed carcinomas at the site of application. Twenty-eight mice developed carcinomas after receiving applications of 2.0 μg (median total dose of 293.5 μg) twice weekly for 75.8 weeks.

Values for two Swiss mice were 5 and 42 carcinomas at the same dose levels and at total doses of 106.2 μg in 57.8 weeks and 255.6 μg in 60.7 weeks. The Swiss mice were more sensitive to the carcinogen than the NMRI mice by a factor of 1.4.

Since only two doses produced effects, a linear relationship of a logarithmic dose-effect curve could not be shown, but the results did suggest that there is a threshold of response.

Lung Carcinomas: In a preliminary study, Hilfrich, et al (143), using intratracheal instillations of BaP suspended in solution without a carrier dust, reported on the induction of tracheal and lung tumors in Syrian golden hamsters.

At three months of age, the hamsters were intratracheally treated once a week with doses of 0.1 to 9.77 mg (increasing by a factor of 2.5) for life. Increased doses decreased survival rate.

Animals receiving the highest doses (3.91 and 9.77 mg) survived three to four weeks, whereas those receiving the lowest doses (0.1 and 0.25 mg) lived more than 30 weeks.

The three- to four-week survival period was not long enough for the induction of neoplasms, although hyperplastic and metaplastic changes were observed. The first neoplasm occurred after four weeks of treatment at the 3.91 mg level. With an increase in survival at the lower dose levels, the incidence of tumors increased.

A dose-response relationship for the induction of squamous cell carcinomas was developed by using intrabronchial pellets containing a graded series of concentrations of carcinogen in cholesterol (144).

Pellets containing 3 mg of material ranging from 0.1 to 100% carcinogen were implanted in the bronchi of rats. Lower concentrations were produced by dilution with cholesterol.

Figure 5.3 shows a plot of the incidence of carcinoma vs carcinogen concentration for 3-methylcholanthrene and BaP. The curves are representative of the response observed with most drug reactions.

The yield of squamous cell carcinomas with 3-methylcholanthrene ranged from 37.9% at 100% carcinogen down to 0% at the lowest level (0.1%). The controls, which contained cholesterol or blank pellets, showed no evidence of spontaneous carcinoma.

These findings represent an overall picture without regard to dose. The squamous cell carcinoma incidence was 20% with 3-methylcholanthrene and 11.6% with BaP.

Squamous metaplasia incidence was high in both treated and control rats. These were preliminary results and the time length of the study was not reported.

Figure 5.3: Relationships of Tumor Yield to Dose, Using a Graded Series of Concentrations of Carcinogen in Cholesterol on an Intrabronchial Pellet

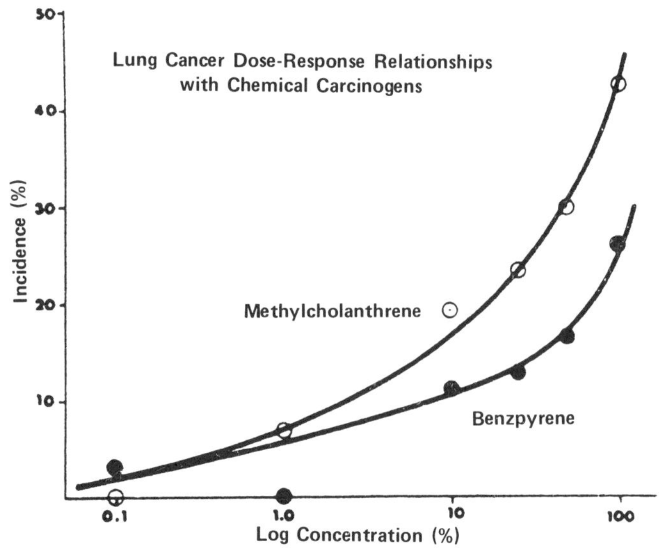

Source: Reference (144) as cited in ORNL/EIS-95

Subcutaneous Tumors: Pfeiffer (145) showed a dose-response relationship between BaP and tumors at the site of subcutaneous injection on the upper back of female NMRI mice. Doses ranging from 3.125 to 100 μg BaP in 0.5 ml tricaprylin were used. The results are shown in Figure 5.4.

Figure 5.4: Rates of Tumors in NMRI Mice After Subcutaneous Injection of Different Doses of BaP in 0.5 ml Tricaprylin

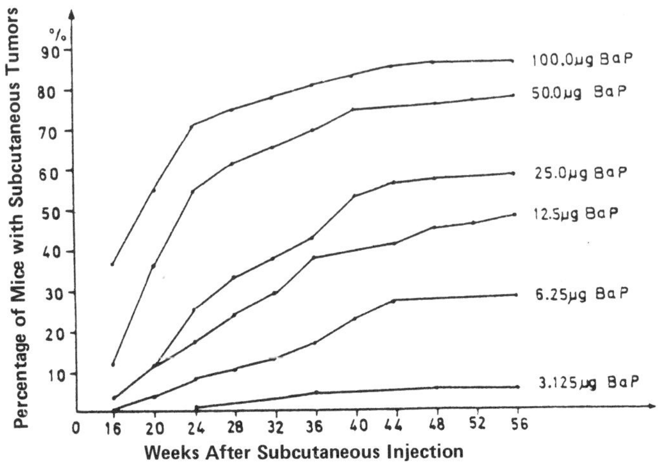

Source: Reference (145) as cited in ORNL/EIS-95

The spontaneous tumor rate for this strain of mice is 0 to 2%. Similar dose-response results were found with dibenz[a,h]anthracene. With the lowest dose of BaP (3.125 μg), tumors developed in 5 of 100 animals within 56 weeks. With the lowest dose of dibenz[a,h]anthracene (2.35 μg), 33 tumors developed.

Human Lung Cancer: Results of a long-term study of mortality among coke-oven workers (146)(147) suggest a dose-response relationship between exposure to coke-oven emissions, which contain PAH, and lung cancer.

Increased risk of lung cancer was related to both time and degree of exposure, as indicated by area of employment. Employment at the top of the ovens, where emission concentrations were highest, was associated with an almost fivefold excess risk of lung cancer. This increased to an almost tenfold risk when only men having five or more years of experience were considered. Lloyd (146) reports that coke-oven workers run a 7.5-fold risk of death from kidney cancers.

Enzyme Induction

The enzyme system primarily responsible for PAH metabolism is a multicomponent, microsomal-bound system of mixed-function oxidases (oxygenases), of which aryl hydrocarbon hydroxylase (AHH) is the most important. It is well established that these enzymes may either inactivate (detoxify) carcinogenic compounds or activate them to more carcinogenic forms (109).

Conney, Miller and Miller (148) describe the oxidative metabolism of PAH by enzymes located in the endoplasmic reticulum. They found that, after pretreatment with 1 mg BaP, liver homogenates from treated rats showed a tenfold greater AHH activity than did similar preparations from control rats. A doubling of enzyme activity was detected as early as three hours after BaP administration.

Wiebel and Gelboin (109) list AHH properties in human lymphocytes from normal subjects. These include (1) a requirement for reduced nicotinamide adenine dinucleotide phosphate (NADPH) and oxygen, (2) inhibition by a mixture of carbon monoxide and oxygen, and (3) inhibition by 7,8-benzoflavone (a potent inhibitor of BaP hydroxylation). These requirements indicate that the AHH of human lymphocytes is similar to that of rat liver or of rodent cells in culture.

Wiebel and Gelboin also reviewed studies demonstrating the relationship between the enzyme system and the biological effects of PAH. The studies suggest that the regulation of this enzyme system may be an important factor in chemical carcinogenesis. BaP hydroxylation is the basis for an extremely sensitive assay of the mixed-function oxidases and is used as an indicator for the presence and inducibility of AHH.

AHH and other mixed-function oxidases have been shown to consist of a thermal cytochrome (P450) and an electron transporting chain. AHH converts PAH to epoxides (114)(116), phenols, dihydrodiols, and quinones (101)(102)(112)(120). Epoxides are metabolized by epoxide hydrase to dihydrodiols (126). Some phenols and dihydrodiols are further metabolized by conjugating enzymes to yield the water-soluble glucuronides or sulfates.

Aryl Hydrocarbon Hydroxylase Inducibility: Studying the role of the AHH enzyme system in chemical carcinogenesis is particularly useful because of two properties: (1) the system is inducible (149)(150), and (2) the enzyme level and inducibility are genetically determined and vary in different species and different strains of mice (109)(150).

AHH is found in 90% of the tissues of the mouse, rat, hamster and monkey. In vivo, the prior administration of a PAH induces the hydroxylase to higher levels in the liver, lungs, gastrointestinal tract, and kidneys of the monkey, hamster and rat. However, the specific activities of the enzyme system in all tissues of the monkey, before and after 3-methylcholanthrene administration, are considerably lower than those observed for the mouse, hamster or rat.

Generally, the monkey is rather resistant to PAH carcinogenesis; this resistance may be related to the low levels of AHH activity. AHH is present in various tissues of six strains of mice; the hepatic enzyme is inducible in the Swiss strains C-57, C3H, and A, but not in strains AKR/N or DBA.

As reported by Nebert and Gelboin (150), the magnitude of AHH induction varied greatly in the tissues and species–from no induction to more than a 100-fold increase in activity. Similar variations in AHH inducibility or activity were found to exist in the human system (151).

In addition to PAH, the enzyme system is also inducible by a variety of other endogenous and exogenous compounds including steroid hormones, barbiturates, insecticides, and whole cigarette smoke (96). Recent studies have shown that inducibility is genetically regulated (152).

Transplacental Induction: The enzyme system can be induced transplacentally. Treatment of the pregnant rat or hamster with PAH or phenobarbital induced the enzyme system in specific fetal tissues as well as in the placenta (150). The extent of enzyme induction was greatest in the adult, less in neonatal tissues, and least in fetal tissues.

There is a high correlation between the placental content and the smoking habits of pregnant women (153)(154). PAH are known constituents of cigarette smoke. The placentas from women who smoked 20 or more cigarettes per day showed a mean AHH activity 5- to 25-fold greater than that found in placentas from women who did not smoke. However, significant AHH activity was found in the placentas of some women who did not smoke, and no measurable AHH activity was found in the placentas of some women who smoked 20 or 30 cigarettes per day (154). No such differences were found by Welch, et al (153).

Tissue Specificity: Although earlier studies indicated that inducibility is an "all-or-none" phenomenon in all tissues of a given mouse strain, Wiebel and Gelboin (109) note a tissue-specific inducibility in certain strains of mice.

In four strains of mice in which hepatic AHH is either not inducible or only slightly inducible, enzyme activity was increased about tenfold in the lungs and two- to fivefold in the kidneys and small intestine after a single intraperitoneal BaA injection (100 mg/kg body weight in 0.2 ml corn oil). AHH activity also increased in the skin of all the strains 16 hours after topical application of 300 μg BaA (Table 5.21).

Table 5.21: Inducibility of AHH in Hepatic and Extrahepatic Tissues from Various Mouse Strains*

	AHH Activity, mols phenolic product/mg protein/30 min**									
	Liver		Lung		Small Intestine		Kidney		Skin***	
Strain	Control	BA	Control	BA	Control	BA	Control	BA	Control	BA
NZB	3,080	2,500	27	300	2.0	8.1	<1	6.2	20	79
NZW	2,090	2,970	37	460	4.7	23.2	3.2	13.4	–	–
AKR	1,810	2,808	47	514	2.7	7.9	<1	6.1	8	72
SJL	2,090	1,790	35	260	2.5	17.2	4.8	10.8	25	106
DBA	920	995	56	350	–	–	2.3	8.2	36	129
C57	1,550	5,760	208	540	–	–	2.4	142.0	63	570

*BaA (100 mg/kg body weight) was injected intraperitoneally in 0.2 ml of corn oil. Control mice received corn oil only. Animals were sacrificed 12 hr later and enzyme activities were determined in homogenates (0.5-1.2 mg protein) of the various tissues.

**Numbers represent the mean values from three animals.

***Sixteen hours after topical application of 300 μg BaA in 0.2 ml acetone or acetone alone to the shaven backs.

Source: Reference (109) as cited in ORNL/EIS-95

Mechanism and Forms: Much is yet to be learned of the mechanisms of detoxification enzyme induction. However, the kinetics of AHH induction at different inducer concentrations were summarized by Wiebel and Gelboin (109), who routinely used BaA as inducer because it is relatively nontoxic.

In cell cultures (1) the inducer was rapidly incorporated into the cell; (2) there was an initial lag of about 30 minutes before enzyme activity increased; (3) induction required continuous protein synthesis; (4) induction required RNA synthesis initially, but not later; and (5) the RNA synthesis was independent of protein synthesis. Currently, there appear to be various levels of AHH induction regulation (at the level of RNA transcription or post-transcription), and their relative importance differs for different cell types and probably for different tissues as well.

Observations by Alvares, Schilling and Kuntzman (155) indicate that PAH are metabolized by more than one form of microsomal hydroxylase, distinguishable by spectral and kinetic properties. The authors observed differences in the kinetics of BaP hydroxylation by hepatic hydroxylase from rats previously treated with either phenobarbital or 3-methylcholanthrene.

Wiebel and Gelboin (109) summarize the relative distribution of these two AHH forms in rats. The first form, which is stimulated by 7,8-benzoflavone, is found in newborn rats and predominates in the liver of male adult rats. This form is induced by phenobarbital. The second form, which is inhibited by 7,8-benzoflavone, is found in extrahepatic tissues, comprises a larger fraction in the liver of female adult rats, and is inducible by PAH in hepatic and extrahepatic tissues.

Both forms are present in any given tissue; however, one form predominates over another. Preliminary observations indicate that there are differences in the profile of BaP metabolites formed by the constitutive and the 3-methylcholanthrene-induced form of AHH.

Rasmussen and Wang (156) treated adult male Sprague-Dawley rats with PAH, phenobarbital, or barbital and studied the induced AHH enzymes in liver and lung in terms of the specific metabolites of BaP that were produced in an in vitro incubation system. The metabolite profile produced by phenobarbital- or barbital-induced enzymes differed from that found with the hydrocarbon-induced liver enzymes.

Only the production of 4,5-dihydro-4,5-dihydroxy-BaP and 3-hydroxy-BaP was increased over the control level when phenobarbital was used as the inducer, whereas the production of all identified metabolites was increased when the hydrocarbons were used as inducers. An exception was benzo[e] pyrene, which was essentially inactive as an inducer for either liver or lung enzymes. Phenobarbital and barbital did not increase enzyme activity in the lung as did BaP, 3-methylcholanthrene, and BaA.

The rats received daily intraperitoneal injections of a solution (8 mg/ml) of sodium phenobarbital or sodium barbital (8 mg/100 g body weight) in NaCl for three days prior to sacrifice. Separate groups of rats received intraperitoneal injections of a solution (4 mg/ml) of BaP or other hydrocarbons (2 mg/100 g body weight) in corn oil 24 hours before sacrifice.

Aryl Hydrocarbon Hydroxylase Induction and Cancer: Results of studies focused on the relationship between AHH inducibility and cancer are far from definitive. Nonetheless, there is some indication in both laboratory animals and humans that greater cancer susceptibility may be associated with greater inducible AHH activity.

Kouri, Ratrie and Whitmire (152) found a close relationship between susceptibility to 3-methylcholanthrene-induced subcutaneous tumors and inducibility of AHH in 14 strains of mice. However, no correlation could be detected between sarcomas evoked by DMBA or BaP and the inducible AHH activity among the same inbred strains.

In a genetic system where AHH inducibility is segregated as a single autosomal dominant gene, mice carrying the inducible allele were five to ten times as sensitive to 3-methylcholanthrene tumorigenesis as their noninducible litter mates. Heterozygotes seemed to have intermediate sensitivity. Of 72 mice bearing 3-methylcholanthrene-induced primary tumors, 68 were AHH inducible.

In backcross experiments, 10 of 12 (83%) mice having tumors were AHH inducible. All tumors were typical fibrosarcomas and were found at the site of subcutaneous injection. In this study hepatic AHH level was assayed.

The susceptibility to tumorigenesis produced by the topical application of BaP and promotion by phorbol ester or by the intraperitoneal administration of large doses of BaP was not correlated with the AHH inducibility in litter mates of C57BL/6N, DBA/2N, and NZW/BLN strains of mice (157). It was concluded that, if this enzyme is necessary for metabolic activation of BaP to the proximal carcinogen, the basal levels of the AHH in skin, lymph nodes, or peritoneal lining are sufficient.

Kellermann, Luyten-Kellermann and Shaw (158)(159) showed that variation in extent of AHH induction by 3-methylcholanthrene in cultured human leukocytes is under genetic control. The normal white population in the United States was divided into three distinct groups of low, intermediate, and high degrees of inducibility. The variation appeared to result from two alleles at a single locus. Phenotype frequencies were 53, 37, and 10%. A total of 161 healthy unrelated adults were sampled.

Kellermann, Shaw and Luyten-Kellermann (160) studied 50 patients having bronchogenic carcinomas; 4% of the patients had a low inducible AHH activity, 66% had a medium, and 30% had a high level of induction. Low, medium, and high AHH-inducibility frequencies in a control group of 85 healthy volunteers were 44.7, 45.9, and 9.4%.

As an additional control group, 46 patients having other types of tumors were randomly selected to demonstrate that low or high inducibility is not simply the result of having a tumor. Percentages of AHH inducibility in this group were 43.5, 45.6, and 10.9%, which were comparable with those of the original control group.

Evaluation of the smoking habits of the patients having lung cancer disclosed that all were heavy smokers. The data indicate that susceptibility to bronchogenic carcinoma is associated with the higher levels of inducible AHH activity. The potential usefulness of lymphocytes in screening human populations for genetic variability or possible effects of PAH was suggested.

Enhancement

In most epidemiological studies, exposure to a combination of chemicals as they appear in the environment or industry is the usual mode of contact, whereas in controlled laboratory studies, animals contact only individual chemicals or, occasionally, two chemicals. Cocarcinogens can markedly increase the effect of carcinogens, but are usually noncarcinogenic or, at most, very weakly carcinogenic.

Impairment of mucociliary transport mechanisms by chemicals could be important in determining the residence time, and thus the concentration, of damaging substances within the respiratory tract. Pulmonary clearance may be impaired by noxious gases and vapors. Reduced clearance of carcinogens has been noted when the carcinogen is attached to particulate matter (161).

Particulates: There is a maximum efficiency of lung (alveolar) deposition of particulates at a size of about 1 and 2 μ; there is a minimum efficiency for a size of about 0.5 μ; the percentage of particle deposition for sizes less than 0.1 μ is just as great as for sizes of greater than 1 μ (162). The bulk of the particulate mass in the atmosphere ranges between 0.1 and 10 μ in size.

PAH are readily adsorbed to very small particles (i.e., those particles having a diameter of less than 0.04 μ), but are not readily eluted from them. As particle size increases, release in the presence of appropriate solvents becomes more rapid and greater in extent (162). Polycyclic organic matter detected in the atmosphere is associated exclusively with particulate matter, especially soot. Most data indicate that more than 75% of the BaP associated with particles is associated with particles less than 5 μ in diameter (163).

Within the limits of 0.01 to 100 μ (mass median diameter), the fraction of the mass of an inhaled particulate cloud deposited in the nose, tracheobronchial, and pulmonary compartments as a function of the mass median diameter is shown in Figure 5.5.

Figure 5.5: Fraction of Mass of Inhaled Particles Deposited in the Three Respiratory Tract Compartments as a Function of Mass Median (Aerodynamic) Diameter

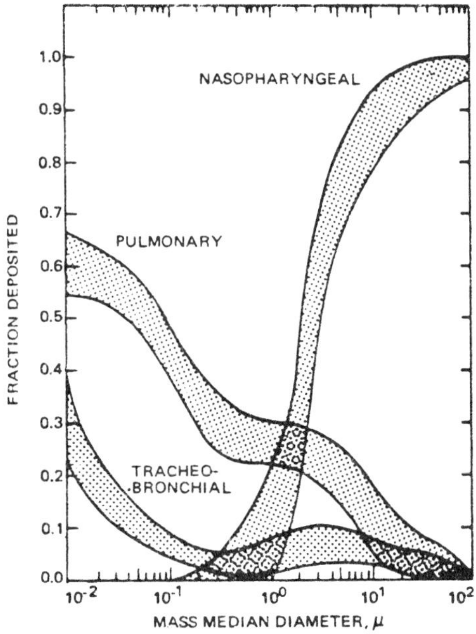

Source: Reference (164) as cited in ORNL/EIS-95

Iron oxide particulate matter appears to have properties that contribute to cancer production. Saffiotti, Cefis and Kolb (165a) first demonstrated that BaP, when attached to iron oxide (Fe_2O_3) particles, induced an unexpected high incidence of bronchogenic carcinomas in hamsters after repeated intratracheal injections. Iron oxide has no extremely irritating or toxic effects.

Lung cancer occurred in up to 100% of the animals in some experiments, and the lung cancers mimicked all the cell types seen in human lung cancers. Dose-response effects were indicated, as was the possibility that a single high dose could induce cancers in the system.

The influence of ferric oxide on BaP carcinogenesis was studied recently by Henry, Port and Kaufman (166). Three different mixtures of BaP and iron oxide with different physical properties were given intratracheally to Syrian golden hamsters.

Hamsters treated with a preparation containing large aggregates of BaP and ferric oxide (95% of the particles less than 3 μ in size) showed an earlier onset and higher incidence of respiratory tract tumors than animals given a mixture containing smaller aggregates. The greatest number of tumors were present in the trachea and the predominant type was the squamous carcinoma.

The third preparation, in which the carcinogen was not attached to the ferric oxide, showed a low tumor incidence similar to that present after intratracheal intubation of BaP in gelatin without a carrier particle. For this system, the physical attachment of BaP to the carrier dust was necessary for a high tumor yield, and particle size was important. The adsorption of BaP onto iron oxide particles in the atmosphere is yet to be shown (167).

The enhanced tumorigenicity of BaP-iron oxide combinations may be due to a reduction in carcinogen clearance rate. Clearance of BaP-hematite particles in the hamster lung following intratracheal instillation was studied by Kennedy and Little (168). BaP proved to be a loosely bound soluble carcinogen, which left the carrier particles and entered the upper airway epithelial cells during clearance of the particles by mucous and ciliary action. Thus, mainly epidermoid carcinomas of the major bronchi and trachea were produced.

Sulfur Dioxide: Laskin, Kuschner and Drew (144) studied the combined effect of sulfur dioxide and BaP on lung tumor development. Rats and hamsters were exposed to atmospheres of fresh air, sulfur dioxide, and/or a BaP-sulfur dioxide mixture. No respiratory tract neoplasms or other significant pathology was found in the hamsters.

The rats developed squamous cell carcinomas histologically similar to those observed in man. Of the 21 rats exposed to both sulfur dioxide and a BaP-sulfur dioxide mixture, 5 developed squamous cell carcinoma of the lung, whereas only two squamous cell carcinomas were detected in 21 rats exposed only to the BaP-sulfur dioxide. No tumors developed in the other atmospheres. The results may have been influenced by the chronic murine pneumonia observed in the rat colony.

Ozone: Although ozone is present in photochemical smog and causes a series of pulmonary disorders, no studies on the possible cocarcinogenic effect of ozone were found. Nettesheim and Schreiber (169) summarize animal studies on the possible effects of ozone.

Ozone has been shown to accelerate lung tumor development in mice. An important effect of ozone in relation to respiratory carcinogenesis is its suppressive effect on AHH activity in the lung. Enzyme activity in the tracheobronchial mucosa of rabbits was inhibited at ozone concentrations as low as 1.4 mg/m^3 (0.75 ppm) (170).

Nitrogen Dioxide: No studies were located on the carcinogenic or cocarcinogenic effect of nitrogen dioxide that resulted in a positive relationship. Kuschner and Laskin (1974), as cited in NIOSH (171), investigated the effects of combined nitrogen dioxide and BaP exposure in rats. Preliminary results indicated the presence of squamous cell carcinoma in the lungs of 1 of 30 rats exposed to 25 ppm nitrogen dioxide with daily 1 hour exposures to combined nitrogen dioxide and

BaP (10 mg/m^3). The concentration of nitrogen dioxide used in this study was 25 times as great as the recommended ceiling level. Concentrations as high as 94.1 mg/m^3 (50 ppm) had no significant effect on AHH activity (170).

Long-Chain Aliphatic Hydrocarbons: The accelerating properties of aliphatic and related hydrocarbons were demonstrated by Horton, Denman and Trossett (172) under three sets of experimental conditions: (1) as solvents for 3-methylcholanthrene or BaP in tests involving repeated application of the carcinogenic solutions to the skin of C3H mice; (2) as preconditioners (without added carcinogen) of the host by their application to the skin prior to any exposure to carcinogens; and (3) as promoting agents following a single application of DMBA. It was observed that

(1) Solutions of polycyclic carcinogens in solvents having long-chain structures, such as n-dodecane and dodecylbenzene, induce skin tumors in mice at much higher rates for a given concentration of carcinogen than do solutions in hydrocarbon solvents of lower molecular weight.

(2) The accelerating hydrocarbons are capable of preconditioning the skin of C3H mice, rendering it more responsive to subsequent applications of a carcinogenic oil.

(3) The accelerating solvents are effective promoters of carcinogenesis initiated by a single application of a carcinogenic material.

The solvents, although primary irritants, were noncarcinogenic to the skin of mice.

In a later study, Bingham and Falk (173), using compounds with long aliphatic chains as cocarcinogens or accelerators, compared the rates of acceleration of cutaneous tumorigenesis in mice. They found a 1,000-fold increase in the enhancement of potency of low concentrations of BaP and BaA when n-dodecane, a low-temperature hydrocarbon carbonization product of coal, was the diluent.

Solutions of BaP, ranging from 0.02 to 0.000002% dissolved in Decalin or a 50:50 mixture of n-dodecane and Decalin, were applied topically (50 mg by volume) to the skin of C3H/He mice three times a week. When BaP was dissolved in Decalin (decahydronaphthalene), no skin tumors developed in any of the mice below the 0.02% concentration. When n-dodecane was the diluent, skin tumors were produced at 0.000002% BaP. Similar results were obtained with BaA.

In a second experiment using various concentrations of BaP and of 1-dodecanol and 1-phenyldodecane, there was a reduction in the time required to induce tumors with increasing concentrations of BaP; however, the cocarcinogen was most effective at the lower level of carcinogen (0.05% BaP in 1-dodecanol as compared with 0.20% BaP). It was impossible to distinguish between the effects of 0.05 and 0.20% BaP when 1-phenyldodecane was the solvent.

Phenols: In inhalation studies using mice, Tye and Stemmer (174) found that coal tars with phenols removed were less potent carcinogens than coal tars containing phenols; the cocarcinogenic potential of phenols was attributed to their irritant properties.

Aldehydes: Aldehydes occur as general pollutants and are important intermediaries in many industrial processes. Several investigators have studied their carcinogenicity and mutagenicity. A 35-week pretreatment of C3H mice with gaseous formaldehyde did not enhance the response to subsequently applied coal tar at 100 mg/m^3 of air (175). However, simultaneous administration of furfural (a homolog of formaldehyde) and BaP did hasten lung tumor development in hamsters (176).

Infections: A possible relationship between respiratory infections and susceptibility to lung cancer has been investigated and debated for years. Various respiratory infections, such as chronic bronchitis, influenza pneumonitis, interstitial pneumonia, and tuberculosis, have been thought to increase the risk of lung cancer in man. Results of numerous clinical, pathological, and epidemiological investigations have been contradictory; laboratory studies have also furnished seemingly contradictory results.

Viral pneumonitis suppressed the development of pulmonary adenomas in mice, whereas chronic murine pneumonia enhanced the development of squamous cell carcinomas (169). Although viral pneumonitis causes extensive pulmonary scarring and consolidation of lung sections, thus reducing the potential amount of target tissue for carcinogens, other microbial effects could result in an enhanced susceptibility of respiratory tract epithelium to carcinogens.

These effects are the prolonged stimulation of cell proliferation in the tracheobronchial epithelium, alteration of the distribution and residence time of particulates, damage to ciliated epithelium, alteration in carcinogen metabolism, and alteration of the immune response to noncross-reacting antigens (169).

Suppression

Two (or more) agents applied simultaneously or sequentially may interact to alter the effect of each other. This may result in a decreased biologic effect, owing to competitive reactions at the sites concerned with biologic activity or to induction or inhibition of enzymes that may detoxify or activate, respectively, chemical carcinogens. There is some evidence that excess dietary vitamin A offers partial protection to epithelial tissues against chemical carcinogens. Several compounds that inhibit AHH activity have been reported.

Vitamin A: The role of vitamin A in controlling the growth and differentiation of mucous epithelia has been known for many years. Vitamin A is necessary for normal growth and differentiation of tracheobronchial epithelium. In the absence of vitamin A, the normal ciliated and mucus-producing cells of this epithelium are replaced by squamous cells, which produce keratin. In hamster organ culture systems, addition of vitamin A or its analogs to the culture induced reversal of keratinization and growth of a new ciliated and mucus-producing epithelium (177).

Current research suggests that ample vitamin A intake imparts some protection to epithelial tissue against chemical carcinogens. This was demonstrated in hamsters by the protection afforded by vitamin A against lung cancer produced by instillation of hydrocarbons and other particles (96). However, whether vitamin A or its analogs are capable of reversing squamous metaplastic or other preneoplastic lesions induced by chemical carcinogens remains unknown.

Saffiotti, et al (165b) investigated the effects of vitamin A on respiratory tract carcinogenesis. Respiratory tract tumors were induced in hamsters by ten intratracheal injections of BaP (3 mg) mixed with ferric oxide (3 mg). One group received stomach-tube feedings of vitamin A palmitate (5,000 IV) twice a week, starting one week after the last BaP treatment. Respiratory tract tumors (all types) developed in 17 of 53 (32%) animals receiving BaP only, but in only 5 of 46 (11%) of animals receiving treatment with vitamin A. The incidence of squamous tumors was markedly lower in the group receiving the vitamin treatment (only one animal in 46 as compared with 11 of 53).

Similarly, squamous metaplasia was seen in 13 cases receiving BaP treatment and in only one case receiving BaP plus vitamin A. The number of forestomach papillomas was also significantly reduced by excess vitamin A. However, animals treated with vitamin A had a faster death rate. The authors suggest that vitamin A has a systemic inhibitory effect on the induction of squamous changes in the columnar mucous epithelium of the respiratory tract.

Cone and Nettesheim (178) observed the suppression of 3-methylcholanthrene-induced squamous metaplasias and early tumors in the respiratory tract of rats by vitamin A, which confirmed the work of Saffiotti, et al (165b). Rats, on a diet free of vitamin A, were given normal or excess levels of retinyl acetate (a vitamin-A compound), by stomach tube several weeks before administration of 3-methylcholanthrene. Development of metaplastic lesions and squamous cell tumors was suppressed at all time intervals.

The study suggests that sufficient vitamin A offers partial protection of respiratory tract epithelium against high doses of 3-methylcholanthrene. Nettesheim and Schreiber (169) report several studies in which tumor development appeared to be enhanced by vitamin A in experimentally prepared tumor systems.

Hill and Shih (179) evaluated 14 vitamin A compounds and analogs and other agents as inhibitors of microsomal mixed-function oxidases from liver and lung tissue of mice and hamsters (Table 5.22). The formation of BaP-X, an unidentified metabolic product of BaP that binds tightly to macromolecules, was strongly inhibited by retinol and several other vitamin A compounds and by butylated hydroxytoluene.

Several of the tested compounds have known anticarcinogenic properties. BaP-X accumulated in large amounts in the presence of cyclohexene oxide, an inhibitor of epoxide hydrase. The formation of tightly bound metabolites of DMBA and 3-methylcholanthrene was also inhibited by the compounds that block production of BaP-X.

By another pathway, BaP is converted to 3-hydroxy-BaP (Figure 5.6); this step was not inhibited by any of the vitamin A compounds. 3-Hydroxy-BaP is further metabolized by a microsomal oxidase to BaP-Y. The disappearance of this compound was inhibited by retinol and some of its derivatives and by butylated hydroxytoluene.

Hill and Shih proposed the scheme in Figure 5.6 as tentative pathways for BaP metabolism. One pathway leads to BaP-X, the formation of which probably proceeds through an epoxide intermediate.

Table 5.22: Inhibition of Liver Mixed-Function Oxidases by Vitamin A Compounds and Analogs and Other Agents

Inhibitor	Cyclophosphamide (K_i)	Nicotine (N-1) (K_i)	BP (B-1) (% inhibition)*,**	DMBA (% inhibition)*,***	MC (% inhibition)*,†	3-Hydroxy-BP (% inhibition)*,††
Retinol	0.12	0.36	38 (31-44)	42 (40, 43)	52 (40, 61)	40 (38, 42)
13-cis-Retinol	0.11	0.26	32 (27-39)	20 (13, 27)	58 (49, 68)	32 (31, 34)
Retinyl acetate	0.12	0.40	37 (24-51)	40 (35, 44)	54 (49, 60)	28 (26, 30)
Retinyl palmitate	0.29	0.90	13 (0-22)	13 (10, 16)	-2 (-5, 0)	0 (3, -3)
Retinal	0.15	>1.0	37 (27-45)	37 (26, 47)	47 (34, 60)	28 (24, 33)
9-cis-Retinal	0.12	>2.0	45 (35-55)	53 (50, 56)	75 (71, 79)	46 (44, 49)
13-cis-Retinal	0.14	0.62	34 (30-36)	34 (33, 36)	56 (46, 66)	27 (26, 28)
Retinoate	>0.5	1.0	4 (0-17)	6 (5, 8)	-2 (-11, 7)	0 (0, 0)
13-cis-Retinoate	>1.0	>2.0	5 (3-13)	4 (3, 4)	-2 (-3, 0)	10 (10, 10)
Methyl retinoate	>1.0	>2.0	11 (0-23)	8 (6, 11)	17 (13, 22)	-3 (-5, -2)
Ethyl retinoate	>0.8	>2.0	10 (0-16)	8 (0, 16)	-3 (-18, 12)	1 (-5, 3)
N-Ethylretinamide	0.25	0.64	25 (18-30)	14 (2, 26)	30 (24, 37)	13 (5, 19)†††
N,N-Diethylretinamide	>1.0	>2.0	26 (19-37)†††	14§	—	10 (6, 13)†††
Ro 8-7699	>0.6	>2.0	9 (0-22)	6 (0, 12)	26 (21, 30)	8 (3, 14)
7,8-Benzoflavone	0.1	>0.2	46 (34-57)	39 (37, 41)	58 (57, 60)	-7 (-3, -15)†††
Butylated hydroxytoluene	0.1	>0.2	76 (67-83)†††	44 (40, 49)	79 (76, 82)	90 (73, 100)†††
Cyclohexene oxide	>2.0	>2.0	-52 (-64, -39)§§	8 (2, 15)	3§	-19 (-18, -20)

*Inhibition of binding to reaction components produced by 50 μM test compound, except cyclohexene oxide which is 2 mM.

**Average value and range for 5 separate determinations; in controls, the product formed ranged from 2.3 to 3.1 nmols.

***Average and individual values for 2 separate determinations; in controls, the product formed was 5.2 and 6.0 nmols.

†Average and individual values for 2 separate determinations; in controls, the product formed was 0.50 and 0.71 nmol.

††Average and individual values for 2 separate determinations; in controls, the substrate used was 1.2 and 1.3 nmol.

†††Three determinations.

§Single determination.

§§Two determinations.

Source: Reference (179) as cited in ORNL/EIS-95

Retinol and some of its derivatives, 7,8-benzoflavone, and butylated hydroxytoluene inhibit this reaction. Cyclohexene oxide apparently inhibits epoxide hydrase to increase the yield of BaP-X.

Figure 5.6: Proposed Scheme for BaP Metabolism

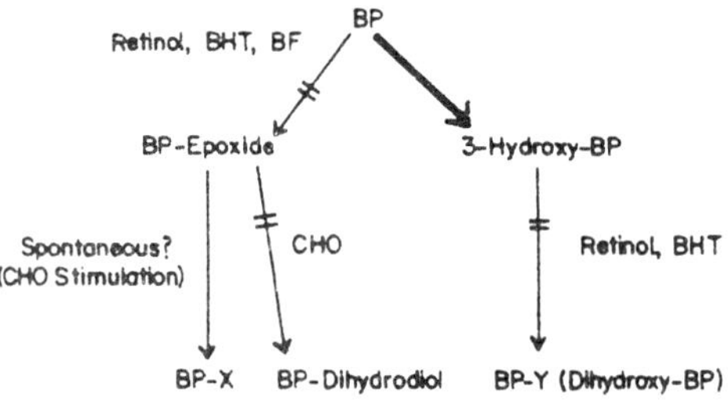

Note: Heavy arrow, reaction occurring both in normal liver and in lungs of mice pretreated with BaA; CHO, cyclohexene oxide; BHT, butylated hydroxytoluene; BF, 7,8-benzoflavone.

Source: Reference (179) as cited in ORNL/EIS-95

Antioxidants: Butylated hydroxytoluene, an antioxidant that serves as a substrate for a microsomal oxidase, strongly inhibited the formation of mammary tumors induced by DMBA (180). Also, butylated hydroxyanisole, butylated hydroxytoluene, and ethoxyquin inhibited chemical carcinogenesis when administered either before or with a carcinogen, suppressing the action of several carcinogens under different experimental conditions.

Inhibition occurred both when the route of administration resulted in direct contact of the carcinogen with the target tissue and when the carcinogen acted at a site remote from that of administration (181).

As an extension of the work with antioxidants, Wattenberg (181) studied the inhibitory effects of disulfiram and several other sulfur-containing compounds in three experimental tumor systems. Disulfiram, dimethyldithiocarbamate, and benzyl thiocyanate, when added to the diet of female Sprague-Dawley rats, inhibited DMBA-induced mammary tumor formation and adrenal necrosis. A single oral dose of disulfiram, given 24 hours before the carcinogen, similarly inhibited DMBA-induced mammary tumor formation.

In the mouse, disulfiram prevented the occurrence of forestomach tumors attributed to BaP in the diet, but did not affect pulmonary adenoma formation in mice given BaP by oral intubation.

7,8-Benzoflavone: 7,8-Benzoflavone inhibited enzymes involved in the activation of DMBA to its carcinogenic form to a greater extent than it inhibited enzymes involved in the detoxification of DMBA (182). However, in the case of the substrate dibenz[a,h]anthracene, 7,8-benzoflavone appeared to inhibit enzymes involved in detoxification to a greater extent than it inhibited activation of dibenz[a,h]anthracene. Data obtained with phenobarbital indicated that dibenz[a,h]anthracene, but not DMBA, has a specific phenobarbital-inducible detoxification system in mouse-skin experiments. AHH in mouse skin epidermis was inducible by the two hydrocarbons and 5,6-benzoflavone.

7,8-Benzoflavone given at the same time as DMBA inhibited the formation of skin tumors in a two-stage system of tumorigenesis and also inhibited the formation of covalently bound complexes of DMBA with epidermal DNA, RNA, and protein. However, 7,8-benzoflavone given at the same time as dibenz[a,h]-anthracene stimulated skin tumor formation, but had little or no effect on the binding of dibenz[a,h]anthracene to epidermal marcromolecules.

Phenobarbital had a small inhibitory effect on DMBA skin tumor initiation and significantly inhibited tumor formation initiated by dibenz[a,h]anthracene.

In a study by Diamond and Gelboin (183), 7,8-benzoflavone inhibited AHH activity and inhibited PAH toxicity. Inhibition of AHH by 7,8-benzoflavone also inhibited tumorigenesis by DMBA in mouse skin (107).

Wattenberg and Leong (184) reported on the inhibition of lung adenoma induction by β-naphthoflavone. β-Naphthoflavone inhibited BaP-induced enzymes.

Antibiotics: The addition of the antibiotics actinomycin D, puromycin, or cycloheximide simultaneously with BaA completely prevented the induction of AHH activity in hamster fetus cell cultures (185). Actinomycin D acts by inhibiting protein synthesis necessary for AHH induction.

Nicotinamide: Nicotinamide inhibited the levels of both the constitutive and induced AHH activity in liver nuclei of C3H/He and DBA/2 mice (186). The enzymes that catalyze the oxidation of cyclophosphamide and nicotine to hydroxy derivatives were also inhibited by the agents that block BaP metabolism in mice and hamster cell cultures (179).

Hormones: The hormonal status of experimental animals is known to influence the appearance and rate of growth of tumors. The rate of growth of mammary tumors in rats was accelerated by pregnancy or by the injection of progesterone, and their appearance was delayed by the administration of estradiol (187).

Booth, Keysell and Sims (119) describe the effect of estradiol on the in vitro metabolism of DMBA and of its hydroxymethyl derivatives by subcellular fractions of female rat liver. The amounts of dihydrodiols, phenolic metabolites, and glutathione conjugates were greatly reduced by the presence of estradiol in such a system.

The steroid probably acted by inhibiting epoxide formation rather than by inhibiting the three types of reaction involved in their further metabolism. This was supported by the finding that estradiol also reduced the total amount of each substrate metabolized.

Estradiol may exert its anticarcinogenic effect by competing for binding sites on the mixed-function oxidase enzyme complex, thus interfering with the metabolism of the hydrocarbon.

Mixed Hydrocarbons: Lacassagne, et al (1934), as cited in Haddow and Weinhouse (188), applied mixtures of hydrocarbons to the skin of mice and found that the weakly carcinogenic hydrocarbons, such as chrysene and dibenz[a,g]-fluorene, inhibited the production of skin tumors by the potent carcinogen 3-methylcholanthrene. A number of subsequent studies by the same authors and others demonstrate the inhibition, by some PAH, of carcinogenesis induced by potent PAH carcinogens.

Falk, Kotin and Thompson (189) studied the effects of a number of hydrocarbons found in polluted urban air and cigarette smoke on sarcoma production by BaP. Benzo[a]fluorene, perylene, perinaphthoxanthrene, benzo[a]carbazole, and chrysene were potent anticarcinogens.

These compounds, when given in the ratio of 0.10 to 0.15 of anticarcinogen to 1.0 of BaP, inhibited sarcomagenesis by more than 70%. Less active were benz[k]fluoranthene, benz[m,n,o]fluoranthene, and 2-naphthol. The inhibition was probably due to a stimulation of enzyme activity capable of detoxifying the carcinogenic agent or was related to an antimetabolite type of competition of the inhibitor and the active carcinogen for a relevant receptor site in the cell.

Selenium: Selenium repressed chemical carcinogenesis in experimental animals (190). Selenium (0.0005%) and vitamin E (0.13%) significantly reduced the number of skin tumors in mice that were induced by DMBA (0.125 mg in 0.25 ml acetone) and 3-methylcholanthrene (0.25 ml of 0.01% solution) in nondietary tumor production experiments. Both selenium and vitamin E reduced the total number of papillomas.

Shamberger, Tytko and Willis (191) also presented data suggesting that selenium could be an environmental factor affecting tumor incidence in man. Selenium may exert an anticarcinogenic effect by acting as an antioxidant.

Personal Hygiene: A high incidence of scrotal cancer among wax pressman employed in the petroleum industry was attributed to contamination of their clothes with wax containing carcinogenic oil (192). Of 77 men who had at least nine years of experience as wax pressman, 11 ultimately developed cancer of the scrotum. While removing wax from filters in the wax processing plant, the workers leaned across a supporting bar, thus heavily contaminating the crotch of their trousers with wax. At this point in the processing, the wax was grossly contaminated with oil.

An industrial hygiene program was instituted that consisted of (1) cancer detection at the time of employee quarterly examinations, (2) prompt treatment by excision of any malignancies or premalignant lesions, and (3) cancer prevention by protecting workers from exposure to carcinogens.

All wax pressmen were issued an impervious rubber or plastic apron, a daily change of clean underwear, and double lockers for clothes changes. In addition, the men were encouraged to clean their own work clothes at least once a week, and a daily shower was enforced. After institution of the industrial hygiene program, no problems among wax pressmen were reported.

An industrial hygiene program was initiated at the Union Carbide Corporation coal hydrogenation pilot plant at Institute, West Virginia, in an effort to reduce employee exposure to potentially cancer-producing oils (193). As part of the control program, a program of employee education was started. The use of protective clothing and the encouragement of personal hygiene procedures was stressed as a first line of defense against employee contamination. The program included a daily change of underwear, outer clothing, and socks, plus a daily inspection of hands, face, and neck. However, operation of the plant was curtailed before the complete clothing program was put into effect.

Prior to closure, small groups of workers were tested for contamination. A skin examination using 2537 Å ultraviolet light in a dark room showed skin contamination after showering. After use of barrier clothing, consisting of pajamas buttoned at the neck and fitted at the arm and leg cuffs, in conjunction with barrier cream on the exposed parts of the body, the men became clean after several days, as determined by ultraviolet light.

Dry cleaning of clothes, followed by a soap and water laundering, was successful in removing most of the pasting oil. Waterless hand soaps were most effective in removing pyrene, a noncarcinogenic material, from the skin. Double lockers and shower facilities were also provided. A drying room and inspection room, equipped with mirrors and ultraviolet lights, were planned.

Selective placement, restriction, and rotation of employees was practiced to some degree, but the burden upon the medical staff and the total number of employees made these practices difficult to accomplish (194). In addition, medical literature does not agree on the type of person (skin type) who should be excluded from exposure to materials that contain suspected or known carcinogens.

Mechanisms

It is now almost generally accepted that PAH must be metabolically activated to induce cancer. However, it is apparent that the ultimate carcinogenic metabolite(s) of PAH has not been positively identified. It is possible that different PAH produce different types of carcinogenic metabolites. Although more work is needed to firmly establish the nature of the ultimate carcinogenic form of PAH, some studies have been aimed at unravelling this complex problem.

Covalent Binding to DNA, RNA, and Protein: There is appreciable evidence that derivatives of chemical carcinogens bind covalently to DNA, RNA, and protein. As reviewed by Miller (195) and Miller and Miller (196), the covalent binding in vivo to proteins was noted in tissues susceptible to their carcinogenic action. Subsequently, covalent bindings of derivatives of chemical carcinogens with both nucleic acids and proteins of tumor-susceptible tissues in vivo and in vitro have been observed. The amounts of these carcinogens that are bound in vivo are of the order of one carcinogen residue per 10^4 to 10^7 monomer residues in the macromolecules.

Although in some cases the correlation between binding and carcinogenicity is low and some noncarcinogenic derivatives bind at appreciable levels, there are no exceptions known in which a chemical carcinogen has failed to exhibit covalent binding to macromolecules in its target tissue in vivo (196). Chemical carcinogens have been shown to be electrophilic compounds (196).

Brookes and Lawley (197) painted a series of PAH on mouse skin and found a strong correlation between the reactivity of PAH with the DNA of the skin and the ability of PAH to induce cancer. This was one of the first suggestions that PAH exercise their carcinogenic activity through a reaction with DNA. Binding to RNA and protein also occurred. An earlier study by Miller (198) had shown that derivatives of BaP bind to protein in the epidermal fraction of mouse skin in vivo.

Wiest and Heidelberger (199), in mouse skin-painting studies, showed an irreversible binding of the carcinogen dibenz[a,h] anthracene, or its metabolites, with the nucleoproteins. This was a covalent binding to cellular constituents in vivo.

Using an in vitro system involving incubation of PAH with either DNA or protein in the presence of a microsomal hydroxylating system, Grover and Sims (200) found that tritiated PAH reacted with the protein or DNA.

Table 5.23 shows the results of incubation of a 50-mg sample of protein or DNA for 1 hour with 5 μg of hydrocarbon. When similar mixtures were incubated without enzyme cofactors (i.e., the parent hydrocarbons were not metabolized), no binding of the hydrocarbons to either DNA or protein was detected.

Table 5.23: Enzyme-Catalyzed Reaction of Tritiated PAH with Protein and DNA in Vitro

Hydrocarbon	Reaction with Protein (mmols/mol)*	Reaction with DNA (μmols/g-atom DNA P)**
Benzo[a] pyrene	0.78	1.41
3-Methylcholanthrene	0.73	0.78
7,12-Dimethylbenz[a] anthracene	0.78	0.64
Dibenz[ah] anthracene	0.95	0.44
Dibenz[ac] anthracene	1.07	0.56
Benz[a] anthracene	0.76	0.70
Pyrene	1.15	0.31
Phenanthrene	0.72	0.05

*Molecular weight of bovine plasma albumin taken as 64,000.
**Calculated on a basis of 8% phosphorus.

Source: Reference (200) as cited in ORNL/EIS-95

A later study by the same authors (201) gave results indicating that K-region epoxides may be the active intermediates involved in the binding of PAH to cellular macromolecules.

At physiological temperature and pH, phenanthrene 9,10-oxide and dibenz[a,h]-anthracene 5,6-oxide reacted with 4-(p-nitrobenzyl)pyridine, nucleic acids, and histone. However, neither the parent compounds (phenanthrene, dibenz-[a,h] anthracene) nor the K-region dihydrodiols derived from these hydrocarbons reacted with p-nitrobenzylpyridine, nucleic acids, or histone.

Some reported data is not in agreement with the concept of an epoxide intermediate. Grilli, Rochi and Prodi (202) found a high level of nonenzymatic binding of PAH to nucleic acids in vitro when thymus DNA was used. The aromatic hydrocarbons were tightly bound and not extractable either by organic solvents or by warming DNA solutions above the melting point and precipitating the DNA again. Microsomal dependent binding occurred between polynucleotides and the hydrocarbons (DMBA, BaP, and 3-methylcholanthrene).

A metabolite of BaP that bound to DNA in vitro without the need for further activation was isolated and preliminarily identified on the basis of mass spectral information and hydration studies as the 4,5 epoxide (203).

To determine whether the K-region epoxide is the PAH derivative that binds to DNA, Baird, et al (1973) compared the DNA-bound products formed by the reaction of the K-region epoxide of 7-methylbenz[a]anthracene with DNA with those of DNA isolated from cells that had been exposed to the parent hydrocarbon. These results suggest that the cellular metabolism of 7-methylbenz[a]-anthracene that results in the covalent binding of this hydrocarbon to DNA in cells may involve more than simple formation of K-region epoxide.

Since the carcinogenic metabolites formed in cells from methyl-substituted PAH may be different from those formed from the nonmethylated hydrocarbons (204), the products formed by the reaction of the K-region epoxide of BaP with DNA in aqueous solution were compared with those found in DNA isolated from cells incubated with BaP in vitro (205).

The hydrocarbon-deoxyribonucleoside products from enzyme digests of mouse embryo cell cultures treated with ^{3}H-BaP were not identical to those found in similar chromatograms of enzyme digests of DNA that had been reacted with BaP 4,5-epoxide in aqueous ethanol solution. This suggests that the metabolic activation of BaP that results in covalent binding to DNA in mouse embryo cells in culture may also be more complex than the simple formation of a K-region epoxide and reaction of that compound with the cellular DNA.

Since tritium-labeled BaP epoxide was not available, it was not possible to determine whether the products formed in the DNA of epoxide-treated cells would be the same as those formed in aqueous solution.

On the basis of the above results, King, Thompson and Brookes (206) proposed a model of the mechanism of BaP metabolism and DNA binding. They suggest that BaP is metabolized by AHH to yield a series of epoxides, both K-region and non-K-region.

These products are substrates for the enzyme epoxide hydrase, which is associated with the AHH complex. By epoxide hydrase action, an intermediate is produced prior to conversion to the dihydrodiols, and this intermediate is responsible for the BaP binding to DNA that occurs in vivo; for example, a diol epoxide of BaP may be the likely metabolite responsible for the observed binding to DNA.

A pathway in which a dihydrodiol epoxide is produced was recently discovered by Sims, et al (207). These investigators noted the formation of a non-K-region epoxide that was converted by epoxide hydrase to a dihydrodiol. The isolated double bond in the dihydrodiol ring was then epoxidized to yield a reactive dihydrodiol epoxide. This metabolite, 7,8-dihydro-7,8-dihydroxybenz[a]pyrene

9,10-oxide (Figure 5.7), appeared to account for a major DNA-bound form of BaP in cultured hamster cells.

Figure 5.7: BaP (a) and 7,8-Dihydro-7,8-Dihydroxybenz[a]pyrene 9,10-Oxide (b)

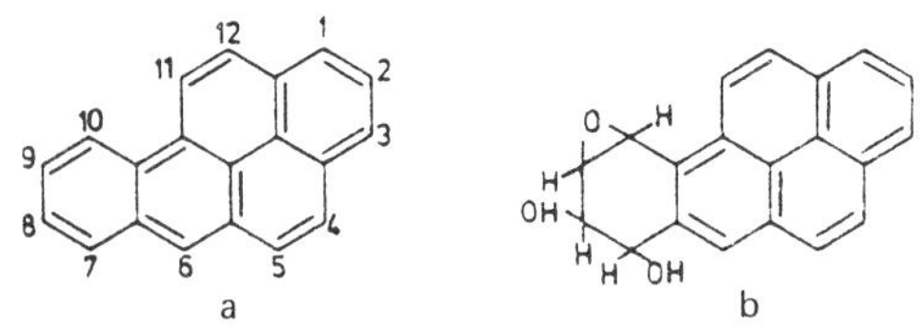

Source: Reference (207) as cited in ORNL/EIS-95

A correlation between oncogenicity and the binding of PAH to macromolecules in man has not been studied. Harris, et al (208) incubated bronchi from three patients with lung cancer and from one patient without lung cancer with several PAH. The bronchial epithelial cells in culture had the capability to bind and activate tritiated DMBA, BaP, 3-methylcholanthrene, and dibenz[a,h]anthracene. Radioactivity was found in both the nuclei and cytoplasm of cells incubated with ^{3}H-DMBA. Ciliated, mucous, and basal cells contained tightly bound radioactivity. Epithelial cells in bronchial glands were overlaid with autoradiographic grains.

Similar results were found with ^{3}H-BaP, ^{3}H-3-methylcholanthrene, and ^{3}H-dibenz[a,h]anthracene. Bound radioactivity was also found in mesenchymal cells, but to a lesser extent. DMBA and BaP bound to DNA more than did dibenz[a,h]anthracene. No information was given on differences of amounts of binding among patients.

DNA-Binding, Suppression, and Enhancement: In the respiratory epithelium, vitamin A deficiency causes basal cell hyperplasia and squamous metaplasia, Genta, et al (209) studied the binding of tritiated BaP (^{3}H-BaP) to DNA in tracheal epithelial cells of normal and vitamin A-deficient hamsters.

When isolated tracheas were incubated in vitro in the presence of ^{3}H-BaP, substantially greater quantities of radioactivity were associated with DNA prepared from vitamin A-deficient tracheas. When 7,8-benzoflavone, an inhibitor of AHH, was added to the system, the binding of ^{3}H-BaP was reduced.

Harris, et al (unpub.), as cited in Genta, et al (209), demonstrated that, in vitamin A-deficient tracheas, there is a higher binding of ^{3}H-BaP to macromolecular structures of basal cells located in squamous metaplastic areas than to basal cells located in normal areas.

In hamster tracheas the binding of BaP to DNA was enhanced if the hamsters had been previously treated intratracheally with BaP plus Fe_2O_3 (210). There was increased binding observed in the tracheas of vitamin A-deficient hamsters (211).

7,8-Benzoflavone given simultaneously with DMBA inhibited the formation of (1) skin tumors in mouse skin epidermis, and (2) covalently bound complexes of DMBA with epidermal DNA, RNA, and protein. However, 7,8-benzoflavone given simultaneously with dibenz[a,h]anthracene stimulated skin tumor formation, but had little effect on the bonding of dibenz[a,h]anthracene to epidermal macromolecules (182).

Metabolites: Results of studies since 1970 have provided information on the possible ultimate carcinogenic metabolite(s) of PAH in vivo. It has been demonstrated by some researchers that K-region epoxides of several PAH are more active than either the parent hydrocarbons or other metabolites in producing malignant formation in cell-culture systems.

More recent studies, however, suggest that diol-epoxides may be involved in the reactions of PAH with the DNA of treated cells and thus be the ultimate carcinogenic metabolite. There is also strong evidence that the carcinogenic metabolites of methyl-substituted PAH may be different from those formed by nonmethylated hydrocarbons.

In 1950, Boyland (139) suggested that epoxides are metabolic intermediates in the conversion of PAH to phenols and trans-dihydrodiols. Since that time, numerous studies have demonstrated that epoxides are intermediates in the microsomal oxidation of several PAH.

Huberman, et al (105) and Grover, et al (113) proposed that epoxides may be the metabolically activated, and possibly the ultimate, carcinogenic form of PAH that do not contain active methyl groups. They demonstrated that K-region (5,6) epoxides of dibenz[a,h]anthracene and BaA were more active in producing malignant transformation of hamster embryo cells than the corresponding hydrocarbons, K-region cis- and trans-dihydrodiols and phenols.

The K-region epoxides of BaA and 3-methylcholanthrene also transformed a clone of C3H-mouse ventral prostate cells, whereas the parent hydrocarbons were not active (113). These results are shown in Tables 5.18 and 5.19. Parent hydrocarbons gave results similar to those of the acetone controls. With either type of derivative, increased concentration of the compounds gave increases in the frequency of transformation.

Epoxides of the K-region of 3-methylcholanthrene, phenanthrene, and chrysene were also more active in producing transformation than were the corresponding hydrocarbons (105).

Marquardt, et al (212) tested the K-region epoxides derived from DMBA and 7-hydroxymethyl-BaA for their ability to induce malignant transformation of a clone of C3H-mouse prostate fibroblasts. In this case the K-region epoxides of both compounds were slightly less active than the parent hydrocarbons in an in vitro transformation. The corresponding hydroxylated derivatives, which were also tested, were without transforming activity.

Addition of α-naphthoflavone, an inhibitor of microsomal mixed-function oxidases, prevented the toxic effects usually produced by DMBA and 7-hydroxymethyl-BaA, but did not inhibit the induction of malignant transformation in cells treated with these compounds. Thus, metabolic activation to epoxides may not be a prerequisite for the oncogenic effect of 7-methyl-substituted PAH.

However, hydroxylation of the 7-methyl group of DMBA may be a necessary first step in carcinogenesis. Liver homogenates from female Sprague-Dawley rats injected subcutaneously with 400 μg BaP were incubated with BaP or 6-methyl-BaP (213). A number of metabolites were identified, including one that was indistinguishable from 6-hydroxymethyl-BaP by either thin-layer chromatography or ultraviolet spectra. This compound was a potent carcinogen when administered by subcutaneous injection to rats.

All rats injected with 100 μg BaP, 6-methyl-BaP, or 6-hydroxymethyl-BaP had sarcomas at the site of injection. No tumors at the other sites were observed. These observations were in accord with the hypothesis that methylation or hydroxymethylation is one of the first steps in metabolic activation of carcinogenic PAH. Earlier studies (214)(215) led the authors to postulate this mechanism for DMBA.

Because some K-region epoxides of DMBA do not exhibit greater carcinogenic potencies than the parent compound, Dipple, Levy and Iype (216) investigated the possible involvement of the ketonic derivatives, 5,6-dihydro-5-oxo-7,12-dimethylbenz[a]anthracene and 5,6-dihydro-6-oxo-7,12-dimethylbenz[a]anthracene. These compounds were inactive in tests for tumor-initiating activity in mouse skin and tumor production in the subcutaneous tissue of the mouse.

A diol-epoxide (Figure 5.7) that binds to DNA might be the ultimate carcinogenic metabolite (207). Huberman, et al (217) and Yang, et al (218) demonstrated the high mutagenic activity and metabolic formation of this compound, which they identified as r-7,t-8-dihydroxy-t-9,10-oxy-7,8,9,10-tetrahydrobenzo-[a]pyrene, and suggested that it may be an ultimate carcinogenic form of BaP.

TERATOGENICITY

Although only a small number of substances have been bioassayed for both teratogenicity and carcinogenicity, a relationship between the two is indicated by the large percentage of compounds that have produced both malformations and cancer.

A comprehensive review by DiPaolo and Kotin (219), shows that the majority of carcinogens exert a teratogenic effect. The rapid growth, lack of differentiation, and reactivity of the cells in the developing embryo suggest that the action of chemical or biological agents will be expressed in alterations at the level of differentiation. Somatic cell division is a prerequisite for both teratogenesis and carcinogenesis. The rapid division and differentiation of fetus cells result in a sensitivity that makes the fetus especially vulnerable to teratogens.

The same compound may have different effects, depending upon the stage of development of the organism at the time of exposure. Thus, a compound found to be carcinogenic to mature cells might be teratogenic to immature, embryonic cells.

In 1973, Shepard (220) edited a catalogue of agents that had been tested for teratogenicity. The catalogue includes a listing of agents along with a short summary of the study including animals studied, doses, route of administration, and bibliographic citation. Both positive and negative results are reported. Several carcinogens, including DMBA and 3-methylcholanthrene, produced teratogenic effects in experimental animals.

Placental Transfer

Shendrikova and Aleksandrov (221) studied the permeability of the rat placenta to DMBA, BaP, and 3-methylcholanthrene. Female rats were given a single dose (200 mg/kg body weight) of one of three compounds (in sunflower-oil suspension) by gastric tube on the twenty-first day of pregnancy.

Fetal tissue content of DMBA reached a maximum of 1.53 to 1.6 μg/g body weight 2 to 3 hours after administration (Figure 5.8). BaP, tested 3 hours after administration, also reached high concentrations (2.77 μg/g); 3-methylcholanthrene was found only in trace quantities. The same ratio of concentrations was also found in the liver of the pregnant rats.

Figure 5.8: Concentration of DMBA

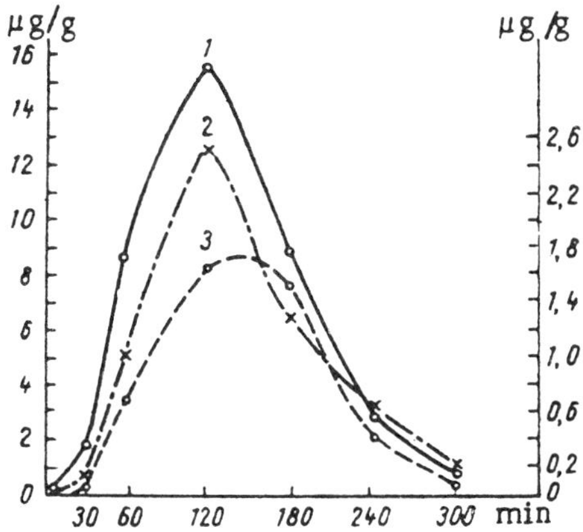

Note: Concentration of DMBA in liver (1) of pregnant rats, placentas (2), and fetuses (3) at various times after administration of the compound in a dose of 200 mg/kg by gastric tube on the 21st day of pregnancy. Abscissa, time after injection of DMBA (in min); ordinate, (left) DMBA concentration (in μg/g) in liver, (right) in fetuses and placentas.

Source: Reference (221) as cited in ORNL/EIS-95

When DMBA was injected intravenously into the animals (25 mg/kg body weight), maximum accumulation occurred in the fetal tissues 1 hour after injection. The period of maximum accumulation was shorter than after intragastric administration (Figure 5.9). The fetal concentration of the compound was roughly of the same magnitude (1.5 to 2 μg/g) when administered by different routes. The similarity in fetal content following administration of doses that differed by an order of magnitude by different routes was explained by the small quantities of DMBA absorbed from the gastrointestinal tract and the consequent delay in reaching its maximal concentration. The differences in concentrations of the different PAH in fetal tissues did not reflect their relative abilities to cross the placental barrier, but were instead due to the differential absorption of these compounds from the gastrointestinal tract (221).

Figure 5.9: Concentration of DMBA in Fetal Tissues

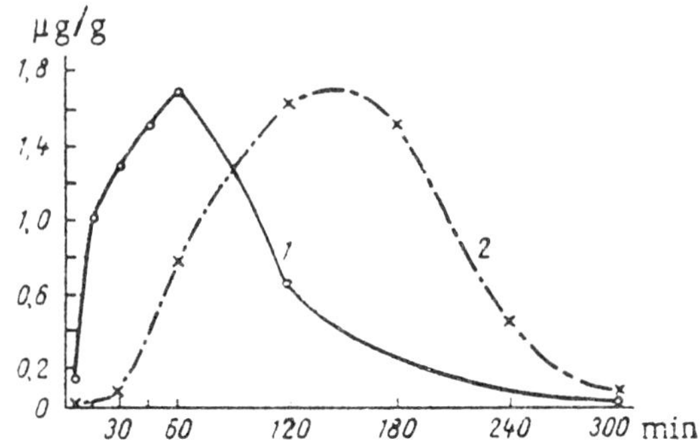

Note: Concentration of DMBA in fetal tissues after intravenous injection of the compound into rats in a dose of 25 mg/kg (1) and after its administration by gastric tube in a dose of 200 mg/kg (2) on the 21st day of pregnancy. Abscissa, time (in min) after administration of DMBA; ordinate, DMBA concentration (in μg/g) in fetal tissues.

Source: Reference (221) as cited in ORNL/EIS-95

Enzyme Activity

Placental enzymes, especially those metabolizing foreign compounds, may be significant factors in the resistance of the developing fetus to potentially toxic or teratogenic substances. The teratogen may be inaccessible to the fetus because the placental enzyme system enzymatically converts the teratogen to nontoxic, more polar derivatives or because the placenta may metabolize an inactive compound to an active teratogen. The functions of the placenta appear to change during the course of gestation.

Treatment of pregnant rats with BaP stimulated AHH activity in the entire implantation site (embryo and decidua), decidua, placenta, fetal membranes, and fetus (222)(223). Wistar rats were treated with an oral dose of BaP (40 mg/kg body weight) dissolved in corn oil 24 hours prior to sacrifice. Maximum induction of AHH activity in the placenta, as measured by the in vitro hydroxylation of BaP, was obtained only from day 15 of gestation, whereas in the fetus, the degree of induction of enzyme activity increased from day 13 to day 18 of gestation (gestational period lasted 21 to 22 days).

Enzyme activity in fetal membranes was higher on days 16, 18 and 20 of gestation than on day 13. Enzyme activity was three to eight times higher in the fetal membranes than in the placenta. In control animals little or no enzyme activity was measured in these tissues. In animals treated on day 8 of gestation, increased AHH was present within 6 hours in the entire implantation site and reached a maximum activity 12 to 18 hours after treatment. By day 10, AHH levels fell back to control levels. In the placenta, following treatment on day 12, AHH levels were increased within 6 hours and remained at the same level until 18 hours after treatment.

By days 14 and 15, the enzyme activity had dropped to half of the highest levels. Following treatment on day 19, AHH levels dropped only slightly from day 19 to day 21. These results demonstrated that, in the rat, the fetal membranes play a more important role for metabolism of BaP than does the placenta. Treatment of the pregnant rat or hamster with either a PAH or phenobarbital induces the enzyme system in specific fetal tissues as well as in the placenta (150). Table 5.24 shows the AHH activity in several tissues of the normal and 3-methylcholanthrene-treated hamster, hamster fetus, and newborn hamster.

Table 5.24: Effect of In Vivo Administration of MC on AHH Activity in Tissues from Pregnant Hamster, Fetus, and Newborn

	AHH Specific Activity (U/mg tissue protein)					
	Mother		Fetus		Newborn	
	Control*	MC*	Control*	MC*	Control**	MC**
Liver	2,430-3,555	3,435-3,945	14-27	174-426	500-620	2,505-2,835
Lung	<6-11	33-48	<6	18-45	21-23	63-98
Small intestine	<6-18	75-83	<6	9-20	<6-17	134-155
Kidney	14-33	162-207	<6	32-33	12-15	132-144
Placenta	<6	15-42	–	–	–	–

*Three experiments conducted.
**Two experiments conducted.

Source: Reference (150) as cited in ORNL/EIS-95

3-Methylcholanthrene in corn oil (100 mg/kg body weight) was administered intraperitoneally to the pregnant hamster, and the animal was sacrificed 24 hours later, on approximately day 15 of gestation. The hydrocarbon was administered by the same route and at the same dosage on the first day of life to the newborn, which were sacrificed 24 hours later. Mothers nursing the neonates did not receive any drugs.

Control levels of AHH were detectable in hamster fetal liver, and the enzyme was induced transplacentally by 10- to 20-fold after treatment of the mother with 3-methylcholanthrene. In the liver, lung, and kidney of the newborn hamster, the 3-methylcholanthrene-induced enzyme was stimulated about 3- to 5-fold, whereas in the small intestine it was more than nine times higher. About a 30-fold increase in AHH activity developed in less than 72 hours between the fetal and neonatal period. Table 5.25 shows the AHH activity in several tissues of the normal and 3-methylcholanthrene-treated pregnant Sprague-Dawley rat and rat fetus. Treatment was the same as for the hamster. The transplacental induction of AHH activity in the fetal rat tissues was similar in magnitude to that found in fetal hamster tissues.

Table 5.25: Effect of In Vivo Administration of MC on AHH Activity in Tissues from Pregnant Rat and Fetus

	AHH Specific Activity (U/mg tissue protein)			
	Mother		Fetus	
	Control*	MC	Control*	MC*
Liver	1,365-1,530	3,165-3,525*	6-9	201-311
Lung	8	159-191*	<6	54-81
Small intestine	<6	605**	<6	9-23
Kidney	6-17	812-942*	<6-6	41-42
Placenta	<6	24-33*	–	–

*Two experiments conducted.
**One experiment conducted.

Source: Reference (150) as cited in ORNL/EIS-95

It is well established that cigarette smoke contains some of the same PAH as environmental contaminants, and several studies have linked levels of placental AHH to smoking habits in pregnant women (153)(154).

Figure 5.10 shows the distribution of AHH in human placentas assayed within 24 hours after delivery (154). A positive relationship was observed between a relatively high level of placental AHH activity and a history of cigarette smoking. The mean AHH activity in the placentas from smoking women was 2- to 25-fold greater than that in the placentas from nonsmokers.

Figure 5.10: Distribution of Human Placental AHH Activity*

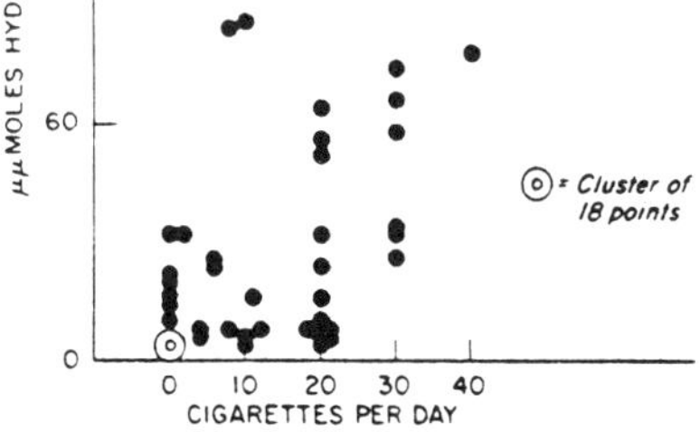

*Note: As a function of cigarette smoking during pregnancy (correlation coefficient r = 0.53; P = 0.001).

Source: Reference (150) as cited in ORNL/EIS-95

The placentas from women who smoked 20 or more cigarettes per day showed a mean AHH activity 5- to 25-fold greater than that found in placentas from women who did not smoke. However, AHH activity was found in the placentas of some women who did not smoke, and no measurable AHH activity was found in the placentas of some women who smoked 20 to 30 cigarettes per day.

The differences may reflect differences in the tar content of cigarettes smoked, method of inhalation, nutritional and medicinal status, or eating habits of the women. On the other hand, they tend to support the studies of Kellermann, Cantrell and Shaw (151), which indicate that genetic differences exist in AHH inducibility by aromatic hydrocarbons in man.

Of the 59 placentas assayed, 35 samples were obtained from women who smoked between 2 and 40 cigarettes daily during gestation, up to and including the day of parturition. The assay of AHH activity was the in vitro enzymatic conversion of BaP to hydroxylated metabolites. Concentration was determined spectrophotofluorometrically (149).

Tumorigenesis

PAH have been shown to produce tumors in animal offspring of mothers treated during pregnancy (224)(225). Female Ha/ICR mice were injected subcutaneously on days 11, 13 and 15 of pregnancy with 0.2 ml of Carbowax-400 containing 4 mg of BaP or vehicle only. Absolute controls received no injections. On day 20 of gestation, the progeny were delivered by Cesarean section and nursed on control mothers.

After weaning at 4 weeks of age, the backs of the mice were shaved, and topical administrations of croton oil (2 drops of 1% croton oil in acetone) were given twice a week. (Croton oils acts as a promotor of epidermal carcinogenesis.)

Mice were sacrificed at 28 weeks of age, and pulmonary adenomas and skin papillomas were noted. In both cases, the tumor incidence and the number of tumors per animal were increased in the progeny of mothers injected with BaP (Tables 5.26 and 5.27). A carcinogenic effect was not noted in other tissues (224). Since PAH are excreted in milk (226), these experiments show the effect of prenatal exposures only.

Table 5.26: Effect of BaP Administration to Pregnant Mice on Pulmonary Adenoma Formation in the Progeny

Experimental Group	Material Administered	No. of Mice	No. of Mice*	Percent of Mice**	No. of Adenomas/ Group	No. of Adenomas/ Mouse	No. of Mice with Multiple Adenomas 3–6	7–9
Absolute controls	Nothing	82	6	6	7	0.09	0	0
Vehicle controls	Carbowax-400	89	10	11.2	13	0.15	1	0
BaP	BaP in Carbowax-400	55	34	61.8	130	2.36	13	6

*Number of mice with adenomas having pulmonary adenomas at 28 weeks of age.
**With adenomas.

Source: Reference (224) as cited in ORNL/EIS-95

Table 5.27: Effect of BaP Administration to Pregnant Mice on Skin Papilloma Formation in the Progeny

Experimental Group	No. of Mice	No. of Mice*	Percent of Mice**	No. of Papillomas/ Group	No. of Papillomas/ Mouse
Absolute controls	82	6	7.3	8	0.10
Vehicle controls	89	8	9.0	9	0.10
BaP	55	13	23.6	18	0.33

*Number of mice having skin papillomas after 24 weeks of topical treatment twice weekly with 1% croton oil; mice were 28 weeks of age.
**With papillomas.

Source: Reference (224) as cited in ORNL/EIS-95

In a more extensive study, Bulay (225), using both BaP and DMBA as carcinogens and using different routes of administration, produced pulmonary adenomas and skin papillomas in Ha/ICR mice in utero. The effects were produced to a greater degree with DMBA, and they could be produced by either intraperitoneal or subcutaneous injections. Although adenomas and papillomas are defined as benign tumors, they may be a precarcinogenic response.

Turusov, et al (227) investigated the effect of a low dose of 3-methylcholanthrene on the offspring of pregnant mice. CF-1 mice received 1 mg of 3-methylcholanthrene dissolved in 0.2 ml of olive oil, by stomach tube, in late pregnancy. Delivery occurred between one and three days after treatment. Additional experimental groups, consisting of offspring of treated and untreated mothers injected with 12.5 μg diethylnitrosamine (DENA) within 24 hours after birth, were started.

For comparison, the authors included results from an earlier study in which 8.4 mg of 3-methylcholanthrene were administered by stomach tube to pregnant females. Tumor incidence is shown in Table 5.28. All mothers that had received 8.4 mg of 3-methylcholanthrene and all of their offspring died, or were killed in poor condition, within 80 weeks, and both mother and offspring bore one or more tumors.

Table 5.28: Tumor Incidence in CF-1 Mice Treated with MC Prenatally and DENA When Newborn

Group	Type of Mice and Treatment	No. at Start (Sex)	Effective No.*	Tumor-bearing Animals No.	Tumor-bearing Animals (%)	No. of Tumors Total	No. of Tumors Per Mouse
1	Pregnant mice treated with 8.4 mg of MC	35 (F)	34	34	100	60	1.8
2	Offspring of pregnant mice treated with 8.4 mg of MC	54 (F)	53	52	98	74	1.4
		45 (M)	45	45	100	65	1.4
3	Pregnant mice treated with 1 mg of MC	38 (F)	35	34	97	56	1.6
4	Offspring of pregnant mice treated with 1 mg of MC	42 (F)	41	41	100	54	1.3
		65 (M)	65	64	98	81	1.2

(continued)

Table 5.28 (continued):

Group	Type of Mice and Treatment	No. at Start (Sex)	Effective No.*	Tumor-bearing Animals No.	(%)	No. of Tumors Total	Per Mouse
5	Offspring (of pregnant mice treated with 1 mg of MC) given 12.5 μg of DENA when newborn	45 (F)	45	45	100	64	1.3
		56 (M)	54	54	100	95	1.8
6	Offspring (of untreated pregnant mice) given 12.5 μg of DENA when newborn	36 (F)	34	33	97	56	1.6
		32 (M)	32	32	100	63	2.0
7	Untreated controls	44 (F)	38	31	82	44	1.2
		49 (M)	47	36	77	52	1.1

Group	Animals with More than One Tumor No.	(%)**	Malignant Lymphoma No.	(%)**	Lung Tumor No.	(%)**	Hepatomas No.	(%)**	Other Tumors No.
1	24	70	20	58	29	85	–	–	11
2	24	45	17	33	49	92	–	–	10
	19	42	16	36	44	98	3	7	2
3	19	54	15	43	29	83	1	3	11
4	11	27	6	15	39	95	–	–	9
	17	26	5	8	63	97	10	15	3
5	17	38	12	27	41	91	7	16	4
	39	72	6	11	53	98	33	61	3
6	20	59	12	35	30	88	12	35	2
	25	78	7	22	25	78	29	90	2
7	10	26	20	53	13	34	–	–	11
	13	28	19	40	24	51	3	6	6

*Number of animals alive at the time of appearance of the first tumor.
**In relation to the effective number.

Reference (227) as cited in ORNL/EIS-95

Although tumor incidence of controls is high (82% in the females and 77% in the males), tumor incidence (nearly 100%) of treated females and offspring occurred earlier, and there were more tumors per mouse in the 3-methylcholanthrene-treated group. More treated mothers bore more than one tumor than did offspring. At the dose level of 8.4 mg, this was mainly due to the occurrence of carcinoma of the forestomach in the treated mothers.

The highest incidence of tumors was observed in the lungs and lymphoreticular tissue, both in mothers and offspring. The same tumor types also existed in untreated controls. The results also showed that prenatal treatment with 3-methylcholanthrene caused a considerable inhibition of liver carcinogenesis induced by diethylnitrosamine.

The average levels of 3-methylcholanthrene in fresh tissue per litter after the administration of 1 mg to pregnant mice were 213 ng/g and 140 ng/g, respectively, 3 and 6 hours after treatment.

Tomatis and Goodall (228) investigated the effect of DMBA on successive generations of mice. They injected pregnant Swiss mice with 400 μg of DMBA during the last period of pregnancy. The percentages of animals having tumors in F-1, F-2, and F-3 generations were 58.6, 52.6, and 30.0, respectively. The percentage of tumors in untreated controls was 23.2. The tumors observed were lung adenomas, mammary carcinomas, granulosa cell tumors, and papillomas of the urinary bladder.

Birth Defects

Currie, et al (103) succeeded in inducing pronounced anomalies of the nervous system, eyes, and skeleton of rat fetuses after treatment with DMBA and its derivative, 7-hydroxymethyl-12-methylbenz[a]anthracene, at various stages of organogenesis.

Using the maximum dose tolerated by pregnant rats (20 mg/kg) at different stages of gestation, Alexandrov (229) succeeded in producing embryotoxic and teratogenic effects with DMBA. An embryotoxic effect was observed after treatment on the fifth, sixth, and ninth day after coitus.

Malformed fetuses were detected after treatment on days 9 and 13 postcoitum. Malformations included hydrocephaly, exencephaly, spina bifida, cleft lip, and cleft maxilla; anophthalmos occurred after treatment on day 9, and ectrodactyly of the forefeet occurred after treatment on day 13.

MUTAGENICITY

The active forms of a large number of known chemical carcinogens are mutagens. The PAH have been shown to bind to DNA. This suggests that malignant transformation can involve an alteration in the genetic constitution of treated cells. PAH carcinogens may therefore produce mutations, some of which may involve the genes that control malignancy (230).

Ames, et al (231) propose that carcinogens that are mutagens cause cancer by somatic mutation in which flat aromatic molecules intercalate in DNA base-pair stacks. This leads, during DNA replication or repair, to an addition or deletion of base pairs in the DNA sequence. These "frameshift" mutagens are more potent if they contain a side chain that covalently reacts with DNA.

Testing Methods

A variety of methods for testing mutagenicity has been developed. Submammalian tests in bacteria, bacteriophage, *Neurospora*, plants, and *Drosophila* help to elucidate basic mechanisms of mutagenesis. However, microorganisms do not metabolize the PAH in the same way as mammals; thus, in vivo mammalian test systems are more relevant for extrapolating data to man. Three in vivo mammalian tests are practical and sensitive: dominant lethal assay, host-mediated assay, and in vivo cytogenetics (96).

Dominant Lethal Assay: The induction of dominant lethal mutations in animals can be assayed with a high degree of sensitivity and practicality after acute, subacute, or chronic administration of test materials by several routes. After administration of a drug, male rodents are mated sequentially with groups of untreated females over the duration of the spermatogenic cycle. Matings within three weeks after a single drug administration represent samplings of sperm exposed during postmeiotic stages, and matings four to eight weeks later represent samplings of sperm exposed during premeiotic and stem cell stages.

The classic form of the dominant lethal assay involves autopsy of females about 13 days after timed matings and enumeration of corpora lutea and total implants. Dominant lethal mutations are directly measured by enumeration of early fetal deaths and indirectly by preimplantation losses. Results are best expressed as early fetal deaths per pregnant female.

Host-Mediated Assay: In the host-mediated assay the chemical being tested is administered to the mammal, which then receives an injection by another route of indicator microorganisms. The microorganisms are later recovered and scored for induction of mutants. This assay determines the influence of in vivo mammalian factors on the activation or detoxification of chemical mutagens.

In Vivo Cytogenic Assay: Rats and Chinese hamsters are most often investigated for structural and numerical chromosomal aberrations in in vivo cytogenetic assays. Cytogenetic effects on metaphase or anaphase preparations of somatic or germinal cells can be studied singly or serially after recovery periods.

Data on mutagenicity testing of air pollutants in in vivo mammalian systems are scarce, and there are no published data on mutagenicity testing by inhalation (96). However, mutagenicity studies using these three systems and specific chemicals have been reported.

In Vitro Assay: A fourth method, the in vitro tissue or cell culture, permits quantitative analysis of genetic variations in cell populations via the plating technique, following treatment of the cells in culture with chemical agents.

Ames, et al (231) describe a test for the detection of carcinogens as mutagens. The test consists of rat or human liver homogenate for metabolism of carcinogens and a set of *Salmonella* histidine mutants for mutagen detection. The carcinogen, liver homogenate, and bacterial tester strain are incubated together on a petri plate. Histidine revertant colonies are counted after a two-day incubation period.

Relationship Between Carcinogenicity and Mutagenicity

Using the in vitro assay method, Chu, Bailiff and Malling (232) tested the mutagenicity, in Chinese hamster cells, of some chemical carcinogens and their related compounds and derivatives (Table 5.29).

The genetic marker assayed in the hamster cells was the change from 8-azaguanine sensitivity to resistance. The results show a direct relation between the degree of carcinogenicity and mutagenicity. Although BaP was not mutagenic in this study, it is usually listed as a known mutagen (233).

Table 5.29: Relative Carcinogenicity and Mutagenicity of Selected Compounds*

Test Compound	Carcinogenicity	Mutagenicity
Benzo[e]pyrene	–	–
Benzo[a]pyrene	+	–
3-Hydroxybenzo[a]pyrene	–	±

(continued)

Table 5.29 (continued):

Test Compound	Carcinogenicity	Mutagenicity
Dibenz[a,c]anthracene	–	–
Dibenz[a,h]anthracene	±	±
7,12-Dimethylbenz[a]anthracene	+++	+++

*Symbols: – not carcinogenic (or mutagenic); ± uncertain or weakly carcinogenic (or mutagenic); + carcinogenic (or mutagenic); +++ strongly carcinogenic (or mutagenic).

Source: Reference (163) as cited in ORNL/EIS-95

Mutagenicity of Alkaline PAH

The recent studies of B.R. Clark, et al indicate that alkaline constituents of the petroleum substitutes are often major contributors to their mutagenicity. In contrast, the mutagenicities of petroleum crudes are determined almost solely by the neutral constituents as would be expected if polycyclic aromatic hydrocarbons were the determinant constituents.

The alkaline mutagens in petroleum substitutes may thus pose a problem unique to the emerging fossil fuels conversion industry. Neutral fractions of petroleum substitutes are, in most cases, the largest contributor to the overall mutagenicity of the whole material. Although neutral fractions usually have a lower specific activity than alkaline fractions, the much greater amounts of neutral substances increase their total contribution relative to that of the alkaline fraction. Components of the neutral fractions which were partially characterized include PAH, multialkylated PAH, and neutral nitrogen pyrrole homologues.

Alkaline fractions are uniquely important in most crude synfuel products. Isolation and analysis of the alkaline mutagens suggests that azaarenes and aromatic primary amines are responsible for the mutagenicity. Polycyclic aromatic primary amines are found to predominate in the mutagenic isolate and are found to exhibit exceptionally high mutagenicities.

Metabolism

A number of studies since 1969 have shown that the chemically nonreactive carcinogens, such as the PAH, must be metabolized by cellular enzymes in order to exert their biological effects (including mutagenicity).

Huberman and Sachs (230) developed a cell-mediated mutagenic assay with carcinogenic hydrocarbons in which Chinese hamster V79 cells, which are susceptible for mutagenesis but do not metabolize PAH, are cocultivated with lethally irradiated rodent cells that can metabolize these compounds.

The marker used for mutagenesis in the V79 cells was the genetic change from susceptibility to resistance to 8-azaguanine. The authors found that the number of mutations was dependent on the number of metabolizing cells and that the carcinogens were not mutagenic when V79 cells were cocultivated with nonmetabolizing cells. Inhibition of the hydrocarbon metabolizing enzymes by 7,8-benzoflavone inhibited mutagenicity.

Cell-mediated mutagenicity was obtained with the carcinogenic hydrocarbons (DMBA, BaP, 3-methylcholanthrene, and 7-methylbenz[a] anthracene). The most potent carcinogen, DMBA, gave the highest frequency of mutations. Pyrene and BaA, which are not carcinogenic, were also not mutagenic. The method detected mutagenicity with 1 μg/ml (230)(235).

E. Huberman and R. Langenbach (234) have shown that the cell-mediated mutation system, with the proper choice of metabolizing cells, can be used to detect the mutagenic activities of different classes of chemical carcinogens. When fibroblastic cells were used as the metabolizing cells, a correlation between the in vivo carcinogenic activity and the in vitro mutagenic activity of 11 aromatic polycyclic hydrocarbons was observed.

When primary liver cells were used as the metabolizing cells, three known liver carcinogens were demonstrated to be mutagenic by the cell-mediated assay, while two noncarcinogenic analogues were not mutagenic. These results from the cell-mediated system suggest that the reactive intermediates of the carcinogens are stable enough to be transferred from the metabolizing cells to the V79 cells.

The cell-mediated mutagenesis system is a simple in vitro assay which may simulate the in vivo situation. This approach could be extended to the cocultivation of cells from other organs or tissues with mutable mammalian cells.

Metabolites

In another study, Huberman, et al (216), using the same cell-mediated assay, identified the major mutagenic intermediate of BaP metabolism as a diol-epoxide. They tested the mutagenicity of BaP and 15 of its derivatives, which included the phenols; the BaP 4,5-epoxide (the K-region epoxide); dihydrodiols; two isomeric 7,8-diol 9,10-epoxides; a 6-methyl derivative; and a 6-hydroxymethyl derivative.

Mutations were characterized by resistance to ouabain or 8-azaguanine. Mutagenesis was tested both in the presence and absence of normal, BaP-metabolizing, golden hamster cells. Little or no mutagenicity was shown by the phenols; 4,5-diols; trans-9,10-diol; and 6-methyl and 6-hydroxymethyl derivatives.

The ($\pm$) 7α,8β-dihydroxy-9α,10α-epoxy-7,8,9,10-tetrahydro-BaP and K-region 4, 5-epoxide exhibited similar and moderate mutagenicity in the absence of BaP-metabolizing cells, but the ($\pm$) 7α,8β-dihydroxy-9β,10β-epoxy-7,8,9,10-tetrahydro-BaP showed a 2,000- and 270-fold higher mutation frequency for ouabain and 8-azaguanine resistance, respectively, than did the K-region 4,5-epoxide.

The trans-7,8-diol was more mutagenic than BaP after metabolism in BaP-metabolizing cells. Mutagenesis by the trans-7,8-diol was inhibited by 7,8-benzoflavone, an inhibitor of mixed-function oxidases. The major metabolite of trans-7,8-diol was isolated and found to be 7α,8β-dihydroxy-9β,10β-epoxy-7,8,9,10-tetrahydro-BaP. The high mutagenic activity of this compound and the demonstration of its metabolic formation (218) suggested that it may be an ultimate carcinogenic form of BaP.

Malaveille, et al (236) examined the mutagenicity of a series of BaP and BaA derivatives towards *Salmonella typhimurium* strain TA 100. BaP, 8-diol and BaA 8,9-diol were more active than the parent compounds in inducing his+ revertant colonies. BaP 9,10-diol was less active than BaP, whereas the K-region diols (BaA 5,6-diol and BaP 4,5-diol) were inactive.

None of the diols were active in the presence of AHH cofactors. The diol-epoxides (BaP 7,8-diol 9,10-oxide and BaP 7,8-diol 10,11-oxide) and the K-region epoxides (BaP 4,5-oxide and BaA 5,6-oxide) were mutagenic without further metabolism. The results support the theory that the non-K-region diol-epoxides are more mutagenic than the simple K-region epoxides.

In this study, dose-response curves for mutagenicity were prepared by using varying amounts of postmitochondrial liver supernatant. In the presence of 150 μl supernatant per plate, an increasing mutagenic response was observed with both BaP and BaA up to a concentration of 45 μM, and BaP was always more active than BaA.

Induced liver enzymes from male Sprague-Dawley rats, treated prior to sacrifice with PAH or phenobarbital, were tested for their ability to activate BaP to a bacterial mutagen (156). The phenobarbital-induced enzymes were much more active than the 3-methylcholanthrene-induced enzymes when low levels (less than 1 mg/ml) of enzyme protein were present in the incubation mixtures. At higher levels of enzyme protein, the 3-methylcholanthrene-induced enzymes were more active.

Addition of an epoxide hydrase inhibitor to the test mixtures with phenobarbital-induced enzymes greatly increased the mutagenic activity, but only slightly increased the activity obtained with the 3-methylcholanthrene-induced enzymes. The bacterial mutagenesis tests were carried out by using reversion of histidine auxotrophs of *Salmonella typhimurium* as indicators of mutagenesis. The authors concluded that their results favored the long-held idea that the 4,5-epoxide (K-region) metabolite is involved in mutagenesis.

REFERENCES

(1) Andelman, J.B., and Suess, M.J., "Polynuclear Aromatic Hydrocarbons in the Water Environment," *Bull. Wld. Hlth. Org.*, 43:479 (1970).
(2) Borneff, J., "Fate of Carcinogens in Aquatic Environment," (1977).
(3) Borneff, J., "Carcinogenic Substances in Water and Soil. Part XV: Interim Results of the Former Investigations," *Arch. Hyg. Bakt.*, (Berl) 148:1 (1964).
(4) Borneff, J., and Kunte, H., "Carcinogenic Substances in Water and Soil. XVI: Evidence of PAH in Water Samples Through Direct Extraction," *Arch. Hyg. Bakt.*, 148:585 (1964).
(5) Borneff, J., and Kunte, H., "Carcinogenic Substances in Water and Soil. XXVI: A Routine Method for the Determination of PAH in Water," *Arch. Hyg. Bakt.*, 153:220 (1969).
(6) Basu, D.K., and Saxena, J., *Analysis of Raw and Drinking Water Samples for Polynuclear Aromatic Hydrocarbons,* EPA P.O. No. CA-7-2999-A, and CA-8-2275-B, Expo. Evalu. Branch, HERL, Cincinnati (1977).
(7) Basu, D.K., and Saxena, J., "Polynuclear Aromatic Hydrocarbons in Selected U.S. Drinking Waters and Their Raw Water Sources," *Environ. Sci. Technol.*, 12:795 (1978).

(8) Borneff, J., and Kunte, H., "Carcinogenic Substances in Water and Soil. XXII: About the Origin and Evaluation of PAH in Water," *Arch. Hyg. Bakt.,* 149:226 (1965).
(9) Dikun, P.P., and Makhinenko, A.I., "Detection of BP in the Schistose Plant Resins, in Its Effluents and in Water Basins After Discharge of Effluents," *Gig. i. Sanit.,* 28:10 (1963).
(10) Fedorenko, Z.P., "The Effect of Biochemical Treatment of Wastewater of a By-Product Coke Plant on the BP Content," *Gig. i. Sanit.,* 29:17 (1964).
(11) Harrison, R.M., et al, "Polynuclear Aromatic Hydrocarbons in Raw, Potable, and Wastewaters," *Water Research*, 9:331 (1975).
(12) Keegan, R.E., "The Trace Fluorometric Determination of Polynuclear Aromatic Hydrocarbons in Natural Water," Ph.D. Thesis, University of New Hampshire. Available from University Microfilms, Ann Arbor, Mich. (1971).
(13) Butenandt, A., and Dannenberg, H., *The Biochemistry of Tumors,* Vol. VI, *(Handbuch der Allgemeine Pathologie),* F. Büchner, et al (editor), Springer Verlag, Berlin (1956).
(14) Buu-Hoi, N.P., *Carcinogenic Materials, Vol. 2, (Medizinische Grundlagen Forschung),* K.F. Bauer (editor), Georg Thieme Verlag, Stuttgart (1959).
(15) World Health Organization, *European Standards for Drinking Water,* 2nd ed., Revised, Geneva (1970).
(16) Stocks, P., *Regional and Local Differences in Cancer Death Rates. Studies on Medical and Population Subjects, No. 1,* Gen. Regis. Off., London (1947).
(17) Diehl, J.S., and Tromp, S.W., *First Report on the Geographical and Geological Distribution of Carcinogens in the Netherlands*, Foundation for the Study of Psychophysics, Leiden (1953).
(18) Tromp, S.W., "Possible Effects of Geophysical and Geochemical Factors on Development and Geographic Distribution of Cancer," *Schweiz. Z. Path.,* 18:929 (1955).
(19) Wynne-Griffith, G., and Davies, R.I., "Cancer and Soils in the County of Anglesey, – A Revised Method of Comparison," *Br. Jour. Cancer,* 8:594 (1954).
(20) Davies, R.I., and Wynne-Griffith, G., "Cancer and Soils in the County of Anglesey," *Br. Jour. Cancer,* 8:56 (1954).
(21) Payne, W.W., and Hueper, W.C., "The Carcinogenic Effects of Single and Repeated Doses of BP," *Am. Ind. Hyg. Assoc. Jour.,* 21:350 (1960).
(22) Shabad, L.M., and Il'nitskii, A.P., "Perspective on the Problem of Carcinogenic Pollution in Water Bodies," *Gig. i. Sanit.,* 35:84 (Russian), Eng. Transt., *Hyg. Sanit.,* 35:268 (1970).
(23) Andelman, J.B., and Snodgrass, J.E., "Incidence and Significance of Polynuclear Aromatic Hydrocarbons in the Water Environment," pp. 69-83 CR, *Critical Reviews in Environmental Control* (1974).
(24) Payer, H.D., et al, "Accumulation of Polycyclic Aromatic Hydrocarbons in Cultivated Microalgae," *Naturwiss.,* 62:536 (1975).
(25) Shabad, L.M., and Smirnov, G.A., "Aircraft Engines as a Source of Carcinogenic Pollution of the Environment (Benzo[a] pyrene Studies)," *Atmos. Environ.,* 6:153 (1972).
(26) Dunn, B.P., and Stich, H.F., "Release of the Carcinogen Benzo[a] Pyrene from Environmentally Contaminated Mussels," *Bull. Environ. Contam. Toxicol,* 15:398 (1976).
(27) Binet, L., and Mallet, L., "Diffusion of PAH in the Living Environment," *Gaz. Hop.* (Paris), 135:1142 (1963), *Chem. Abstr.,* 60:2282c (1964).
(28) Lo, M., and Sandi, E., "Polycyclic Aromatic Hydrocarbons (Polynuclears) in Foods," *Residue Reviews,* Vol. 69, (Gunther and Gunther, editors), Springer-Verlag (1978).
(29) Lee, R.F., et al, "Uptake, Metabolism and Discharge of Polycyclic Aromatic Hydrocarbons by Marine Fish," *Mar. Biol.,* 17:201 (1972).
(30) Cahnmann, H.J., and Kuratsun, M., "Determination of Polycyclic Aromatic Hydrocarbons in Oysters Collected in Polluted Water," *Anal. Chem.,* 29:1312 (1957).
(31) Guerrero, H., et al, "High-Pressure Liquid Chromatography of Benzo[a] pyrene and Benzo[g,h,i] perylene in Oil-Contaminated Shellfish," *Jour. Assoc. Off. Anal. Chem.,* 59:989 (1976).
(32) Hetteche, H.O., "Plant Waxes as Collectors of PCAH in the Air of Polluted Areas," *Staub*, 31:72 (1971).

(33) Kolar, L.R., et al, "Contamination of Soil, Agricultural Crops, and Vegetables by 3,4-Benzopyrene in the Vicinity of Ceska Budejovice," *Cesu. Hyg.,* 20:135 (1975).

(34) Swallow, W.H., "Survey of Polycyclic Aromatic Hydrocarbons in Selected Foods and Food Additives Available in New Zealand," *New Zealand Jour. Sci.,* 19:407 (1976).

(35) Martin, R.J., and Duggan, R.E., "Pesticide Residues in Total Diet Samples (III)," *Pest. Monit. Jour.,* 1:111 (1968).

(36) Howard, J.W., et al, "Extraction and Estimation of Polycyclic Aromatic Hydrocarbons in Vegetable Oils," *Jour. Assoc. Off. Anal. Chem.,* 49:1236 (1966).

(37) Biernoth, G., and Rost, H.E., "The Occurrence of PAH in Coconut Oil and Their Removal," *Chem. Ind.,* 45:2002 (1967).

(38) Thorsteinsson, T., "Polycyclic Hydrocarbons in Commercially and Home-Smoked Food in Iceland," *Cancer,* 23:455 (1969).

(39) Dungal, N., "Can Smoked Food Be Carcinogenic?" *Acta Unio Intern. Contra. Cancrum,* 17:365 (1961).

(40) Howard, J.W., et al, "Extraction and Estimation of PAH in Smoked Foods. Part I. General Method," *Jour. Assoc. Off. Anal. Chem.,* 49:595 (1966).

(41) Malanoski, A.J., et al, "Survey of Polycyclic Aromatic Hydrocarbons in Smoked Foods," *Jour. Assoc. Off. Anal. Chem.,* 51:114 (1968).

(42) Masuda, Y., and Kuratsune, M., "Polycyclic Aromatic Hydrocarbons in Smoked Fish," *Katsuobuski, GANN,* 62:27 (1971).

(43) Zitko, V., "Aromatic Hydrocarbons in Aquatic Fauna," *Bull. Environ. Contam. Toxicol.,* 14:621 (1975).

(44) Bjørseth, A., *Analysis of Polycyclic Aromatic Hydrocarbons in Environmental Samples by Glass Capillary Gas Chromatography,* Vol. 3, *Carcinogenesis, Polynuclear Aromatic Hydrocarbons,* Jones and Freudenthal (editors), Raven Press, New York (1978).

(45) Howard, J.W., and Fazio, T., "A Review of Polycyclic Aromatic Hydrocarbons in Foods," *Jour. Agri. Food Chem.,* 17:527 (1969).

(46) Grimmer, G., and Hildebrandt, A., "Content of Polycyclic Hydrocarbons in Crude Vegetable Oils," *Chem. Ind.,* 45:2000 (1967).

(47) Lijinsky, W., and Shubik, P., "The Detection of Polycyclic Aromatic Hydrocarbons in Liquid Smoke and Some Foods," *Toxicol. Appl. Pharmacol.,* 7:337 (1965).

(48) Lijinsky, W., and Ross, A.E., "Production of Carcinogenic Polynuclear Hydrocarbons in the Cooking of Food," *Food Cosmet. Toxicol.,* 5:343 (1967).

(49) Howard, J.W., et al, "Extraction and Estimation of Polycyclic Aromatic Hydrocarbons in Total Diet Composites," *Jour. Assoc. Off. Anal. Chem.,* 51:122 (1968).

(50) Draudt, H.N., "The Meat Smoking Process: A Review," *Food Technol.,* 17:85 (1963).

(51) Shiraishi, Y., et al, "Determination of Polycyclic Aromatic Hydrocarbons in Foods. IV. 3,4-Benzopyrene in Fish and Shellfish," *Jour. Food Hyg. Soc. Japan,* 16:178 (1975).

(52) Lee, R.F., et al, "Fate of Petroleum Hydrocarbons Taken Up from Food and Water by the Blue Crab Calinectes Sapidus," *Marine Biol.,* 37:363 (1976).

(53) Veith, G.D., et al, "Measuring and Estimating the Bioconcentration Factors of Chemicals in Fish," *Jour. Fish. Res. Board Can.,* 36:1040 (1979).

(54) Hansch, C., and Leo, A.J., *Substituent Constants for Correlation Analysis in Chemistry and Biology,* Wiley-Interscience, New York (1979).

(55) Ahorkas, J.T., et al, "Metabolism of Polycyclic Hydrocarbons by a Highly Active Aryl Hydrocarbon Hydroxylase System in the Liver of a Trout Species," *Biochem. Biophys. Res. Comm.,* 63:635 (1975).

(56) Graf, W., et al, "Levels of Carcinogenic Polycyclic Aromatic Hydrocarbons in Human and Animal Tissues," *Zbl. Bakt. Hyg. I. Abst. Orig. B.,* 161:85 (1975).

(57) Radding, S.G., et al, *The Environmental Fate of Selected Polynuclear Aromatic Hydrocarbons,* EPA-560/5-750-009 (1976).

(58) Lu, P.Y., et al, "The Environmental Fate of Three Carcinogens: Benzo[a]pyrene, Benzidine, and Vinyl Chloride Evaluated in Laboratory Model Ecosystems," *Arch. Environ. Contam. Toxicol.,* 6:129 (1977).

(59) Lijinsky, W., and Shubik, P., "PH Carcinogens in Cooked Meat and Smoked Food," *Industr. Med. Surg.*, 34:152 (1965).
(60) Panalaks, T., "Determination and Identification of Polycyclic Aromatic Hydrocarbons in Smoked and Charcoal-Broiled Food Products by High Pressure Liquid Chromatography," *Jour. Environ. Sci. Health*, 11:299 (1976).
(61) Frethein, K., "Carcinogenic Polycyclic Aromatic Hydrocarbons in Norwegian Smoked Meat," *Jour. Agri. Food Chem.*, 24:976 (1976).
(62) Howard, J.W., et al, "Extraction and Estimation of Polycyclic Aromatic Hydrocarbons in Smoked Foods. II. Benzo[a]pyrene," *Jour. Assoc. Off. Anal. Chem.*, 49:611 (1966).
(63) Bailey, E.J., and Dungal, N., "Polycyclic Hydrocarbons in Iceland Smoked Food," *Br. Jour. Cancer*, 12:348 (1958).
(64) Filipovic, J., and Toth, L., "Polycyclische Kohlenwasserstoffe in Geräucherten Jugoslavischen Fleischwaren," *Fleischwirtschaft*, 51:1323 (1971).
(65) Toth, L., and Blass, W., "Einfluss der Räuchertechnologie auf den Gehalt von Geräucherten Fleischwaren an Cancerogenen Kohlenwasserstoffen," *Fleischwirt.*, 21:1121 (1972).
(66) Rhee, K.S., and Bratzier, L.J., "Benzo[a]pyrene in Smoked Meat Products," *Jour. Food Sci.*, 35:146 (1970).
(67) Simon, S., et al, "Effect of Cellulose Casing on Absorption of Polycyclic Hydrocarbons in Wood Smoke by Absorbents," *Jour. Agric. Food Chem.*, 17:1128 (1969).
(68) Youngblood, W.W., and Blumer, M., "Polycyclic Aromatic Hydrocarbons in the Environment: Homologous Series in Soils and Recent Marine Sediments," *Geochem. Cosmochin. Acta.*, 39:1303 (1975).
(69) White, R.H., et al, "Determination of Polycyclic Aromatic Hydrocarbons in Liquid Smoke Flavors," *Jour. Agri. Food Chem.*, 19:143 (1971).
(70) Shiraishi, Y., et al, "Determination of Polycyclic Aromatic Hydrocarbons in Foods. III. 3,4-Benzpyrene in Vegetables," *Jour. Food Hyg. Soc. Japan*, 15:18 (1974).
(71) Siddiqui, I., and Wagner, K.H., "Determination of 3,4-Benzpyrene and 3,4-Benzofluoranthene in Rain Water, Groundwater, and Wheat," *Chemosphere*, 1:83 (1972).
(72) International Agency for Research on Cancer (IARC), *Evaluation of the Carcinogenic Risk of Chemicals to Man: Certain Polycyclic Aromatic Hydrocarbons and Heterocyclic Compounds*, Vol. 3 (1973).
(73) Shiraishi, Y., et al, "Determination of Polycyclic Aromatic Hydrocarbons in Foods. II. 3,4-Benzopyrene in Japanese Foods," *Jour. Food Hyg. Soc. Japan*, 14:173 (1973).
(74) Graf, W., and Nowak, W., "Promotion of Growth in Lower and Higher Plants by Carcinogenic Polycyclic Aromatics," *Arch. Hyg. Bakt.*, 150:513 (1966).
(75) Kuratsune, M., "Benzo[a]pyrene Content in Certain Pyrogenic Materials," *Jour. Natl. Cancer Inst.*, 16:1485 (1956).
(76) Fabian, B., "Carcinogenic Substances in Edible Fat and Oil. Part VI: Further Investigations on Margarine and Chocolate," *Arch. Hyg.* (Berl), 153:21 (1965).
(77) Nishimura, K., and Masuda, M., "Minor Constituents of Whisky Fusel Oils. I. Basic, Phenolic and Lactonic Compounds," *Jour. Food Sci.*, 36:819 (1971).
(78) Vitzthum, O.G., et al., "New Volatile Constituents of Black Tea Aroma," *Jour. Agri. Food Chem.*, 23:999 (1975).
(79) Kuratsune, M., and Hueper, W.C., "Polycyclic Aromatic Hydrocarbons in Roasted Coffee," *Jour. Natl. Cancer Inst.*, 24:463 (1960).
(80) Kuratsune, M., and Hueper, W.C., "Polycyclic Aromatic Hydrocarbons in Coffee Soots," *Jour. Natl. Cancer Inst.*, 20:37 (1958).
(81) Grimmer, G., "Detection and Occurrence of Polycyclic Hydrocarbons in Yeast Cultured on Mineral Oils," *Dtsch. Lebensm. Rundsch.*, 70:394 (1974).
(82) Kertész-Sáringer, M., and Morlin, Z., "On the Occurrence of Polycyclic Aromatic Hydrocarbons in the Urban Area of Budapest," *Atmos. Environ.*, 9:831 (1975).
(83) Gordon, R.J., "Distribution of Airborne Polycyclic Aromatic Hydrocarbons Throughout Los Angeles," *Environ. Sci. Technol.*, 10:370 (1976).

(84) Gordon, R.J., and Bryan, R.J., "Patterns of Airborne Polynuclear Hydrocarbon Concentrations at Four Los Angeles Sites," *Environ. Sci. Technol.,* 7:1050 (1973).
(85) Sawicki, E., et al, "Polynuclear Aromatic Hydrocarbon Composition of the Atmosphere in Some Large American Cities," *Am. Ind. Hyg. Assoc. Jour.,* 23:137 (1962).
(86) U.S. EPA, *Special Report: Trends in Concentrations of Benzene-Soluble Suspended Particulate Fraction and Benzo[a]pyrene,* EPA-450/2-74-022 (1974).
(87) Greinke, R.A., and Lewis, I.C., "Development of a Gas Chromatographic-Ultraviolet Absorption Spectrometric Method for Monitoring Petroleum Pitch Volatiles in the Environment," *Anal. Chem.,* 47:2151 (1975).
(88) Bartle, K.D., et al, "High-Resolution GLC Profiles of Urban Air Pollutant Polynuclear Aromatic Hydrocarbons," *Intern. Jour. Environ. Anal. Chem.,* 3:349 (1974).
(89) Lunde, G., and Bjørseth, "Polycyclic Aromatic Hydrocarbons in Long-Range Transported Aerosols," *Nature,* 268:518 (1977).
(90) Hoffman, D., and Wynder, E.L., *Organic Particulate Pollutants–Chemical Analysis and Bioassays for Carcinogenicity, Air Pollution,* Vol. II, 3rd ed., Stern (editor), Academic Press, New York (1977).
(91) Santodonato, J., et al, *Health Assessment Document for Polycyclic Organic Matter,* U.S. EPA, Washington, D.C. (1978).
(92) Faoro, R.B., and Manning, J.A., "Trends in Benzo[a]pyrene," (1978).
(93) Krstulovic, A.M., et al, "Distribution of Some Atmospheric Polynuclear Aromatic Hydrocarbons," *Amer. Lab.* (1977).
(94) Fox, M.A., and Staley, S.W., "Determination of Polycyclic Aromatic Hydrocarbons in Atmospheric Particulate Matter by High Pressure Liquid Chromatography Coupled with Fluorescene Techniques," *Anal. Chem.,* 48:992 (1976).
(95) Colucci, J.M., and Begeman, C.R., *Polynuclear Aromatic Hydrocarbons and Other Pollutants in Los Angeles Air.* Vol 2. *Proceedings of the International Clean Air Congress,* Academic Press (1971).
(96) National Academy of Sciences (NAS), *Particulate Polycyclic Organic Matter,* National Research Council Committee on Biologic Effects of Atmospheric Pollutants, Washington, D.C. (1972).
(97) Gerarde, H.W., *Toxicology and Biochemistry of Aromatic Hydrocarbons,* Elsevier Publishing Company, New York (1960).
(98) Shubik, P., and Della Porta, G., "Carcinogenesis and Acute Intoxication with Large Doses of Polycyclic Hydrocarbons," *Arch. Pathol.,* 64:691-703 (1957).
(99) Stowell, R.E., and Maas, L.C., "Effects of Massive Doses of Methylcholanthrene on Epidermal Carcinogenesis," *Cancer Res.,* 6:121-27 (1946).
(100) Huggins, C., and Morii, S., "Selective Adrenal Necrosis and Apoplexy Induced by 7, 12-Dimethylbenz[a]anthracene," *J. Exp. Med.,* 114(4):741 (1961).
(101) Boyland, E., and Sims, P., "Metabolism of Polycyclic Compounds: The Metabolism of 7,12-Dimethylbenz[a]anthracene by Rat-Liver Homogenates," *Biochem. J.,* 95(3):780-87 (1965).
(102) Boyland, E., Sims, P., and Huggins, C., "Induction of Adrenal Damage and Cancer with Metabolites of 7,12-Dimethylbenz[a]anthracene," *Nature,* 207:816-17 (1965).
(103) Currie, A.R., Bird, C.C., Crawford, A.M., and Sims, P., "Embryopathic Effects of 7,12-Dimethylbenz[a]anthracene and Its Hydroxymethyl Derivatives in the Sprague-Dawley Rat," *Nature,* 226:911-14 (1970).
(104) Grover, P.L., Sims, P., Huberman, E., Marquardt, H., Kuroki, T., and Heidelberger, C., "In Vitro Transformation of Rodent Cells by K-Region Derivatives of Polycyclic Hydrocarbons," *Proc. Natl. Acad. Sci.,* 68(6):1098-1101 (1971).
(105) Huberman, E., Kuroki, T., Marquardt, H., Selkirk, J.K., Heidelberger, C., Grover, P.L., and Sims, P., "Transformation of Hamster Embryo Cells by Epoxides and Other Derivatives of Polycyclic Hydrocarbons," *Cancer Res.,* 32:1391-96 (1972).
(106) Lubet, R.A., Brown, D.Q., and Kouri, R.E., "The Role of 3-OH Benzo[a]pyrene in Mediating Benzo[a]pyrene Induced Toxicity and Transformation in Cell Culture," *Res. Commun. Chem. Pathol. Pharmacol.,* 6(3):929-42 (1973).

(107) Gelboin, H.V., Wiebel, F., and Diamond, L., "Dimethylbenzanthracene Tumorigenesis and Aryl Hydrocarbon Hydroxylase in Mouse Skin: Inhibition by 7,8-Benzoflavone," *Science,* 170:169-71 (1970).

(108) Williams, R.T., "The Metabolism of Aromatic Hydrocarbons," *Detoxication Mechanisms: The Metabolism and Detoxication of Drugs, Toxic Substances and Other Organic Compounds,* 2nd ed., John Wiley and Sons, Inc., New York (1959).

(109) Wiebel, F.J., and Gelboin, H.V., "Enzyme Induction and Polycyclic Hydrocarbon Metabolism in Cell Culture, Experimental Animals and Man," *Chemical Carcinogenesis Essays. Proceedings of a Workshop on Approaches to Assess the Significance of Experimental Chemical Carcinogenesis Data for Man, Organized by IARC and the Catholic Univ. of Louvain, Brussels, Belgium, Dec. 10-12, 1973,* Publ. No. 10, Lyon: International Agency for Research on Cancer (1974).

(110) Waterfall, J.F., and Sims, P., "Epoxy Derivatives of Aromatic Polycyclic Hydrocarbons. The Preparation and Metabolism of Epoxides Related to Benzo[a]pyrene and to 7,8- and 9,10-Dihydrobenzo[a]pyrene," *Biochem. J.,* 128:265-77 (1972).

(111) Sims, P., "The Metabolism of 7- and 12-Methylbenz[a]anthracene and Their Derivatives," *Biochem. J.,* 105:591-98 (1967).

(112) Sims, P., "The Metabolism of Some Aromatic Hydrocarbons by Mouse Embryo Cell Culture," *Biochem. Pharmacol.,* 19:285-97 (1970).

(113) Grover, P.L., Hewer, A., and Sims, P., "Epoxides as Microsomal Metabolites of Polycyclic Hydrocarbons," *FEBS Lett.,* 18:76-80 (1971).

(114) Selkirk, J.K., Huberman, E., and Heidelberger, C., "An Epoxide Is an Intermediate in the Microsomal Metabolism of the Chemical Carcinogen, Dibenz[a,h]anthracene," *Biochem. Biophys. Res. Commun.,* 43(5):1010-16 (1971).

(115) Grover, P.L., Hewer, A., and Sims, P., "K-Region Epoxides of Polycyclic Hydrocarbons: Formation and Further Metabolism of Benz[a]anthracene 5,6-oxide by Human Lung Preparations," *FEBS Lett.,* 34(1):63-68 (1973).

(116) Jerina, D.M., Kaubisch, N., and Daly, J.W., "Arene Oxides as Intermediate in the Metabolism of Aromatic Substrates: Alkyl and Oxygen Migrations During Isomerization of Alkylated Arene Oxides," *Proc. Natl. Acad. Sci.,* 68:2545 (1971).

(117) Sims, P., "Epoxy Derivatives of Aromatic Polycyclic Hydrocarbons. The preparation and Metabolism of Epoxides Related to 7,12-Dimethylbenz[a]anthracene," *Biochem. J.,* 131:405-13 (1973).

(118) Booth, J., Keysell, G.R., Pal, K., and Sims, P., "The Metabolism of Polycyclic Hydrocarbons by Cultured Human Lymphocytes," *FEBS Lett.,* 43(3):341-44 (1974).

(119) Booth, J., Keysell, G.R., and Sims, P., "Effect of Oestradiol on the In Vitro Metabolism of 7,12-Dimethylbenz[a]anthracene and its Hydroxymethyl Derivatives," *Biochem. Pharmacol.,* 23:735-44 (1974).

(120) Sims, P., "Qualitative and Quantitative Studies on the Metabolism of a Series of Aromatic Hydrocarbons by Rat-Liver Preparations," *Biochem. Pharmacol.,* 19:795-818 (1970).

(121) Wang, I.Y., Creasy, R.K., and Crocker, T.T., "Effect of Cigarette-Smoking on Metabolism of Benzo[a]pyrene by Human Placental Microsomes," *Proc. Am. Assoc. Cancer Res.,* 15:20 (1974).

(122) Huberman, E., and Sachs, L., "Metabolism of the Carcinogenic Hydrocarbon Benzo[a]pyrene in Human Fibroblast and Epithelial Cells," *Int. J. Cancer,* 11:412-18 (1973).

(123) Pal, K., Grover, P.L., and Sims, P., "The Metabolism of Carcinogenic Polycyclic Hydrocarbons by Tissues of the Respiratory Tract," *Biochem. Soc. Trans.,* 3:174-75 (1975).

(124) Keysell, G.R., Booth, J., and Sims, P., "The Metabolic Formation of Water-Soluble Derivatives from Dihydrodiols of Polycyclic Hydrocarbons," *Br. J. Cancer* 31(2):266 (1975).

(125) Kotin, P., Falk, H.L., and Miller, A., "Effect of Carbon Tetrachloride Intoxication on Metabolism of Benzo[a]pyrene in Rats and Mice," *J. Natl. Cancer Inst.,* 28:725-45 (1962).

(126) Nebert, D.W., "Genetic and Environmental Factors Affecting Placental and Fetal Metabolism of Xenobiotics," *The Placenta: Biological and Clinical Aspects, Harold C.*

Mack Symposium on the Physiology and Pathology of Reproduction, 1972, S. Moghissi and E.S.E. Hafez (editors), pp. 207-37 (1974).

(127) Peacock, P.R., "Evidence Regarding the Mechanism of Elimination of 1,2-Benzpyrene, 1,2,5,6-Dibenzanthracene, and Anthracene from the Bloodstream of Injected Animals," *Br. J. Exp. Pathol.,* 17:164-72 (1936).

(128) Kotin, P., Falk, G.L., and Busser, R., "Distribution, Retention, and Elimination of ^{14}C-3,4-Benzopyrene after Administration to Mice and Rats," *J. Natl. Cancer Inst.,* 23(3):541-55 (1959).

(129) Heidelberger, C., and Jones, H.B., "The Distribution of Radioactivity in the Mouse Following Administration of Dibenzanthracene Labelled in the 9 and 10 Positions with Carbon-14," *Cancer,* 1:252-60 (1948).

(130) Heidelberger, C., and Weiss, S.M., "The Distribution of Radioactivity in Mice Following Administration of 3,4-Benzpyrene-5-^{14}C," *Cancer Res.,* 11:885-91 (1951).

(131) Levine, W.G., "Hepatic Uptake, Metabolism, and Biliary Excretion of 7,12-Dimethylbenzanthracene in the Rat," *Drug Metabol. Disposition,* 2(2):169-77 (1974).

(132) Harper, K.H., "The Metabolism of Pyrene," *Br. J. Cancer,* 11:499-507 (1957).

(133) Hartwell, J.L., *Survey of Compounds Which Have Been Tested for Carcinogenic Activity,* 2nd ed. Public Health Service Pub. 149. U.S. Government Printing Office Washington, D.C. (1951).

(134) Shubik, P., and Hartwell, J.L., *Survey of Compounds Which Have Been Tested for Carcinogenic Activity.* Public Health Service Pub. 149, Suppl. 1. Government Printing Office, Washington, D.C. (1957).

(135) Shubik, P., and Hartwell, J.L., *Survey of Compounds Which Have Been Tested for Carcinogenic Activity.* Public Health Service Pub. 149, Suppl. 2. Government Printing Office, Washington, D.C. (1969).

(136) Thompson, J.I., and Company, *Survey of Compounds Which Have Been Tested for Carcinogenic Activity.* Public Health Service Pub. 149. Government Printing Office, Washington, D.C. (1971).

(137) Tractor/Jitco, *Survey of Compounds Which Have Been Tested for Carcinogenic Activity (1961-1967).* Public Health Service Pub. 149 (1973).

(138) Tractor/Jitco, *Survey of Compounds Which Have Been Tested for Carcinogenic Activity (1970-1971).* Public Health Service Pub. 149 (1973).

(139) Boyland, E., "The Biological Significance of Metabolism of Polycyclic Compounds," *Symp. Biochem. Soc.,* 5:40-54 (1950).

(140) Adamson, R.H., Cooper, R.W., and O'Gara, R.W., "Carcinogen Induced Tumors in Primitive Primates," *J. Natl. Cancer Inst.,* 45:555-59 (1970).

(141) Cottini, G.B., and Mazzone, G.B., "Effects of 3,4-Benzpyrene on Human Skin," *Am. J. Cancer,* 37:186-95 (1939).

(142) Schmidt, K.G., Schmahl, D., and Misfeld, J., "Investigations on the Carcinogenic Burden by Air Pollution in Man. VI. Experimental Investigations to Determine a Dose-Response Relationship and to Estimate a Threshold Dose of Benzo[a]pyrene in the Skin of Two Different Mouse Strains," *Zbl. Bakt. Hyg. I. Abt. Orig. B.,* 158:62-68 (1973).

(143) Hilfrich, J., Bresch, H., Misfeld, J., and Mohr, U., "Investigation on the Carcinogenic Burden by Air Pollution in Man. V. Tumors of the Respiratory Tract in Syrian Golden Hamsters After Intratracheal Instillation of Benzo[a]pyrene," *Zbl. Bakt. Hyg. I. Abt. Orig. B.,* 158:59-61 (1973).

(144) Laskin, S., Kuschner, M., and Drew, R.T., "Studies in Pulmonary Carcinogenesis," *Inhalation Carcinogenesis, Proceedings of a Biology Division, Oak Ridge National Laboratory, Conference, Gatlinburg, Tenn., Oct. 8-11, 1969,* AEC Symp. Ser. 18, M.G. Hanna, P.Nettesheim, and J.R. Gilbert (editors), pp. 321-51. Conf.-691001. (1970).

(145) Pfeiffer, E.H., "Investigations on the Carcinogenic Burden by Air Pollution in Man. VII. Studies on the Oncogenetic Interaction of Polycyclic Aromatic Hydrocarbons," *Zbl. Bakt. Hyg. I. Abt. Orig. B.,* 158:69-83 (1973).

(146) Lloyd, J.W., "Long-Term Mortality Study of Steelworkers. V. Respiratory Cancer in Coke Plant Workers," *J. Occup. Med.,* 13(2):53-67 (1971).

(147) Redmond, C.K., Ciocci, A., Lloyd, J.W., and Rush, H.W., "Long-Term Mortality Study of Steelworkers. VI. Mortality from Malignant Neoplasms Among Coke Oven Workers," *J. Occup. Med.,* 14(8):621-29 (1972).

(148) Conney, A.H., Miller, E.G., and Miller, J.A., "Substrate-Induced Synthesis and Other Properties of Benzpyrene Hydroxylase in Rat Liver," *J. Biol. Chem.,* 228:753-66 (1957).

(149) Nebert, D.W., and Gelboin, H.V., "Substrate-Inducible Microsomal Aryl Hydroxylase in Mammalian Cell Culture. I. Assay and Properties of Induced Enzyme," *J. Biol. Chem.,* 243(23):6242-49 (1968).

(150) Nebert, D.W., and Gelboin, H.V., "The In Vivo and In Vitro Induction of Aryl Hydrocarbon and Hydroxylase in Mammalian Cells of Different Species, Tissues, Strains, and Developmental and Hormonal States," *Arch. Biochem. Biophys.,* 134:76-89 (1969).

(151) Kellermann, G., Cantrell, E., and Shaw, C.R., "Variations in Extent of Aryl Hydrocarbon Hydroxylase Induction in Cultured Human Lymphocytes," *Cancer Res.,* 33:1654-56 (1973).

(152) Kouri, R.E., Ratrie, H., and Whitmire, C.E., "Evidence of a Genetic Relationship Between Susceptibility to 3-Methylcholanthrene-Induced Subcutaneous Tumors and Inducibility of Aryl Hydrocarbon Hydroxylase," *J. Natl. Cancer Inst.,* 51(1):197-200 (1973).

(153) Welch, R.M., Harrison, Y.E., Conney, A.H., Poppers, P.J., and Finster, M., "Cigarette Smoking: Stimulatory Effect on Metabolism of 3,4-Benzpyrene by Enzymes in Human Placenta," *Science,* 160:541-42 (1968).

(154) Nebert, D.W., Winker, J., and Gelboin, H.V., "Aryl Hydrocarbon Hydroxylase Activity in Human Placenta from Cigarette Smoking and Nonsmoking Women," *Cancer Res.,* 29:1763-69 (1969).

(155) Alvares, A.P., Schilling, G.R., and Kuntzman, R., "Differences in the Kinetics of Benzpyrene Hydroxylation by Hepatic Drug-Metabolizing Enzymes from Phenobarbital- and 3-Methylcholanthrene-Treated Rats," *Biochem. Biophys. Res. Commun.,* 30:588-93 (1968).

(156) Rasmussen, R.E., and Wang, I.Y., "Dependence of Specific Metabolism of Benzo[a]pyrene on the Inducer of Hydroxylase Activity," *Cancer Res.,* 34:2290-95 (1974).

(157) Benedict, W.F., Considine, N., and Nebert, D.W., "Genetic Differences in Aryl Hydrocarbon Hydroxylase Induction and Benzo[a]pyrene-Produced Tumorigenesis in the Mouse," *Mol. Pharmacol.,* 6:266-77 (1973).

(158) Kellermann, G., Luyten-Kellermann, M., and Shaw, C.R., "Genetic Variation in Aryl Hydrocarbon Hydroxylase in Human Lymphocytes," *Am. J. Human Genet.,* 25:327-31 (1973).

(159) Kellermann, G., Luyten-Kellermann, M., and Shaw, C.R., "Metabolism of Polycyclic Aromatic Hydrocarbons in Cultured Human Leukocytes Under Genetic Control," *Humangenetik,* 20:257-63 (1973).

(160) Kellermann, G., Shaw, C.R., and Luyten-Kellermann, M., "Aryl Hydrocarbon Hydroxylase Inducibility and Bronchogenic Carcinoma," *New Engl. J. Med.,* 289(18):934-36 (1973).

(161) Henry, M.C., and Kaufman, D.G., "Clearance of Benzo[a]pyrene from Hamster Lung After Administration on Coated Particles," *J. Natl. Cancer Inst.,* 51:1961-64 (1973).

(162) National Air Pollution Control Administration (NAPCA), *Air Quality Criteria for Particulate Matter.* AP-49. NAPCA, Washington, D.C. (1969).

(163) Environmental Protection Agency (EPA), *Scientific and Technical Assessment Report on Particulate Polycyclic Organic Matter (PPOM).* EPA-600/6-75-001 (1975).

(164) Task Group on Lung Dynamics, "Deposition and Retention Models for Internal Dosimetry of the Human Respiratory Tract," *Health Phys.,* 12:173-207 (1966).

(165a) Saffiotti, U., Cefis, F., and Kolb, L.H., "Bronchogenic Carcinoma Induction by Particulate Carcinogens," *Proc. Am. Assoc. Cancer Res.,* 5(1):59 (1964).

(165b) Saffiotti, U., Montesano, R., Sellakumar, A.R., and Borg, S.A., "Experimental Cancer of the Lung. Inhibition by Vitamin A of the Induction of Tracheobronchial Squamous Metaplasia and Squamous Cell Tumors," *Cancer,* 20(5):857-64 (1967).

(166) Henry, M.C., Port, C.D., and Kaufman, D.G., "Importance of Physical Properties of Benzo[a]pyrene-Ferric Oxide Mixtures in Lung Tumor Induction," *Cancer Res.,* 35:207-17 (1975).

(167) Olsen, D., and Haynes, J.L., *Preliminary Air Pollution Survey of Organic Carcinogens. A Literature Review,* National Air Pollution Control Administration Pub. APTD-69-43 (1969).

(168) Kennedy, A.R., and Little, J.B., "The Transport and Localization of Benzo[a]pyrene-Hematite and Hematite-^{210}Po in the Hamster Lung Following Intratracheal Instillation," *Cancer Res.,* 34:1344-52 (1974).

(169) Nettesheim, P., and Schreiber, H., "Advances in Experimental Lung Cancer Research," *Handbuch d. allg. Pathologic Bd: VI/7.* Heidelberg: Springer Verlag (1975).

(170) Palmer, M.S., Exley, R.W., and Coffin, D.L., "Influence of Pollutant Gases on Benzpyrene Hydroxylase Activity," *Arch. Environ. Health,* 25:439-42 (1972).

(171) National Institute for Occupational Safety and Health (NIOSH), *Criteria for a Recommended Standard: Occupational Exposure to Oxides of Nitrogen (Nitrogen Dioxide and Nitric Oxide),* HEW Publ. No. (NIOSH) 76-149, Cincinnati, Ohio (1976).

(172) Horton, A.W., Denman, D.T., and Trosset, R.P., "Carcinogenesis of the Skin. II. The Accelerating Properties of Aliphatic and Related Hydrocarbons," *Cancer Res.,* 17:758-66 (1957).

(173) Bingham, E., and Falk, H.L., "Environmental Carcinogens: The Modifying Effect of Cocarcinogens on the Threshold Response," *Arch. Environ. Health,* 19:779-83 (1969).

(174) Tye, R., and Stemmer, K.L., "Experimental Carcinogenesis of the Lung. II. Influence of Phenols in the Production of Carcinoma," *J. Natl. Cancer Inst.,* 39:175-86 (1967).

(175) Horton, A.W., Tye, R., and Stemmer, K.L., "Experimental Carcinogenesis of the Lung. Inhalation of Gaseous Formaldehyde or an Aerosol of Coal Tar by C3H Mice," *J. Natl. Cancer Inst.,* 30(1-6):31-43 (1963).

(176) Feron, V.J., "Respiratory Tract Tumors in Hamsters After Intratracheal Instillations of Benzo[a]pyrene Alone and with Furfural," *Cancer Res.,* 32:28-36 (1970).

(177) Sporn, M.B., Clamon, G.H., Smith, J.M., Dunlop, N.M., Newton, D.L., and Saffiotti, U., "The Reversal of Keratinized Squamous Metaplastic Lesions of Vitamin A Deficiency in Tracheobronchial Epithelium by Vitamin A and Vitamin A Analogs in Organ Culture: A Model System for Anticarcinogenesis Studies," *Experimental Lung Cancer. Carcinogenesis and Bioassays,* E. Karbe and J.F. Park (editors), New York: Springer-Verlag (1974).

(178) Cone, M.V., and Nettesheim, P., "Effects of Vitamin A on 3-Methylcholanthrene-Induced Squamous Metaplasias and Early Tumors in the Respiratory Tracts of Rats," *J. Natl. Cancer Inst.,* 50(6):1599-1604 (1973).

(179) Hill, D.L., and Shih, T.W., "Vitamin A Compounds and Analogs as Inhibitors of Mixed-Function Oxidases that Metabolize Carcinogenic Polycyclic Hydrocarbons and Other Compounds," *Cancer Res.,* 34:564-70 (1974).

(180) Wattenberg, L.W., "Inhibition of Carcinogenic and Toxic Effects of Polycyclic Hydrocarbons by Phenolic Antioxidants and Ethoxyquin," *J. Natl. Cancer Inst.,* 48:1425-30 (1972).

(181) Wattenberg, L.W., "Inhibition of Carcinogenic and Toxic Effects of Polycyclic Hydrocarbons by Several Sulfur-Containing Compounds," *J. Natl. Cancer Inst.,* 52(5):1583-87 (1974).

(182) Bowden, G.T., Slaga, T.J., Shapas, B.G., and Boutwell, R.K., "The Role of Aryl Hydrocarbon Hydroxylase in Skin Tumor Initiation by 7,12-Dimethylbenz[a]anthracene and 1,2,5,6-Dibenzanthracene Using DNA Binding and Thymidine-^{3}H Incorporation into DNA as Criteria," *Cancer Res.,* 34:2634-42 (1974).

(183) Diamond, L., and Gelboin, H.V., "Alpha-Naphthoflavone: An Inhibitor of Hydrocarbon Cytotoxicity and Microsomal Hydroxylase," *Science,* 166:1023-25 (1969).

(184) Wattenberg, L.W., and Leong, J.L., "Inhibition of the Carcinogenic Action of Benzo[a]pyrene by Flavones," *Cancer Res.,* 30:1922-25 (1970).

(185) Nebert, D.W., and Gelboin, H.V., "Substrate-Inducible Microsomal Aryl Hydroxylase in Mammalian Cell Culture. II. Cellular Responses During Enzyme Induction," *J. Biol. Chem.,* 243(23):6250-61 (1968).

(186) Watanabe, M., Ariji, F., Takusagawa, K., and Konno, K., "Aryl Hydrocarbon Hydroxylase in Liver Nuclei of C3H/He and DBA/2 Mice," *Gann,* 66:399-406 (1972).
(187) Huggins, C., Moon, R.C., and Morii, S., "Extinction of Experimental Mammary Cancer. I. Estradiol-17β and Progesterone," *Proc. Natl. Acad. Sci. USA,* 48:379-86 (1962).
(188) Haddow, A., and Weinhouse, S. (editors), *Advan. Cancer Res.,* 10:1-260 (1967).
(189) Falk, H.L., Kotin, P., and Thompson, S., "Inhibition of Carcinogenesis: The Effect of Polycyclic Hydrocarbons and Related Compounds," *Arch. Environ. Health,* 9:169 (1964).
(190) Shamberger, R.J., "Relationship of Selenium to Cancer. I. Inhibitory Effect of Selenium on Carcinogenesis," *J. Natl. Cancer Inst.,* 44(4):931-36 (1970).
(191) Shamberger, R.J., Tytko, S., and Willis, C., "Antioxidants and Cancer. II. Selenium Distribution and Human Cancer Mortality in the United States, Canada, and New Zealand," *Seventh Annual Conference on Trace Substances in Environmental Health, Memorial Union, University of Missouri-Columbia, Mo., June 12-14, 1973,* D.D. Hemphill (editor) (1973).
(192) Eckardt, R.E., "Cancer Prevention in the Petroleum Industry," *Int. J. Cancer,* 2:656-61 (1967).
(193) Ketcham, N.H., and Norton, W., "The Hazards to Health in the Hydrogenation of Coal. III. The Industrial Hygiene Studies," *Arch. Environ. Health* 1:194-207 (1960).
(194) Sexton, R.J., "The Hazards to Health in the Hydrogenation of Coal. IV. The Control Program and the Clinical Effects," *Arch. Environ. Health,* 1:208-31 (1960).
(195) Miller, J.A., "Carcinogenesis by Chemicals: An Overview," *Cancer Res.,* 30:559-76 (1970).
(196) Miller, J.A., and Miller, E.C., "The Metabolic Activation of Chemical Carcinogens to Reactive Electrophiles," Univ. Wisconsin Medical Center, Madison, Wis. (1975). Presented at Symposium, m-RNA: The Relation of Structure to Function, Gatlinburg, Tenn. (April, 1976).
(197) Brookes, P., and Lawley, P.D., "Evidence of the Binding of Polynuclear Aromatic Hydrocarbons to the Nucleic Acids of Mouse Skin: Relation Between Carcinogenic Power of Hydrocarbons and Their Binding to Deoxyribonucleic Acid," *Nature,* 202:781-84 (1964).
(198) Miller, E.C., "Studies on the Formation of Protein-Bound Derivatives of 3,4-Benzpyrene in the Epidermal Fraction of Mouse Skin," *Cancer Res.,* 11:100-108 (1951).
(199) Wiest, W.G., and Heidelberger, C., "The Interaction of Carcinogenic Hydrocarbons with Tissue Constituents. II. 1,2,5,6-Dibenzanthracene-9,10-^{14}C in Skin," *Cancer Res.,* 13:250 (1953).
(200) Grover, P.L., and Sims, P., "Enzyme-Catalyzed Reactions of Polycyclic Hydrocarbons with Deoxyribonucleic Acid and Protein In Vitro," *Biochem. J.,* 110:159-60 (1968).
(201) Grover, P.L., and Sims, P., "Interactions of the K-Region Epoxides of Phenanthrene and Dibenz[a,h]anthracene with Nucleic Acids and Histone," *Biochem. Pharmacol.,* 19:2251-59 (1970).
(202) Grilli, S., Rocchi, P., and Prodi, G., "Nonenzymatic and Microsome-Dependent Binding of Polycyclic Hydrocarbons to DNA and Polynucleotides," *Chem. Biol. Interactions,* 11:351-63 (1975).
(203) Wang, I.Y., Rasmussen, R.E., and Crocker, T.T., "Isolation and Characterization of an Active DNA-Binding Metabolite of Benzo[a]pyrene from Hamster Liver Microsomal Incubation System," *Biochem. Biophys. Res. Comm.,* 49(4):1142-50 (1972).
(204) Dipple, A., Lawley, P.D., and Brookes, P., "Theory of Tumor Initiation by Chemical Carcinogens: Dependence of Activity on Structure of Ultimate Carcinogen," *Europ. J. Cancer,* 4:493-506 (1968).
(205) Baird, W.M., Harvey, R.G., and Brookes, P., "Comparison of the Cellular DNA-Bound Products of Benzo[a]pyrene with the Products Formed by the Reaction of Benzo[a]pyrene-4,5-Oxide with DNA," *Cancer Res.,* 35:54-57 (1975).
(206) King, H.W.S., Thompson, M.H., and Brookes, P., "The Benzo[a]pyrene Deoxyribonucleoside Products Isolated from DNA after Metabolism of Benzo[a]pyrene by Rat Liver Microsomes in the Presence of DNA," *Cancer Res.,* 34:1263-69 (1975).

(207) Sims, P., Grover, P.L., Swaisland, A., Pal, K., and Hewer, A., "Metabolic Activation of Benzo[a]pyrene Proceeds by a Diol-Epoxide," *Nature,* 252:326-27 (1974).
(208) Harris, C.C., Genta, V.M., Frank, A.L., Kaufman, D.G., Barrett, L.A., McDowell, E.M., and Trump, B.F., "Carcinogenic Polynuclear Hydrocarbons Bind to Macromolecules in Cultured Human Bronchi," *Nature,* 252:68-69 (1974).
(209) Genta, V.M., Kaufman, D.G., Harris, C.C., Smith, J.M., Sporn, M.B., and Saffiotti, U., "Vitamin A Deficiency Enhances Binding of Benzo[a]pyrene to Tracheal Epithelial DNA," *Nature,* 247:48-49 (1974).
(210) Kaufman, D.G., Genta, V.M., Harris, C.C., Smith, J.M., Sporn, M.B., and Saffiotti, U., "Binding of ^{3}H-Labeled Benzo[a]pyrene to DNA in Hamster Tracheal Epithelial Cells," *Cancer Res.,* 33:2837-41 (1973).
(211) Kaufman, D.G., Genta, V.M., and Harris, C.C., "Studies on Carcinogen Binding In Vitro in Isolated Hamster Tracheas," *Experimental Lung Cancer. Carcinogenesis and Bioassays,* E. Karbe and J.F. Park (editors), pp. 564-74. New York: Springer-Verlag (1974).
(212) Marquardt, H., Sondergren, J.E., Sims, P., and Grover, P.L., "Maligant Transformation In Vitro of Mouse Fibroblasts by 7,12-Dimethylbenz[a]anthracene and 7-Hydroxymethylbenz[a]anthracene and by their K-Region Derivatives," *Int. J. Cancer,* 13:304-10 (1974).
(213) Flesher, J.W., and Sydnor, K.L., "Possible Role of 6-Hydroxymethylbenzo[a]pyrene as a Proximate Carcinogen of Benzo[a]pyrene and 6-Methylbenzo[a]pyrene," *Int. J. Cancer,* 11:433-37 (1973).
(214) Flesher, J.W., and Sydnor, K.L., "Comparative Studies on Distribution of DMBA-3H and Hydroxy-DMBA-3H and their Carcinogenic Activity," *Int. J. Cancer,* 5:253-59 (1970).
(215) Flesher, J.W., and Sydnor, K.L., "Carcinogenicity of Derivatives of 7,12-Dimethylbenz-[a]anthracene," *Cancer Res.,* 31:1951-54 (1971).
(216) Dipple, A., Levy, L.S., and Iype, P.T., "Investigation of the Possible Involvement of Ketonic Derivatives of Polycyclic Hydrocarbons in Hydrocarbon Carcinogenesis," *Cancer Res.,* 35:652-57 (1975).
(217) Huberman, E., Sachs, L., Yang, S.K., and Gelboin, H.V., "Identification of Mutagenic Metabolites of Benzo[a]pyrene in Mammalian Cells," *Proc. Natl. Acad. Sci. USA,* 73(2):607-11 (1976).
(218) Yang, S.K., McCort, D.W., Roller, P.P., and Gelboin, H.V., "Enzymatic Conversion of Benzo[a]pyrene Leading Predominantly to the Diol-Epoxide r-7,t-8-Dihydroxy-t-9,10-Oxy-7,8,9,10-Tetrahydrobenzo[a]pyrene Through a Single Enantiomer of r-7, t-8-Dihydroxy-7,8-Dihydrobenzo[a]pyrene," *Proc. Natl. Acad. Sci. USA,* 73(8):2594-98 (1976).
(219) DiPaolo, J.A., and Kotin, P., "Teratogenesis-Oncogenesis: A Study of Possible Relationships," *Arch. Path.,* 81:3-23 (1966).
(220) Shepard, T.H., *Catalog of Teratogenic Agents,* John Hopkins Univ. Press, Baltimore, Md. (1973).
(221) Shendrikova, I.A., and Aleksandrov, V.A., "Comparative Penetration of Polycyclic Hydrocarbons Through the Rat Placenta into the Fetus," *Bull. Exp. Biol. Med.,* 77(2):169-71 (1974).
(222) Schlede, E., and Merker, H.J., "Effect of Benzo[a]pyrene Treatment on the Benzo-[a]pyrene Hydroxylase Activity in Maternal Liver, Placenta, and Fetus of the Rat During Day 13 to Day 18 of Gestation," *Naunyn-Schmiedeberg's Arch. Pharmacol.,* 272:80-100 (1972).
(223) Schlede, E., and Merker, H.J., "Benzo[a]pyrene Hydroxylase Activity in the Whole Implantation Site, Decidua, Placenta and Fetal Membranes of the Rat," *Naunyn-Schmiedeberg's Arch. Pharmacol.,* 282:59-73 (1974).
(224) Bulay, O.M., and Wattenburg, L.W., "Carcinogenic Effects of Subcutaneous Administration of Benzo[a]pyrene During Pregnancy on the Progeny," *Prof. Soc. Exp. Biol. Med.,* 135:84-86 (1968).
(225) Bulay, O.M., "The Study of Development of Lung and Skin Tumors in Mice Exposed in Utero to Polycyclic Hydrocarbons," *Acta. Med. Turc.,* 7:3-38 (1970).

(226) Shay, H., Friedmann, B., Gruenstein, M., and Weinhouse, S., "Mammary Excretion of 20-Methylcholanthrene," *Cancer Res.,* 10:797 (1950).

(227) Turusov, V., Tomatis, L., Guibbert, D., Duperray, B., and Pacheco, H., "The Effect of Prenatal Exposure of Mice to Methylcholanthrene Combined with the Neonatal Administration of Diethylnitrosamine," *Transplacental Carcinogenesis: Proceedings of a Meeting Held at the Medizinsche Hochschule, Hanover, Fed. Rep. Germany, Oct. 6-7, 1971,* L. Tomatis, U. Mohr, and W. Davis (editors), Lyon: International Agency for Research on Cancer (1973).

(228) Tomatis, L., and Goodall, C.M., "The Occurrence of Tumors in F-1, F-2, and F-3 Descendants of Pregnant Mice Injected with 7,12-Dimethylbenz[a] anthracene," *Int. J. Cancer,* 4:219-25 (1969).

(229) Alexandrov, V.A., "Embryotoxic and Teratogenic Effects of Chemical Carcinogens," *Transplacental Carcinogenesis. Proceedings of a Meeting Held at the Medizinsche Hochschule, Hanover, Fed. Rep. of Germany, Oct. 6-7, 1971,* L. Thomatis, U. Mohr, and W. Davis (editors), pp. 112-16, Lyon: International Agency for Research on Cancer (1973).

(230) Huberman, E., and Sachs, L., "Cell-Mediated Mutagenesis of Mammalian Cells with Chemical Carcinogens," *Int. J. Cancer,* 13:326-33 (1974).

(231) Ames, B.N., Durston, W.E., Yamasaki, E., and Lee, F.D., "Carcinogens are Mutagens: A Simple Test System Combining Liver Homogenates for Activation and Bacteria for Detection," *Proc. Natl. Acad. Sci. USA,* 70:2281-85 (1973).

(232) Chu, E.H.Y., Bailiff, E.G., and Malling, H.V., "Mutagenicity of Chemical Carcinogenesis in Mammalian Cells," *Proceedings of 10th International Cancer Congress,* Houston, Tex., Medical Arts Publishing Co. (1970).

(233) Hueper, W.C., "Public Health Hazards from Environmental Chemical Carcinogens, Mutagens and Teratogens," *Health Phys.,* 21:689-707 (1971).

(234) Huberman, E., and Langenbach, R., *Cell-Mediated Mutagenesis by Chemical Carcinogens,* NTIS CONF-780227-4 (1978).

(235) Huberman, E., "Correlation Between Mutagenicity and Carcinogenicity: Mammalian Cell Transformation and Cell-Mediated Mutagenesis by Carcinogenic Polycyclic Hydrocarbons," *Mutat. Res.,* 29(2):285-91 (1975).

(236) Malaveille, C., Bartsch, H., Grover, P.L., and Sims, P., "Mutagenicity of Non-K-Region Diols and Diol-Epoxides of Benz[a] anthracene and Benzo[a] pyrene in *S. Typhimurium* TA 100," *Biochem. Biophys. Res. Commun.,* 66(2):693-700 (1975).

HEALTH PROTECTION

The information in this chapter is based on *Synthetic Fossil Fuel Technologies: Health Problems and Intersociety Cooperation*, NTIS CONF-790415-3, 1979, prepared by R.B. Gammage and J.E. Turner of Oak Ridge National Laboratory for the U.S. Department of Energy.

Ambient Water Quality Criteria for Polynuclear Aromatic Hydrocarbons, EPA Report 440/5-80-069, October 1980, prepared by the U.S. Environmental Protection Agency.

ACTIVITIES OF SOCIETIES AND FEDERAL AGENCIES

Faced with the health problems posed by synthetic fuel technologies, a number of organizations with professional interests and legal responsibilities are taking action. It is generally recognized that the cooperation of various local and federal agencies with mutual and sometimes conflicting requirements is essential. The scientific and technical problems are enormous. The resources available for their solution must be used in the optimum manner with a minimum of duplication to obtain the most benefit for human health.

Federal Agencies

NIOSH: The National Institute of Occupational Safety and Health, in accordance with the Occupational Safety and Health Act of 1970, has research efforts in four areas for occupational health hazards associated with coal gasification. These activities include,

> industrial hygiene characterization of gasification plants; a comprehensive series of recommendations for gasification plant development based upon existing knowledge; development of a health maintenance and health surveillance protocol for gasification plant employees; and a program to evaluate and recommend engineering and process design alterations, work place monitoring systems, safe work practices, and protective equipment to prevent disease in workers employed in gasification plants. (2)

EPA: The Environmental Protection Agency has recognized the need to control exposures at acceptable levels. They have established a list of toxic substances associated with related aqueous effluents (4). These have been put into a number of chemical categories, one of which contains sixteen polynuclear aromatic (PNA) hydrocarbons that are to be monitored in effluents. The EPA's Industrial Environmental Research Laboratory (IERL) at Research Triangle Park has sponsored a series of Symposia on Environmental Aspects of Fuel Conversion Technology (3). These activities are directed toward furthering sampling and assessment work.

The evaluation of the Kosovo Lurgi Gasification Plant in Yugoslavia is the largest and most comprehensive of IERL's commercial activities dealing with data aquisition.

DOE: The Department of Energy instigated the development of a Balanced Program Plan (5) identifying specific needs for biomedical and environmental research and development. Comprehensive information overview documents have also been assembled on the environmental, health, and control aspects of coal conversion (6). In the area of low-Btu gasifiers in industry, environmental and health plans have been written for the first two units (Duluth, Minnesota and Pike County, Kentucky).

Health problems are addressed with respect to issues, requirements, and tasks. For example, one of the issues for the University of Minnesota Gasification Facility is whether the occupational work force is being adequately protected. One of the requirements is to establish the level and nature of exposure in a worker's environment at each stage of gasification.

Four specific tasks are to (a) select, install, and operate inplant area monitors for selected pollutants (PNA compounds, CO, H_2S, etc.), (b) identify and monitor potential sources of fugitive emissions or leaks, (c) review and select personal monitors for selected pollutants (especially for coal oil or coal tar separation operations), and (d) correlate information for consistency, uncertainties, and definition of industrial hygiene needs. The Department of Energy works extensively through its national laboratories and energy research centers. From among these broad programs one at Brookhaven National Laboratory (BNL) and one at Oak Ridge National Laboratory (ORNL) may be cited.

IRLG: The Interagency Regulatory Liaison Group was formed in August, 1977, by the heads of four agencies: the Consumer Product Safety Commission, EPA, the Food and Drug Administration, and OSHA. The information uncovered by this group clearly illustrates the potential for duplication of effort and the need for interorganizational liaison.

An early survey revealed that these four agencies alone had some 300 ongoing projects at a cost of $39 million concerned with control of toxic substances (8). The agencies agreed to pool and to share information and to develop cooperative efforts to optimize health-protection programs. One of their working groups has recently drawn up development plans (9) for 24 toxic entities (e.g., asbestos, cadmium, coke-oven emissions, nitrosamines, polychlorinated biphenyls, radiation, and polyvinyl chloride).

For each, the report summarizes the agencies' authority to regulate the adverse biological effects, opportunities to avoid conflicts and duplication, economic ramifications of regulations, and recommendations for future action. The agencies will also establish a joint computerized information system on toxic substances that will contain all regulations and pertinent legal decisions.

Regulation of coke-oven emission sources concerns both OSHA and EPA and presents a potential area of conflict. On October 22, 1976, OSHA promulgated a new standard for coke-oven emissions of 150 μg of benzene-soluble particulates per cubic meter of air, averaged over 8 hr. The standard was to be phased in over a period of 3 years.

The EPA is concerned about emission sources, such as wet-coal charging operations, battery stacks, and quench towers, that could impact public health. A background information document is being assembled that includes health risk assessment and selection of a surrogate, such as BaP, that can be related to visible emissions, the general physical appearance of some of the coke-oven emissions. The Occupational Safety and Health Administration is concerned that if EPA proceeds with separate rule making, its own standard may be preempted because of the way Section 4(b)(1) of the Occupational Safety and Health Act of 1970 is drafted (9).

National Laboratories

Brookhaven National Laboratory (BNL): The Working Group on Assessing Industrial Hygiene Monitoring Needs for the Coal Conversion and Oil Shale Industries is an active organization with a clearly defined function of evaluating instrumentation and monitoring needs. The group, which is chaired by Otto White, Jr., of BNL, "was formed as an initial effort to avoid a traditional problem associated with new industries, i.e., not recognizing health hazards concurrent with the development of the technology." (10)

In its preliminary report, the group identifies research and development requirements for assessing potentially hazardous exposures. They classify these as being (a) immediately hazardous to life and health (e.g., O_2 deficiency, explosive coal dust); (b) high risk, but not immediately hazardous (e.g., some organics, metals, fibers, and dust); and (c) moderate risk and not immediately hazardous. The working group met again in November, 1978. A more comprehensive report is being prepared.

Oak Ridge National Laboratory (ORNL): The ORNL Life Sciences Synthetic Fuels programs at ORNL provide examples of other interdisciplinary work in which the national laboratories are participating. These programs represent a very broad approach, addressing the whole gamut of problems: chemical characterization, health and safety, biological effects, environmental transport and ecological effects, control technology, and assessment and siting.

Some eight laboratory divisions participate in this program. A Life Sciences Committee on Synthetic Fossil Fuels has succeeded in formulating human and environmental health assessment documents specific to the low-Btu coal gasifier at the University of Minnesota at Duluth (7) and is in the process of doing the same for the low-Btu gasifier in Pike County, Kentucky, and the H-Coal liquefaction plant at Catlettsburg, Kentucky. Symposia and workshops are also organized (11)(12).

Numerous computerized data bases, such as that of the Environmental Mutagen Information Center (EMIC) exist at ORNL.

Professional Societies

ACS: The American Chemical Society's present meeting represents an important activity of the Society's Division of Chemical Health and Safety. The subject of the meeting, "Safe Handling of Chemical Carcinogens, Mutagens, and Teratogens: The Chemists' Viewpoint," attests to the recognition of the importance of the problems that have been described.

In being cosponsored by the Chemical Society of Japan and with the participation of the Royal Australian Chemical Institute, the New Zealand Institute of Chemistry, and the Chemical Institute of Canada, this meeting is indeed international in scope. Meetings of this kind afford important opportunities for inter-society cooperation, liaison, and exchange of ideas and information.

International Commission for Protection Against Environmental Mutagens and Carcinogens (ICPEMC): The commission was founded in January, 1977, sponsored by the Institut de la Vie and affiliated with the International Association of Environmental Mutagen Societies. It has recruited some 20 members from universities, research laboratories, and industry. The primary objective of ICPEMC is "to identify and promote scientific principles in the fields of environmental mutagenesis, carcinogenesis, and genetic toxicity." The commission will develop statements to serve as a basis for guides and regulations to prevent or minimize deleterious effects of chemicals in man.

The ICPEMC has formed five committees to review specific areas. These areas are (a) short-term screening systems, (b) relationships between carcinogens and mutagens, (c) preparation of a registry of national regulatory principles, (d) risk estimate procedures, and (e) epidemiological approaches. The ICPEMC has also formed task groups to generate its priorities and statements. The Commission Secretary is P.H.M. Lohman, Medical Biological Laboratory TNO, P.O. Box 45, 2280 AA Rijswijk, Netherlands.

The formation of ICPEMC is reminiscent of the International Commission on Radiological Protection, the foremost international organization in the field of radiation protection standards. Such commissions can play a vital role as a scientific reviewing body, free of governmental and political objectives.

AIHA: The American Industrial Hygiene Association has several standing technical committees in areas related to problems that arise in synthetic fossil fuel technology. The AIHA also has a technical committee on Workplace Environmental Exposure Levels.

HPS: The Health Physics Society was founded as a professional organization dedicated to the protection of man from the potential harmful effects of ionizing radiation while realizing its benefits for mankind. While its principal theme has been radiation, the society has provided a forum for discussion of the potential hazards from nonradiation pollutants.

Its 1974 Midyear Symposium on Population Exposures included a session on exposure to nonradioactive pollutants (14). The journal *Health Physics* has also published discussions of the role of this society. It has been pointed out that the membership of the HPS constitutes a significant national resource for addressing problems from human exposure to chemical insults (15). The lack of input from the health physics profession in the formulation of programs to attack problems of chemical exposures is regrettable (16).

In another effort, the authors have been active as members of a committee on Chemical Pollutants formed recently in the East Tennessee Chapter of the Health Physics Society. This committee requested that the Board of Directors of the parent HPS bring to the attention of the membership the question of HPS activity in the field of chemical health hazards. While the HPS did not overtly assume an expanded role as a matter of stated policy, it has not discouraged members from presenting their views at its meetings and in its journals.

In this regard, one avenue of approach to chemical hazards is the application of data that has been acquired from radiation. For example, cell survival data are being used as a basis for comparing damage from chemical insults with damage from radiation (13).

EMS: The Environmental Mutagen Society has polled its members on the proposal of the Environmental Mutagen Information Center (EMIC) at Oak Ridge National Laboratory to build a current computerized information file. Often, data on chemicals, such as mutagenesis assays, require considerable time to reach publication in the open literature. The Environmental Mutagen Society (EMS) is considering forming a new data file to accept such information from investigators as soon as it is released. Restrictions on the use of any data in the file would be determined by the submitter. The existence and use of this file would in no way alter normal publication procedures.

Committee 17 of the EMS has published an article (17) describing methods (principally screening systems) for detecting environmental mutagens, ways of assessing the potential health impact of environmental mutagens, and regulatory principles for preventing environmental mutagenesis. In this report, it is pointed out that a direct comparison can be made between mutagenic effects produced by chemicals and by radiation, for which a large body of data exists.

In analogy with the radiation unit, rem (roentgen-equivalent-man), the rec (rem-equivalent-chemical) is defined as the dose (average concentration multiplied by exposure time) that produces an amount of genetic damage equivalent to that produced by 1 rem of radiation. This committee recommends "specific limits for the environmental distribution of mutagenic agents, including both ionizing radiations and chemical compounds, such that the resulting genetic damage does not exceed a 12.5% increase over the spontaneous mutation background." (17)

SOEH: The Society for Occupational and Environmental Health was founded in 1972 to identify physical, chemical, and other hazards in the occupational and general environments and to propose actions to reduce their danger. The SOEH urges a multidisciplinary approach to the key problems in all their facets. It has sponsored a number of conferences and workshops on specific issues, such as beryllium, vinyl and polyvinyl chlorides, and lead and arsenic.

ASTM: The American Society of Testing and Materials E-34 Committee is working actively. Thirteen task forces are currently developing draft standards for occupational exposure to a wide variety of hazardous chemicals, potentially dangerous procedures, and safety equipment. Polynuclear aromatic hydrocarbons are not included. Consideration, however, is being given to formation of a task force on PNA hydrocarbons that could limit itself initially to a few of the more seemingly soluble problems such as those mentioned in the section "Volatile PNA Vapors and Contaminated Surfaces." Such a task force would provide the focus for discussion and definition of the various problems and deciding where practical solutions are desirable or possible.

Summary of Organizational Activities

While by no means exhaustive, this survey of synthetic fuel health problems and some examples of the activities of various organizations is revealing. The efforts of government agencies sometimes involve conflicting requirements of the promotor and the protector in the same institution, as do cooperative efforts between government and industry.

In contrast to the role of the International Commission on Radiation Protection (ICRP) or the National Council on Radiation Protection and Measurement (NCRP) in radiation protection, there is no neutral international or national scientific body to which conflicts and disputes can be taken. Perhaps the newly organized ICPEMC might succeed in filling this gap.

Scientific societies are active to various degrees themselves and through intersociety liaisons. However, no society seems to be taking a clear lead. Perhaps an ASTM task group comes closest to providing a vehicle for a concerted program. It is believed that all must work together to define the problems in their broadest context and to begin developing standards. Some of the easier tasks would appear to be agreeing on guidelines for the volatile PNA hydrocarbons and surface contamination.

Table 6.1: Selected Examples of Interorganizational, Interdisciplinary, and Intersociety Liaison on Health Problems Related to Synthetic Fossil Fuel Technologies

Organization	Example of Activity	Reference
NIOSH	Conducts research in four areas concerned with coal gasification; develops criteria documents	(1)(2)
EPA	Classifies hazardous substances; sponsors symposia on environmental aspects of synfuel technologies	(3)(4)
DOE	Balanced Program Plan; develops information overviews; develops coal gasification facility plans	(5)(6)(7)
IRLG	Four agencies pool and share information	(8)(9)
BNL	Maintains a working group on monitoring needs for coal conversion and oil shale	(10)
ORNL	Life Sciences Synthetic Fuels Program; EMIC; sponsors symposia and workshops	(5)(6)(7)(11)(12)
ACS	Division of Chemical Health and Safety; international cosponsorship of meetings	
ICPEMC	Develops statements as basis for guides and regulations concerning mutagenic and carcinogenic chemicals	
AIHA	Maintains several standing technical committees	
HPS	Holds sessions on hazards of nonradioactive pollutants included in some meetings; compares chemical and radiation cell survival data	(13)(14)(15)(16)

(continued)

Table 6.1: (continued)

Organization	Example of Activity	Reference
EMS	Polled members on formation of current data file on chemical hazards by EMIC; reports on environmental mutagenic hazards	(17)
SOEH	Sponsors symposia and workshops on key problems	
ASTM	Task forces currently developing standards for occupational exposure to a variety of chemicals	

Source: NTIS CONF-790415-3

CRITERION FORMULATION

Existing Guidelines and Standards

There have been few attempts to develop exposure standards for PAH, either individually or as a class. In the occupational setting, a Federal standard has been promulgated for coke oven emissions, based primarily on the presumed effects of the carcinogenic PAH contained in the mixture as measured by the benzene soluble fraction of total particulate matter.

Similarly, the American Conference of Governmental Industrial Hygienists recommends a workplace exposure limit for coal tar pitch volatiles, based on the benzene-soluble fraction containing carcinogenic PAH. The National Institute for Occupational Safety and Health has also recommended a workplace standard for coal tar products (coal tar, creosote, and coal tar pitch), based on measurements of the cyclohexane extractable fraction. These standards are summarized in Table 6.2.

Table 6.2

Substance	Exposure Limit	Agency
Coke oven emissions	150 $\mu g/m^3$, 8 hr*	U.S. Occupational Safety and Health Administration
Coal tar products	0.1 mg/m^3, 10 hr*	U.S. National Institute for Occupational Safety and Health
Coal tar pitch of volatiles	0.2 mg/m^3**, 8 hr*	American Conference of Governmental Industrial Hygienists

*Time-weighted average.
**Benzene soluble fraction.

Source: EPA 440/5-80-069

A drinking water standard for PAH as a class has been developed. The 1970 World Health Organization European Standards for Drinking Water recommends a concentration of PAH not to exceed 0.2 $\mu g/\ell$. This recommended standard is based on the composite analysis of six PAH in drinking water: (a) fluoranthene, (b) benzo[a]pyrene, (c) benzo[g,h,i]perylene, (d) benzo[b]fluoranthene, (f) benzo[k]fluoranthene, and (g) indeno[1,3,-cd]pyrene.

The designation of these six PAH for analytical monitoring of drinking water was not made on the basis of potential health effects or bioassay data on these compounds (18). Thus, it should not be assumed that these six compounds have special significance in determining the likelihood of adverse health effects resulting from absorption of any particular PAH. They are, instead, considered to be useful indicators for the presence of PAH pollutants.

Borneff and Kunte (18) found that PAH were present in groundwater at concentrations up to 50 ng/ℓ, and in drinking water at concentrations up to 100 ng/ℓ. Based on these data they suggested that water containing more than 200 ng/ℓ should be rejected. However, as data from a number of U.S. cities indicate, levels of PAH in raw and finished waters are typically much less than the 0.2 μg/ℓ criterion.

Current Levels of Exposure

Considerable data has been presented (see previous chapter) which may be used to calculate an estimate of human exposure to PAH by all routes of entry to the body. However, quantitative estimates of human exposure to PAH require numerous assumptions concerning principal routes of exposure, extent of absorption, conformity of human lifestyle, and lack of geographic-, sex-, and age-specific variables. Nevertheless, by working with estimates developed for PAH as a class, it is possible through certain extrapolations to arrive at an admittedly crude estimate of PAH exposure.

Unfortunately, there are no environmental monitoring data available for most of the PAH which are specified under the Consent Decree in NRDC vs Train. By far the most widely monitored PAH in the environment is BaP; data on BaP levels in food, air, and water are often used as a measure of total PAH.

Among the PAH routinely monitored in water, four compounds are included in the Consent Decree list: BaP, IP, BbFL, and BjFL. In addition, levels of FL and BPR have been routinely determined in water, as recommended by the World Health Organization.

The reported estimated average concentrations of BaP, carcinogenic PAH (BaP, BjFL, and IP), and total PAH in drinking water are 0.55 ng/ℓ, 2.1 ng/ℓ, and 13.5 ng/ℓ, respectively (19). Thus, assuming that a human consumes 2 ℓ of water per day, the daily intake of PAH via drinking water would be:

$$0.55 \text{ ng/}\ell \times 2\ \ell\text{/day} = 1.1 \text{ ng/day (BaP)}$$
$$2.1 \text{ ng/}\ell \times 2\ \ell\text{/day} = 4.2 \text{ ng/day (carcinogenic PAH)}$$
$$13.5 \text{ ng/}\ell \times 2\ \ell\text{/day} = 27.0 \text{ ng/day (total PAH)}$$

Borneff (20) estimates that the daily dietary intake of PAH is about 8 to 11 μg/day. As a check on this estimate, PAH intake may be calculated based on reported concentrations in various foods and the per capita estimates of food consumption by the International Commission on Radiological Protection (21).

Taking a range of 1.0 to 10.0 ppb as a typical concentration for PAH in various foods, and 1,600 g/day as the total daily food consumption by man from all types of foods (i.e., fruits, vegetables, cereals, dairy products, etc.), the intake of PAH from the diet would be in the range of 1.6 to 16.0 μg/day. An estimate of BaP ingestion from the diet may be similarly derived.

Using 0.1 to 1.0 ppb as the range of BaP concentration in various foods, total daily BaP intake would be .16 to 1.6 μg/day.

Ambient air is reported to contain average levels of 0.5 ng/m^3, 2.0 ng/m^3, and 10.9 ng/m^3 for BaP, carcinogenic PAH, and total PAH, respectively. Taking the range of 15 to 23 m^3 as the average amount of air inhaled by a human each day results in an estimated intake of 0.005 to 0.0115 μg/day, 0.03 to 0.046 μg/day, and 0.164 to 0.251 μg/day for BaP, carcinogenic PAH, and total PAH respectively.

In summary, a crude estimate of total daily exposure to PAH would be as follows.

Table 6.3: Estimate of Human Exposure to PAH from Various Media

Source	BaP	Carcinogenic PAH*	Total PAH
		(μg/day)	
Water	0.0011	0.0042	0.027
Food	0.16-1.6	-	1.6-16
Air	0.005-0.0115	0.03-0.046	0.164-0.251
Total	0.166-1.6	-	1.6-16

*Total of BaP, BjFL and IP; no data are available for food.

Source: EPA 400/5-80-069

Two important factors are not taken into account in this estimate. First, it is known that tobacco smoking can contribute greatly to PAH exposure in man. Exposure to BaP from smoking one pack of cigarettes per day was shown to be 0.4 μg/day (22). Second, the possibility for dermal absorption of PAH is assumed to contribute only a negligible amount to the total exposure. Only in certain occupational situations is dermal exposure expected to be quantitatively important.

Special Groups at Risk

An area of considerable uncertainty with regard to the carcinogenic hazard of PAH to man involves the relationship between aryl hydrocarbon hydroxylase (AHH) activity and cancer risk. Genetic variation in AHH inducibility has been implicated as a determining factor for susceptibility to lung and laryngeal cancer (23)(24). It was suggested that the extent of AHH inducibility in lymphocytes was correlated with increasing susceptibility to lung cancer formation.

Paigen, et al (25) have examined the question of genetic susceptibility to cancer, and concluded that epidemiologic evidence supports this hypothesis. Moreover, they were able to show that AHH inducibility in lymphocytes segregates in the human population as a genetic trait. However, their studies failed to find a correlation between this inducibility and presumed cancer susceptibility, either among healthy relatives of cancer patients or in patients who had their cancer surgically removed. It is noteworthy that previous investigations on AHH inducibility were conducted in persons with active cancer.

Recent studies with other human tissues (liver and placenta) have provided important new data concerning the carcinogen-metabolizing capacity of man and its implications for cancer susceptibility. Conney, et al (26) examined individual differences in the metabolism of drugs and carcinogens in human tissues, and have identified drugs which may serve as model substrates to provide an indirect index of carcinogen metabolism for man.

The rates for antipyrene, hexobarbital, and zoxazolamine hydroxylation in human autopsy livers were highly, but not perfectly, correlated with the rates of BaP metabolism. In human placenta, an almost perfect correlation was found between zoxazolamine hydroxylase activity and BaP hydroxylase activity (27). Thus, metabolism of BaP and zoxazolamine by human placenta occurs by the same enzyme system(s) or by different enzyme systems under the same regulatory control (28). BaP and zoxazolamine hydroxylase activities were also shown to be significantly enhanced in placentas obtained from women who smoked cigarettes.

The lack of perfect correlations for the hepatic metabolism of BaP and certain drugs in many subjects indicated the presence of several monooxygenases in human liver which catalyze the oxidative metabolism of these compounds. Furthermore, large inter-individual differences exist in the capacity of humans to metabolize foreign chamicals both in vitro and in vivo.

Further studies showed that 7,8-benzoflavone markedly stimulated the hydroxylation of BaP, antipyrene, and zoxazolamine in human liver samples, but with a wide variation in magnitude among different samples. These results suggested the presence of multiple monooxygenases or cytochrome P-450 in the different liver samples (29). Moreover, 7,8-benzoflavone did not affect the hydroxylation of coumarin or hexobarbital, thereby indicating the existence of different monooxygenases for metabolism of these substrates.

Multiple forms of cytochrome P-450 have been shown in the livers of rats, rabbits, and mice, but not thus far in humans (28). More important, however, MCA is a potent inducer of BaP hydroxylase activity in rats but does not stimulate antipyrene hydroxylase, clearly suggesting that metabolism of PAH in rodents may be regulated by different enzyme systems than in humans (28).

In contrast to the apparent multiplicity of cytochrome P-450 dependent enzyme systems for the oxidative metabolism of PAH in man, a single epoxide hydrase with broad substrate specificity may be present in human liver (26)(30). Because the hydration of arene oxides may lead to the formation of dihydrodiol carcinogen precursors, the capacity of different humans to metabolize epoxides may affect cancer susceptibility. It is not known, however, if enhanced dihydrodiol formation would increase cancer risk or decrease cancer risk.

Thomas and Slaga (31) did not obtain a correlation of AHH induction with skin-tumor-inducing ability in mice for a series of unsubstituted hydrocarbons. Nevertheless, the highest AHH enzyme activity was found in the epidermal layer of the skin, which is the major point of contact with many environmental chemicals. These results may be interpreted to indicate that a chemical carcinogen may not necessarily induce its own bioactivation, but instead can be transformed into a reactive intermediate by virtue of increased AHH activity stimulated by other noncarcinogenic compounds.

Due consideration must also be given to the fact that, in addition to the initiation of resting cells by a chemical carcinogen, a promotion phase involving cell proliferation is also involved in skin carcinogenesis (32). Therefore, although certain aromatic hydrocarbons are effective enzyme inducers, their bioactivated metabolites may function only as an initiator having no promoting ability. A potent complete carcinogen, however, will be transformed not only into a powerful tumor initiator but will also be able to interact with cellular membranes, alter genetic expression, and ultimately cause irreversible cell proliferation.

These observations raise certain doubts concerning the validity and/or reliability of equating enzyme inducibility with carcinogenic potential for chemical agents. Further reinforcement of this opinion has been provided by Shulte-Hermann (33) who showed that cell proliferation is not a direct result of enzyme induction, even though both processes are normally coupled.

The further possibility that the genetics of AHH inducibility is organ-dependent rather than strain-dependent in animals has important implications for evaluating susceptibility to PAH-induced cancers (34). Most significant is the demonstration that pulmonary AHH may be inducible in all strains of mice, regardless of the inducibility of hepatic AHH. Since the respiratory epithelium represents a primary portal of entry for PAH, AHH activity which is induced in this tissue may bear importantly on susceptibility to malignancy.

Enzyme induction by PAH is not limited to AHH. Owens (35) recently demonstrated that MCA can induce hepatic UDP-glucuronosyltransferase activity in certain inbred strains of mice. This enzyme catalyzes the conjugation and excretion of PAH substrates after they have first been oxygenated by AHH. The induction of this transferase activity and that of AHH was apparently regulated by a single genetic locus.

However, transferase inducibility does not depend on AHH levels, but rather is stoichiometrically related to the concentration of a specific and common cytosolic receptor regulating both enzyme induction processes. Owens further demonstrated that AHH activity can be fully induced in certain mouse strains (e.g., by 2,3,7,8-tetrachlorodibenzo-p-dioxin) without greatly enhancing the transferase activity.

Earlier studies had established that chrysene and chlorpromazine were potent inducers of AHH activity while having little effect on transferase activity (36) (37). Subsequent exposure to carcinogenic PAH (i.e., MCA) could lead to maximal oxidative metabolism but little transferase-catalyzed removal of metabolites by glucuronic acid conjugation.

This situation would be exacerbated by the fact that metabolites of MCA are incapable of further inducing the transferase activity. This effect may have considerable toxicologic significance in that highly reactive epoxides of PAH formed by the action of AHH under these circumstances may not be adequately removed by glucuronidation. Thus, one must consider the total exposure of all environmental agents and their possible effect on critical enzymatic processes before attempting to assess the toxicologic impact of exposure to a specific PAH.

In summary, there is a need to further explore the relative effects of enzyme induction on the metabolic activation of chemicals to toxic products, versus metabolism of chemicals via detoxification pathways, when considering the possibility of special groups at risk.

Basis and Derivation of Criterion

The presently available data base is inadequate to support the derivation of individual criteria for each of the PAH as specified under the Consent Decree. This problem arises primarily from the diversity of test systems and bioassay conditions employed for determining carcinogenic potential of individual PAH in experimental animals. Furthermore, it is not possible to estimate the intake via water of individual PAH, except for those compounds which have been selected by the World Health Organization for environmental monitoring.

Therefore, an approach to criterion development is adopted by EPA with the objective of deriving criteria for individual carcinogenic PAH, which will lead to effective control of PAH as a class. This approach is attractive in that it recognizes the fact that environmental exposures to PAH invariably occur by contact with complex, undefined, PAH mixtures. The attempt to develop a drinking water criterion for PAH as a class is hindered by several gaps in the scientific data base:

(a) The PAH class is composed of numerous compounds having diverse biological effects and varying carcinogenic potential. A representative PAH mixture, has not been defined.

(b) The common practice of using data derived from studies with BaP to make generalizations concerning the effects of environmental PAH may not be scientifically sound.

(c) No chronic animal toxicity studies exist involving oral exposure to PAH mixtures.

(d) No direct human data exist concerning the effects of exposure to defined PAH mixtures.

However, assuming that the development of a criterion must proceed despite these obstacles, certain approaches may be taken to circumvent deficiencies in the data base. The choice of an appropriate animal bioassay from which to derive data for application to the human cancer risk assessment should be guided by several considerations. Primary emphasis must be placed on appropriate animal studies which:

(a) include sufficient numbers of animals for statistically reliable results;

(b) involve long-term low-level exposures to PAH;

(c) include a proper control group; and

(d) achieve positive dose-related carcinogenic response.

Because there are no studies available regarding chronic oral exposure to PAH mixtures, it is necessary to derive a criterion based upon data involving exposure to a single compound.

Two studies can be selected, one involving BaP ingestion (38) and one involving DBA ingestion (39). Both compounds are recognized as animal carcinogens, and both are known to be environmental contaminants to which humans are exposed.

Presently, there is no way to quantitate the potential human health risks incurred by the interaction of PAH, either among themselves or with other agents, (e.g., tumor initiators, promoters, inhibitors) in the environment. In addition, it is known that PAH commonly produce tumors at the site of contact (i.e., forestomach tumors by oral exposure to BaP; lung tumors by intratracheal administration; skin tumors by dermal application). Thus, consideration of the extent of absorption may not always be necessary in the case of carcinogenic PAH, and will in fact result in underestimation of actual risk if only distant target sites are considered.

The water quality criterion for BaP derived using the linearized multistage model, as described in the Human Health Methodology Appendices to the October 1980 *Federal Register* is 28 ng/ℓ.

The water quality criterion for BaP is based on the experiment reported by Neal and Rigdon (38), in which benzo[a]pyrene at doses ranging between 1 and 250 ppm in the diet was fed to strain CFW mice for approximately 110 days. Stomach tumors, which were mostly squamous cell papillomas but some carcinomas, appeared with an incidence statistically higher than controls at several doses. The extrapolation was based on the following parameters.

Dose (mg/kg/day)	Incidence* (No. Responding/No. Tested)
0.0	0/289
0.13	0/25
1.3	0/24
2.6	1/23
3.9	0/37
5.2	1/40
5.85	4/40
6.5	24/34
13.0	19/23
32.5	66/73

le = 110 days w = 0.034 kg
Le = 183 days R = 30 ℓ/kg
L = 630 days

*The incidences at the highest three doses were not used in the extrapolation due to lack of fit of the multistage model.

With these parameters, the carcinogenic potency factor for humans, q_1, is 11.53 $(mg/kg/day)^{-1}$. The result is that the water concentration of BaP should be less than 28 ng/ℓ in order to keep the individual lifetime risk below 10^{-5}. It is recognized that numerous carcinogenic PAH other than BaP are found in water. However, there is probably little need to derive criteria for all such PAH, since efforts to reduce BaP levels to within acceptable limits will result in the reduction of all PAH.

For the sake of comparison, a water quality criterion for DBA was calculated using the procedure developed by Mantel and Bryan (40). As opposed to the linearized multistage model, which is logistic and defines acceptable risk as 1/100,000, the Mantel and Bryan model is probablistic and defines acceptable risk as 1/100,000,000. Furthermore, the Mantel and Bryan model is concerned with the maximum tumor incidence in treated animals at the 99% confidence level versus the 95% confidence level in the linearized multistage model. Using the Mantel and Bryan approach with DBA, the resultant water quality criterion is 13.3 ng/ℓ.

Under the Consent Decree in NRDC vs Train, criteria are to state "recommended maximum permissible concentrations (including where appropriate, zero) consistent with the protection of aquatic organisms, human health, and recreational activities." BaP is a known animal carcinogen. Because there is no recognized safe concentration for a human carcinogen, the recommended concentration in water for maximum protection of human health is zero.

Because attaining a zero concentration level may be infeasible in some cases and in order to assist the Agency and states in the possible future development of water quality regulations, the concentrations of BaP corresponding to several incremental lifetime cancer risk levels have been estimated.

A cancer risk level provides an estimate of the additional incidence of cancer that may be expected in an exposed population. A risk of 10^{-5} for example, indicates a probability of one additional case of cancer for every 100,000 people exposed, a risk of 10^{-6} indicates one additional case of cancer for every million people exposed, and so forth.

PAH are widely distributed in the environment as evidenced by their detection in sediments, soils, air, surface waters, and plant and animal tissues. The ecological impact of these chemicals, however, is uncertain. Numerous studies show that despite their high lipid solubility, PAH show little tendency for bioconcentration in the fatty tissues of animals or man. This observation is not unexpected, in light of convincing evidence to show that PAH are rapidly and extensively metabolized.

Lu, et al (41) have published the only available study regarding the bioconcentration and biomagnification of a PAH in model ecosystem environments. They reported that the bioconcentration of BaP, expressed as concentration in mosquito fish/concentration in water was zero. This was apparently due to the fact that the fish metabolized the BaP about as rapidly as it was absorbed.

On the other hand, in a 33 day terrestrial-aquatic model ecosystem study, BaP showed a small degree of biomagnification which probably resulted from food chain transfer. In this case the biomagnification factor for mosquito fish was 30. Based on the results of Lu, et al a bioconcentration (BCF) factor of 30 was employed for the purpose of calculating a water quality criterion.

In the *Federal Register* notice of availability of draft ambient water quality criteria, EPA stated that it is considering setting criteria for BaP at an interim target risk level of 10^{-5}, 10^{-6}, or 10^{-7} as shown in the following Table 6.4.

Table 6.4: BaP

Exposure Assumptions (per day)	Risk Levels and Corresponding Criteria ....(ng/ℓ)..............			
	0	10^{-7}	10^{-6}	10^{-5}
2 ℓ of drinking water and consumption of 6.5 g fish and shellfish	0	0.28	2.8	28.0
Consumption of fish and shellfish only.		3.11	31.0	311.0

Source: EPA 440/5-80-069

The criteria are calculated by applying a linearized multistage model. Since the extrapolation model is linear at low doses, the additional lifetime risk is directly proportional to the water concentration. Therefore, water concentrations corresponding to other risk levels can be derived by multiplying or dividing one of the risk levels and corresponding water concentrations shown in the table by factors such as 10, 100, 1,000, and so forth.

Approximately 9% of the PAH exposure, assumed to be BaP, results from the consumption of aquatic organisms which exhibit an average bioconcentration potential of 30 fold based on the work of Lu, et al (41). The remaining 91% of PAH exposure results from drinking water.

Concentration levels were derived assuming a lifetime exposure to various amounts of PAH:

(a) occurring from the consumption of both drinking water and aquatic life grown in water containing the corresponding PAH concentrations, and

(b) occurring solely from the consumption of aquatic life grown in the waters containing the corresponding PAH concentrations.

Because data indicating other sources of exposure and the concentration to total body burden are inadequate for quantitative use, the criterion reflects the increment to risks associated with ambient water exposure only.

The water quality criterion for carcinogenic PAH compounds is based on the assumption that each compound is as potent as BaP and that the carcinogenic effect of the compounds is proportional to the sum of their concentrations. Based on an oral feeding study of BaP in mice, the concentration of BaP estimated to result in a lifetime risk of 10^{-5} is 28 ng/ℓ. Therefore, with the assumption above, the sum of the concentrations of all carcinogenic PAH compounds should be less than 28 ng/ℓ in order to keep the lifetime cancer risk below 10^{-5}.

REFERENCES

(1) *Criteria for a Recommended Standard–Occupational Exposure to Coal Tar Products*, DHEW (NIOSH), Publication No. 78-107 (1977).

(2) Young, R.J., McKay, W.J., and Evans, J.M., "Coal Gasification and Occupational Health," *Am. Ind. Hyg. J.* 39, 985-997 (1978).

(3) Kingsbury, G.L., "Development of Multimedia Environmental Goals for Pollutants from Fuel Conversion Processes," *Symp. Proc.: Environmental Aspects of Fuel Conversion Technol., III*, EPA-600/7-78-063, pp. 53-75 (1978).

(4) Williamson, J.E., Strup, P.E., and Jones, P.W., "The Quantitative Analysis of Selected PAH's in Aqueous Effluents by High Performance Liquid Chromatography," *Proc. Third Intl. Symp. on Polynuclear Aromatic Hydrocarbons*, Columbus, Ohio (October 25-27, 1978).

(5) *Balanced Program Plan, Vol. IV, Coal Conversion*, ORNL-5123 (1976).

(6) Braunstein, H.M., Copenhaver, E.D., and Pfuderer, H.A., *Environmental, Health, and Control Aspects of Coal Conversion: An Information Overview*, ORNL/EIS-94 (Vol. 1) and ORNL/EIS-94 (Vol. 2)(1977).

(7) Cowser, K.E., editor, *Proposed Environmental and Health Program for University of Minnesota Gasification Facility,* Oak Ridge National Laboratory, Oak Ridge, Tennessee (January 23, 1978).

(8) "Regulating Hazardous Substances," *Environ. Sci. Technol.* 13, No. 2, 146-148 (1979).

(9) Interagency Regulatory Liaison Group (CPSC, EPA, FDA, OSHA), *Hazardous Substances* (December 1, 1978).

(10) White, Otto, *Working Group on Assessing Industrial Hygiene Needs for the Coal Conversion and Oil Shale Industries*, Report BNL-24925, Brookhaven National Laboratory, Upton, New York, 11973 (August 1978).

(11) Gammage, R.B., *Proc. of the Second ORNL Workshop on Exposure to Polynuclear Aromatic Hydrocarbons in Coal Conversion Processes* (March 9-11, 1977) CONF-770361 (1977).

(12) *Symposium on Potential Health and Environmental Effects of Synthetic Fossil Fuel Technologies* (September 25-28, 1978) CONF-780903 (1979).

(13) Griffin, D., Jones, T.D., and Walsh, P.J., "Chemical Cytotoxicity: A Cancer Promotor," *Proc. of the Third Intl. Symp. on Polynuclear Aromatic Hydrocarbons*, Columbus, Ohio (October 25-27, 1978).

(14) *Symposium on Population Exposures, Eighth Midyear Topical Symposium of the Health Physics Society*, NTIS CONF-741018 (October 21-24, 1977).

(15) Turner, J.E., Shank, K.E., Loebl, A.S., and Easterly, C.E., "The Role of the Health Physicist in the Development of Safe Energy Sources," *Health Phys.* 34, 189-190 (1978).

(16) Gammage, R.B., Thorngate, J.H., Parkinson, W.W., Hawthorne, A.R., and Vo-Dinh, T., "On the Desirability of the Health Physics Society Assuming Responsibilities in Nonnuclear and Nonradiation Fields," *Health Phys.* 35, 711-714 (1978).

(17) "Environmental Mutagenic Hazards," *Science* 187, 503-514 (1975).

(18) Borneff, J. and Kunte, H., "Carcinogenic Substances in Water and Soil, XXIV: A Routine Method for the Determination of PAH in Water," *Arch. Hyg.* (Berl) 153:220 (1969).

(19) Basu, D.K. and Saxena, J., *Analysis of Raw and Drinking Water Samples for Polynuclear Aromatic Hydrocarbons*, EPA P.O. No. CA-7-2999-A and CA-8-2275-B, Expo. Evalu. Branch, HERL, Cincinnati (1977).

(20) Borneff, J., "Fate of Carcinogens in Aquatic Environment" (1977).

(21) International Commission on Radiological Protection No. 23, *Report of the Task Group on Reference Man*, Pergamon Press (1974).

(22) National Academy of Sciences, *Biological Effects of Atmospheric Pollutants: Particulate Polycyclic Organic Matter*, Washington, D.C. (1972).

(23) Kellerman, G., et al, "Aryl Hydroxylase Inducibility and Bronchogenic Carcinoma," *New England Jour. Med.* 289:934 (1973).

(24) Kellerman, G., et al, "Genetic Variation of Aryl Hydrocarbon Hydroxylase in Human Lymphocytes," *Am. Jour. Hum. Genet.* 25:327 (1973).

(25) Paigen, B., et al, "Human Aryl Hydrocarbon Hydroxylase and Cancer Risk," *Carcinogenesis* Vol. 3, *Polynuclear Aromatic Hydrocarbons*, Jones, P.W. and Frendenthal, R.I., (editors), Raven Press (1978).

(26) Conney, A.H., et al, "Use of Drugs in the Evaluation of Carcinogen Metabolism in Man," *Screening Tests in Chemical Carcinogenesis*, IARC No. 12, Montesano, R. and Tomatis, L., (editors) Lyon, France (1976).

(27) Kapitulnik, J., et al, "Comparison of the Hydroxylation of Zoxazolamine and Benzo-[a] pyrene," *Clin. Pharmacol. and Therap.* 20:557 (1976).

(28) Kapitulnik, J., et al, "Benzo[a] pyrene 7,8-dihydrodiol Is More Carcinogenic than Benzo-[a] pyrene in Newborn Mice," *Nature* 266:378 (1977).

(29) Kapitulnik, J., et al, "Activation of Monooxygenases in Human Liver by 7,8-Benzo-flavone," *Clin. Pharmol. and Therap.* 22:475 (1977).

(30) Kapitulnik, J., et al, "Comparative Metabolism of Benzo[a] pyrene and Drugs in Human Liver," *Clin. Pharmol. and Therap.* 21:166 (1977).

(31) Thomas, S. and Slaga, T.J., "Mouse Epidermal Aryl Hydrocarbon Hydroxylase," *J. Invest. Dermatol.*, 66:108 (1976).

(32) Yuspa, S.H., et al, "Cutaneous Carcinogenesis: Past, Present and Future," *J. Invest. Dermatol.* 67:199 (1976).

(33) Shulte-Herman, R., "Stimulation of Liver Growth and Mixed Function Oxidase by α-Hexachlorocyclohexane: Separation of Inductive Pathways," *Microsomes and Drug Oxidations*, V. Ulrich, editor, Pergamon Press, New York (1977).

(34) Kouri, R.E., et al, "Studies on Pulmonary Aryl Hydrocarbon Hydroxylase Activity in Inbred Strains of Mice," *Chem.-Biol. Interactions* 13:317 (1976).

(35) Owens, I.S., "Genetic Regulation of UDP-Glucuronosyltransferase Induction by Polycyclic Aromatic Compounds in Mice," *Jour. Biol. Chem.* 252:2827 (1977).

(36) Aitio, A., "Different Elimination and Effect on Mixed Function Oxidase of 20-Methyl-cholanthrene After Intragastric and Intraperitoneal Administration," *Res. Commun. Chem. Path. Pharmacol.* 9:701 (1974).

(37) Aitio, A., "Effect of Chrysene and Carbon Tetrachloride Administration on Rat Hepatic Microsomal Monoxygenase and UDP-Glucuronosyltransferase Activity," *FEBS Lett.* 42:46 (1974).

(38) Rigdon, R.H. and Neal, J., "Effect of Feeding Benzo[a] pyrene on Fertility, Embryos, and Young Mice," *Jour. Natl. Cancer Inst.* 34:297 (1965).

(39) Snell, K.C. and Stewart, H.L., "Induction of Pulmonary Adenomatosis in DBA/2 Mice by the Oral Administration of Dibenz[a,h] anthracene," *Acta. Un. Int. Canc.* 19:692 (1962).

(40) Mantel, N. and Bryan, R.W., "Safety Testing of Carcinogenic Agents," *Jour. Natl. Cancer Inst.* 27:455 (1961).

(41) Lu, A.Y.H., et al, *Carcinogenesis* Vol. 1 p. 116, Freudenthal, R.I. and Jones, P.W., editors, Raven Press, New York (1976).

ANALYTICAL METHODS

INTRODUCTION

The information in this section is based on:

> *Measurement of Polycyclic Organic Matter for Environmental Assessment,* EPA Report 600/7-79-191, August 1979, prepared by P.L. Levins, C.E. Rechsteiner, Jr. and J.L. Stauffer of Arthur D. Little Inc. for the U.S. Environmental Protection Agency.

The interest in POM has given rise to a host of qualitative and quantitative procedures for POM analysis. Of the various procedures, some are intended to give a single quantitative value for a specific compound, such as for benzo[a]pyrene, while others attempt to give a composite value indicative of the total quantity of POM present in a sample. The main difference between most specific and general analytical techniques is a separation of the species of interest prior to detection.

Prior to the detection of POM, some suitable procedure should be used to isolate POM from interferences and to concentrate the sample. Historically, the techniques of thin layer chromatography and paper chromatography were used to isolate specific POM for subsequent identification. Current practice utilizes column chromatography to separate POM from non-POM followed by either high-pressure liquid chromatography or gas chromatography for resolving specific POM isomers from one another. The use of these procedures, in conjunction with POM analysis, has been reviewed extensively (1). The use of gas chromatography (GC) when combined with mass spectrometry is a powerful tool for the detection and quantification of POM ranging up to the dibenzopyrenes in molecular weight.

Fluorescence Methods

Fluorescence methods were among the very first methods used for the detection of POM. The work in the early 1930's by J.W. Cook and coworkers (2)(3) made extensive use of fluorescence spectroscopy to identify benzo[a]pyrene as a carcinogenic constituent of coal tar. Various workers (4)(5)(6) have employed fluorescence spectroscopy to follow the metabolic reactions of specific POM compounds during hydroxylation. The analysis of specific POM has also been undertaken with fluorescence methods. Benz[a]anthracene (7), benzo[a]pyrene (8), and indeno[1,2,3-cd]pyrene (9) are just a few examples of specific POM which have been analyzed with fluorescence spectroscopy. Fluorescence techniques have been used by Sawicki (10) and others to examine a wide range of POM containing samples.

Fluorescence methods are capable of measuring subnanogram quantities of individual POM species, but tend to be fairly nonselective. The normal spectra obtained from POM and related species tend to be intense and lack resolution in fluid media. To determine levels of specific compounds, thorough separation of the sample constituents is needed to minimize spectral interferences. Efforts to overcome this difficulty have been directed towards low temperature techniques.

The fluorescence of various POM at temperatures of 20°K or less no longer consists of a series of broad, smooth peaks found in room temperature fluid media, but contains many discrete peaks of varying intensity. The unique nature of the spectra obtained with this method is demonstrated by the fact that the six isomeric monomethylchrysenes can be distinguished from one another by their low temperature fluorescence spectra (11). However, identification of an unknown species from its spectrum is not an easy task. Compilation of reference spectra and rapid methods of searching reference spectra are not as well developed or as widespread as those often used in infrared or mass spectroscopy.

Furthermore, it may be difficult to recognize spectra, even of pure compounds, because of phenomena such as saturation at high concentrations. Also, certain classes of POM, such as nitrogen containing heterocyclics, exhibit low efficiency towards fluorescence and may not be detected in small quantities.

Work at Oak Ridge National Laboratories (12) has applied synchronous fluorescence and phosphorescence to PAH analysis. With the synchronous technique, the normally broad fluorescence spectra of the PAH species is simplified to a series of distinct and better resolved peaks. The improvement in the distinguishing features of the different PAH spectra enhances the specificity of these luminescence techniques, and the potential application of these methods to provide more specific information than total POM measurements needs to be further explored.

Ultraviolet Absorption Spectroscopy

Ultraviolet absorption spectroscopy (UV) is another analytical detection method which has been widely used. For many years, the preferred approach for POM analysis was to use a combination of liquid chromatography (LC) and thin layer chromatography (TLC) to isolate specific POM followed by UV techniques for species identification and quantification (13). Benz[a]anthracene (14), benzo[a]pyrene (15), and indeno[1,2,3-cd]pyrene (9) are typical examples of POM which may be detected.

Compilations of UV data for numerous POM are found in a number of sources (16)(17). As with fluorescence measurements, the individual spectra for various

POM are unique, although portions of spectra for different compounds may be the same; for example, both chrysene and 4,5-methylene chrysene have similar absorption bands at 361 nm and both benzo[a]pyrene and benzo[ghi]perylene have similar bands at 382 nm. The possibility of spectral overlap requires complete separation of sample components to insure accurate measurement of component levels, as with fluorescence techniques. Also, the overall sensitivity of measurements by UV methods is somewhat lower than for fluorescence methods (one-tenth to one-thousandth that of fluorescence). Hence, the use of UV measurements for POM analysis has declined, being replaced with the more sensitive fluorescence methods and the highly sensitive and specific GC/MS methods.

Gas Chromatographic Methods

Gas chromatographic methods have shown the most rapid development for POM analyses in recent years. The first useful application of GC techniques to POM analysis dates to 1965, with methods being found in the *Journal of Chromatography* (18) and *Analytical Chemistry* (19). The instrumentation for a GC separation is relatively inexpensive and samples may be analyzed conveniently with high speed and good reproducibility.

Detection of the GC effluent may be by any of a wide range of analytical procedures. Since UV methods have been widely used for POM detection and the UV spectra are specific for different POM, methods which couple GC for separation and UV for detection have been developed. Other detectors, such as flame ionization detection (FID), electron capture detection (ECD), mass spectrometer (MS) detection or infrared (IR) detection have been coupled with GC separation for sensitive and specific analyses.

Numerous GC column packings have been developed to enhance the separation of specific POM. Methyl phenyl silicones, such as SE-52, OV-17, or SP2250, and carborane methyl- and methyl phenyl silicones, such as Dexsil 300 or Dexsil 400, have been widely applied to POM separations.

These GC phases will resolve groups of POM from one another, 3-ring POM from 4-ring, 4-ring POM from 5-ring, etc., but the resolution of specific isomers is not possible in every case.

Two developments greatly improved the specificity of GC analyses of POM. Capillary columns have been known for their inherently high resolution since the work of Golay in 1958 (20). However, extensive use of capillary columns was delayed until the development of efficient injection systems and reproducible coating of the stationary phase on the capillary tube. The high resolution obtainable by capillary GC techniques allows separation of selected isomers unresolved with normal packed columns (see e.g., Lee, et al (21) for an extensive list of POM relative retention indices). However, capillary columns will only tolerate small sample volumes. Thus, the detector for capillary GC methods must have high sensitivity such as flame ionization detectors (FID), electron capture detector (ECD), or a mass spectrometer (MS).

Of these, the last detector system, a mass spectrometer, allows confirmation of species identification and, through the use of computerized data handling procedures, greatly enhances the specificity of the analytical measurement.

The other advance in GC technique which has affected POM analysis is the liquid crystal GC column phase [for example, N,N'-bis(p-phenylbenzylidene) α,α'-bi-p-toluidine (BPhBT)]. In 1975, Janini (22) and coworkers first applied nematic liquid crystals to the separation of polycyclic aromatic hydrocarbons. These columns resolve geometric isomers of different POM, but the observed elution order of the isomers is frequently different from that observed for the general purpose silicone columns. For example, the normal elution order benzo[e]pyrene, benzo[a]pyrene, pyrene observed on OV-17 and Dexsil-300 and -400 columns is different from that of a liquid crystal column such as SP-301 (BPhBT) (benzo[a]pyrene, perylene, and benzo[a]pyrene).

To obtain the best separation of POM, isothermal column operation is required which limits the use of these columns with complex samples. When coupled to a selective detector, e.g., a mass spectrometer, which may be adjusted to minimize the interferences due to other non-POM species in a complex sample, liquid crystal columns are highly useful for determining quantities of narrow groups of specific POM.

The power of the combined GC/MS system for POM analyses makes such systems the method of choice for specific POM analyses.

Miscellaneous Methods

The intense interest in POM analysis has brought forth a large variety of special techniques for specific analyses. Infrared (23) and Raman spectroscopic techniques, nuclear magnetic resonance (24) methods, polarography (25) and potentiometric titration (26) are examples of the types of alternative techniques which have been applied to POM analysis. For the most part, these methods tend to be prone to numerous interferences in real samples, and offer no significant advantage over UV, fluorescence or GC methods discussed previously.

NEMATIC LIQUID CRYSTALS FOR PNA SEPARATIONS

The information in this section is based on:

> "Improved Chromatography of PNA Components Using Nematic Liquid Crystals" prepared by W.L. Zielinski, Jr. and M. Janini of National Cancer Institute, Frederick Cancer Research Center, *Proceedings of the Second ORNL Workshop on Exposure to Polynuclear Aromatic Hydrocarbons in Coal Conversion Processes. I Medical Surveillance; II Industrial Experiences, Personnel Protection and Monitoring,* NTIS CONF-770361, December 1977, compiled by B. Gammage of Oak Ridge National Laboratory.

Many of the activities in air pollution monitoring on the national scene are concerned with the quantitative assay of the carcinogenic PNA, benzo[a]pyrene (BaP). Its quantitation by GLC, however, has long been a problem due to the inherent lack of suitable resolution of BaP from its geometric isomer, benzo[e]-pyrene (BeP). Some success in the separation of this important isomeric pair has been obtained by GLC on Dexsil (27), but a substantially improved resolution was best obtained by HPLC (28)-(30). While HPCL is actively pursued today for analysis of PNA components, it is clear that in a general sense, GLC offers the greatest potential for PNA analysis due to its capabilities of covering a broad PNA molecular-weight range from two- to six-ring PNA or higher. This potential, however, can only be approached pending the availability of GLC stationary phases that can offer unique capabilities for the selective separation of geometric isomers of the same ring class.

PNA Shape Differences

While PNA geometric isomers may have similar volatilities and degrees of interaction with a given GLC stationary phase, and hence have similar retentions upon GLC analysis, it can be noted by examination of their planar structures that they differ in molecular shape. It has been known for some years that liquid crystals used as GLC stationary phases can provide selective separations of aromatic geometric isomers. It was perhaps unfortunate that most of these activities centered on separations of aromatic geometric isomers of low molecular weight.

Several years ago, a liquid crystal was utilized that exhibited a mesomorphic phase at sufficiently high column temperatures for separations of PNA components (22), including separation of the so-called benzpyrene fraction components superior to that obtainable by HPLC analysis. The liquid crystal employed in this study, BMBT [N,N'-bis(p-methoxybenzylidene)-α,α'-bi-p-toluidine], was obtained from a commercial chemical supplier and was used for the separation of three-ring PNA at 210°C and five-ring PNA (representing the so-called benzypyrene fraction) at 260°C.

Application was also shown for the temperature programming (210° to 255°C at 4° per min) of 16 PNA components ranging from two to five rings, as well as for the substantially improved resolution of sterol epimers of the cholestane and androstane classes (31). This event was perhaps even more notable for its illustration that an analytical dimension for separations based on differences in solute molecular shape could be attained in GLC analysis of solutes of practical importance.

Development of New Liquid Crystals for PNA Separations

One specific limitation that has been reported (22) in the use of this liquid crystal (BMBT) for routine analysis was its bleed characteristic when operated at elevated temperatures for prolonged periods. In seeking to improve GLC operational stability, higher molecular weight liquid crystals of the same structural class were prepared synthetically and characterized. In the past year, there has been reported GLC evaluations of three synthetic liquid crystals: namely, the n-butoxy (32) and the n-hexoxy and phenyl (33) homologs of BMBT (termed respectively BBBT, BHxBT, and BPhBT). While BBBT showed substantial improvement in thermal stability over BMBT (32), BHxBT and BPhBT exhibited superior stability characteristics.

To verify the thermal stability of these liquid phases, the theoretical plate count of chrysene was determined and was found to be reproducible to within 5% over 100 hr of continuous operation at 275°C, using three different glass columns packed with 2.5% (w/w) each of BPhBT, BHxBT, and 1:1 BHxBT/BPhBT on 100/120 mesh HP Chromosorb W. Operation of the BPhBT and the 1:1 BPhBT/BHxBT columns at 275°C beyond 100 hr resulted in a gradual deterioration of performance as witnessed by some loss of retention and resolution. Yet, it has been independently confirmed (34)(35) that BPhBT column life can be extended to months when certain precautions are taken. A comparison of transition temperatures for these new liquid crystals and those of BMBT were determined by differential scanning calorimetry, and are given as follows:

Table 7.1: Transition Temperatures (°C)

Liquid Crystal	Solid-N	Solid-S	S-S	S-N	N-1	ΔN
BMBT	181				320	139
BBBT		159		188	303	115
BHxBT		127	203	229	276	47
BPhBT	257				403	146

Source: NTIS CONF-770361

Utility of GLC Liquid Crystals in PNA Analysis

If one examines the trends in transition temperatures going from BMBT (methoxy) to BBBT (n-butoxy) to BHxBT (n-hexoxy), it is apparent that the nematic range decreases with increasing alkoxy chain length. Actually, in such a series of continually increasing chain lengths of the alkyl tails (36), the chemical progresses through properties of pure nematic, nematic and smectic, and smectic until, ultimately, all liquid crystalline properties disappear. In such a series only certain liquid crystals may be appropriate for GLC analytical purposes.

While we have prepared all normal chain members of the liquid crystalline series represented by N,N'-bis(p-alkoxybenzylidene)-α,α'-bi-p-toluidine from ethoxy through octoxy, we have elected to report only on the n-butoxy (BBBT) and n-hexoxy (BHxBT) analogs. The selection of these was based both upon knowledge of changes in liquid crystalline properties with increasing alkoxy-chain length and upon considerations for diminishing column bleed by increasing the molecular weight.

Some specific comments regarding the transition-temperature data might be appropriate at this point. Both BBBT and BHxBT melt into a smectic (S) phase, with the latter having two smectic phases [termed Smectic I (127° to 203°C) and Smectic II (203° to 229°C)]. The smectic phase in a liquid crystal represents a closer packing of the liquid crystal molecules than does the packing in the case of a nematic phase. As a result of this phenomenon, the oddity of increased PNA retention with increasing column temperature is observed as the liquid crystal changes from the smectic phase to a nematic phase (32). This oddity is attributed to a higher degree of solute penetrability into a nematic phase than is possible in a smectic phase.

Nonetheless, it should be pointed out (and perhaps stressed to some extent) that GLC separations on liquid crystals are best achieved using a nematic-phase operating temperature. If one then considers the utility of liquid crystals for GLC, one characteristic that could enhance capabilities for programmed temperature analysis of a wide molecular weight range of solutes in a given sample lies in the temperature range of the nematic phase (ΔN) in the liquid crystal employed. If the ΔN values for the four liquid crystals listed in the table are examined it is apparent that BHxBT would not be as conducive to programmed temperature analysis as BMBT, BBBT, and BPhBT. However, it is noted that BMBT shows evidences of substantially higher column-bleed characteristics than do the other liquid crystals listed. On the other hand, the most stable of these (BPhBT) does not become a nematic until 257°C (its melting point), and may not be as appropriate as BMBT and BBBT for the analysis of two- or three-ring PNA. Thus, the various liquid crystals reported for high-temperature PNA analysis offer different desirable traits from an analysis standpoint.

In summary, it is suggested that isothermal GLC analyses of PNA can best be done on BMBT and BBBT for two- or three-ring PNA, on BBBT and BHxBT for four-ring PNA (including benz[a]anthracene (33), on BHxBT and BPhBT for five-ring PNA (the benzpyrene fraction, including benzo[a]- and benzo[e]pyrenes), and on BPhBT for five- to seven-ring PNA (33). Isothermal baseline separation of four- and five-ring PNA was also reported on a 1:1 BHxBT/BPhBT column at 265°C with elution of the last five-ring PNA (benzo[a]pyrene) in 7 min (33).

While BHxBT and BPhBT were both employed in a GLC-MS system for PNA separations and detection (33) the latter of these was observed to be the more stable and was employed in a method describing the analysis of benzo[a]pyrene in cigarette smoke condensate (37) with an expressed minimum detection limit for this carcinogen of less than 10 ng using a flame-ionization detector. BPhBT was also shown to be effective recently for the separation of PNA components in coal-tar pitch (34).

PNA IN LIQUID FUEL

The information in this section is based on:

> *Quantitative Analysis of Polynuclear Aromatic Hydrocarbons in Liquid Fuels,* EPA Report 600/2-80-069, April 1980, prepared by Radian Corporation for the U.S. Environmental Protection Agency.

In order to correlate polynuclear aromatic hydrocarbon (PNA) levels in combustion source exhaust emissions with the concentrations of PNA in various liquid hydrocarbon fuels, the concentration of twelve specified PNA in various test fuels was determined. The PNA of interest include:

Phenanthrene	Triphenylene
Anthracene	Benzo[a]pyrene
Fluoranthene	Benzo[e]pyrene
Pyrene	Benzo[g,h,i]perylene
Benz[a]anthracene	Coronene
Chrysene	Anthanthrene

The approach employed for the analysis of the specified PNA in each sample was combined gas chromatography-mass spectrometry (GC-MS) with operation in the selected ion monitoring mode. A preliminary isolation/concentration scheme was employed to remove the PNA from the hydrocarbon fuel matrix. Isomeric PNA were chromatographically resolved by utilizing liquid crystal liquid phases.

A total of ten test fuels were analyzed. These fuels are listed in Table 7.2. (As a control, Couch No. 5 was tested as 9A and 9B.)

Table 7.2: Test Fuel Description

No.	Test Fuel Designation	Uses	Reference
1	Jet A high sulfur	Commercial jet aircraft	(38)
2	Jet A low sulfur	Commercial jet aircraft	(38)
3	EM-238-F	Diesel-powered passenger cars and trucks	(39)(40)
4	EM-239-F	Diesel-powered passenger cars and trucks	(39)(40)
5	EM-240-F	Diesel-powered passenger cars and trucks	(39)(40)
6	EM-241-F	Diesel-powered passenger cars and trucks	(39)-(41)
7	EM-242-F	Diesel-powered passenger cars and trucks	(39)-(41)
8	Couch No. 2	Diesel-powered passenger cars and trucks	(39)-(41)
9 (A&B)	Couch No. 5	Diesel-powered passenger cars and trucks	(39)-(41)

Source: EPA-600/2-80-069

Isolation and Concentration of PNA from Liquid Fuels

An extraction scheme was developed to isolate and concentrate the PNA present in liquid fuels. The procedure employed was a modification of the method described by Matsushita (42) and is illustrated in Figure 7.1. The fuel sample (500 ml) was extracted three times with 300 ml of dimethyl sulfoxide (DMSO). An equal volume of cold 20% HCl was then added to the combined DMSO extracts. The resulting DMSO-HCl mixture was extracted three times with 900 ml of cyclohexane. The cyclohexane extracts were combined and washed successively first with 200 ml of 5% sodium hydroxide followed by two 200 ml portions of distilled water. This extract was then concentrated by removal of the cyclohexane by distillation. The extract was placed in a round bottom flask in a heating mantle.

A three-ball Snyder column was placed on top of the flask as a distillation column. Distillation was conducted at atmospheric pressure until all cyclohexane had been removed.

Initially, it was anticipated that a large concentration factor (500-1,000) would be required in order to achieve the desired sensitivity (1 μg/gal). Therefore, it was anticipated that 500 ml of sample would be extracted and the solvent concentrated to 1 ml. With this procedure, PNA at 1 μg/gal in the fuel would be at a concentration of 130 ng/ml in the concentrated extract.

However, this concentration factor could not readily be achieved in the samples. Typically, the extraction and concentration steps removed 90 to 95% of the fuel. The remaining 5 to 10% consisted of aromatic compounds which were not removed by the procedure. For the first two samples analyzed, Jet A Fuel-Low Sulfur and Jet A-High Sulfur Fuel, vacuum distillation was employed to provide a higher concentration factor.

The PNA were present in much higher levels than anticipated, frequently in the ppm range. Also, the GC-MS analysis technique was shown to be sensitive to the low ppb range with only moderate concentration. Therefore, the vacuum distillation step was omitted from the isolation scheme for the remaining samples.

Figure 7.1: Isolation and Concentration Scheme for PNA in Liquid Fuels

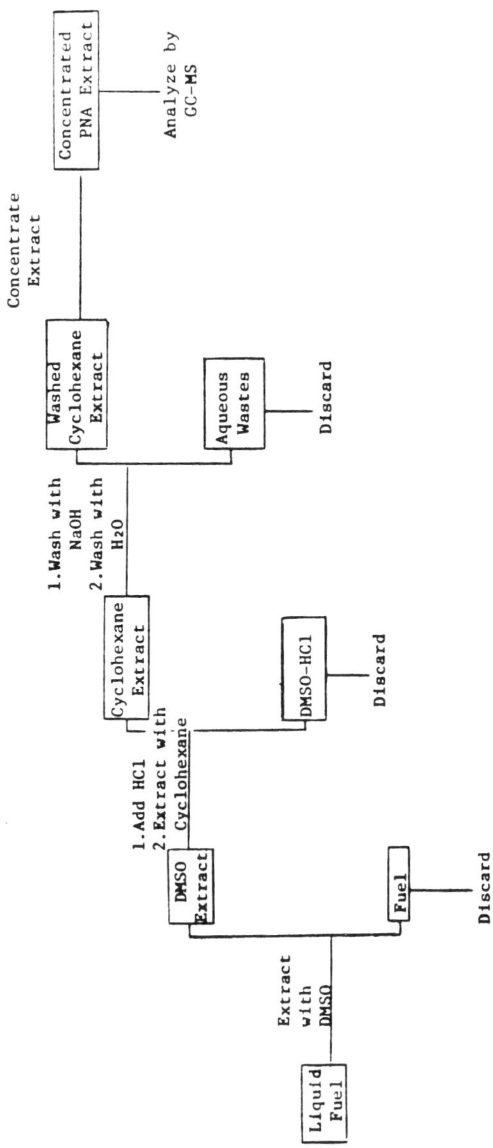

Source: EPA 600/2-80-069

All samples were extracted in duplicate. Each sample was spiked prior to extraction with either 9-bromoanthracene, d_{10}-anthracene, or indenopyrene. The spikes were, for the most part, added to the fuel to the 10 ppb level. Two Couch No. 5 samples (9A and 9B) were spiked to a 1 ppb level.

Identification of PNA

The liquid fuel concentrates were analyzed by combined gas chromatography/mass spectrometry (GC-MS) using a Hewlett-Packard 5982A GC-MS system.

The technique of selected ion monitoring (SIM) was employed for all analyses. In this technique, the intensity of key ions was monitored during the chromatographic separation. Identification of the desired compounds was based on the appearance of these key ions at known gas chromatographic retention times. A second criterion for identification was also employed during the analysis. The relative intensity of the key ions for each compound had to be within 20% of the intensity found from the analysis of a standard. Table 7.3 lists the PNA determined in this program along with the key ions and their relative intensities employed in the analysis.

Table 7.3: Key Ions for the SIM Analysis of PNA in Liquid Fuels

Compound	M_1	Relative Intensity (%)	M_2	Relative Intensity (%)	M_3	Relative Intensity (%)
Phenanthrene	178	100	89	17	176	15
Anthracene	178	100	89	16	176	15
Fluoranthene	202	100	101	23	100	14
Pyrene	202	100	101	26	100	17
Benz[a]anthracene	228	100	229	19	226	19
Triphenylene	228	100	229	19	226	23
Chrysene	228	100	229	19	226	20
Benzo[e]pyrene	252	100	126	23	250	16
Benzo[a]pyrene	252	100	126	23	250	16
Benzo[g,h,i]perylene	276	100	138	37	274	27
Anthanthrene	276	100	138	37	274	27
Coronene	300	100	150	64	149	46

Source: EPA-600/2-80-069

The twelve PNA of interest consist of six series of isomers. Other isomeric species of the desired compounds exist. For example, there are six isomers of benzo[a]pyrene but only two (benzo[a]pyrene and benzo[e]pyrene) were of interest in this program. In order to accurately measure each compound, it was imperative that each isomeric species be resolved.

A liquid crystal liquid phase, N,N'-bis(p-methoxy-benzylidene)-α,α'-bi-p-toluidine (BMBT), was studied for the chromatographic separation of the PNA isomers. This packing, which had been reported by Janini (22), provides excellent resolution of certain PNA isomers.

Two additional liquid crystals recently developed (32)(33) were also studied.

These liquid phases, N,N'-bis(p-butoxy-benzylidene)-α,α'-bi-p-toluidine (BBBT) and N,N'-bis(p-phenyl-benzylidine)-α,α'-bi-p-toluidine (BPhBT), also provided excellent separation of certain PNA isomers. These phases are commercially available from Supelco.

Each of the three liquid crystals have different characteristics. While both BBBT and BMBT are satisfactory for 3 to 7 ring PNA, BBBT has less column bleed at elevated temperatures. Chromatographic resolution on liquid crystals is highly dependent on temperature, with the best resolution achieved with the liquid crystal in the nematic state. BMBT and BBBT are nematic in the temperature range of 190° to 300°C, while BPhBT is nematic in the temperature range of 260° to 400°C. Therefore, BPhBT is generally limited to 5 to 7 ring compounds because of its temperature range. Early in the study it was determined that slight variations in the temperature programming conditions could cause drastic changes in resolution. Because of this temperature dependence of the liquid crystals, it was very difficult to resolve all of the PNA isomers in one analysis as illustrated in Figure 7.2.

Changes in the temperature programming will easily resolve either phenanthrene and anthracene (m/e 178) or chrysene, triphenylene and benz[a] anthracene (m/e 228). However, careful selection of the chromatographic conditions was required in order to separate all of these isomers in one analysis.

Therefore, it was decided to employ separate GC analyses for the different isomeric species. The benefits from this approach were: the instrument gain could be adjusted to obtain maximum sensitivity for the high molecular weight PNA by allowing measurement of the more abundant low molecular weight species separately; the best chromatographic column for each group of isomers was employed, thus improving the separation; temperature control was not critical, and the overall analysis time was often shortened. Table 7.4 summarizes the chromatographic conditions employed for all analyses.

Quantification of each PNA identified was achieved by examination of the stored mass spectral data. For each compound, the areas under the intensity profiles for each key ion were calculated employing the Hewlett-Packard data system. The total mass of each compound injected was determined from a calibration curve.

Results

The results from the analyses of the ten fuel samples are presented in Table 7.5. All samples were analyzed in duplicate, and the results shown are the average from the two determinations. The two Couch No. 5 samples (9A and 9B) were submitted as blind duplicates. The pooled standard deviation of each compound was calculated by taking the square root of the average of all the standard deviations squared.

Anthanthrene and coronene were not detected in any of the samples analyzed. Although the detection limit for each PNA in the samples varied, in general, the detection limit of the method employed was approximately 5 μg/gal (1 ppb). The remaining ten PNA were found in levels ranging from 6 μg/gal (1.6 ppb) to 3.1 g/gal (810,000 ppb) of fuel. In general, the concentration decreased with increasing molecular weight of the PNA.

Figure 7.2: Selected Ion Chromatograms from the Analysis of a Standard PNA Mixture on BBBT

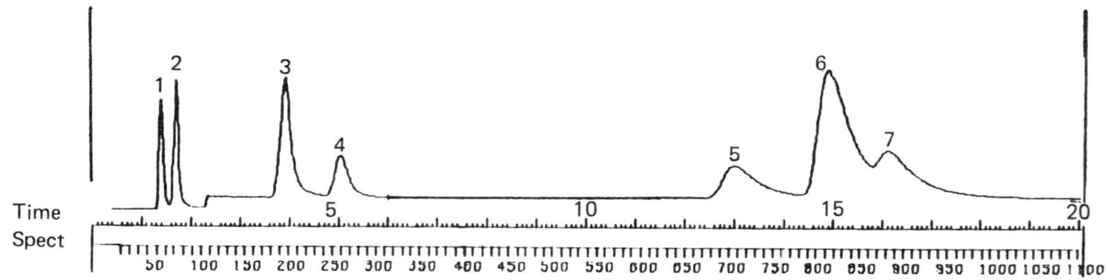

GC Conditions: 220° for 8 minutes
then 8°/minute to 290°C

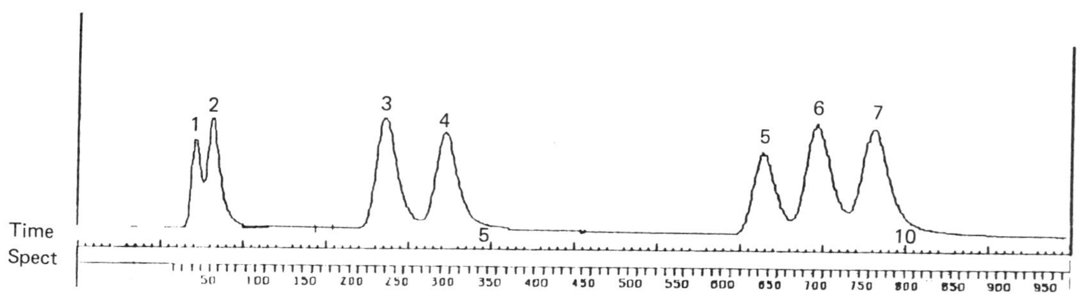

GC Conditions: 240° for 4 minutes
then 8°/minute to 280°C

PNA:

(1) Phenanthrene
(2) Anthracene
(3) Fluoranthene
(4) Pyrene
(5) Triphenylene
(6) Benz[a]anthracene
(7) Chrysene

Source: EPA-600/2-80-069

Table 7.4: GC-MS Operating Conditions for Liquid Fuel Extracts

Column (glass)	Packing (%)	Carrier Gas	Flow Rate (ml/min)	Initial Temperature (°C)	Initial Time (minutes)	Final Temperature (°C)	Program Rate (°C/min)	Samples Analyzed	Isomers Resolved
6' x ¼"	1.5 BBBT*	He	30	200	4	280	8	Jet A fuel, low sulfur Jet A fuel, high sulfur	All
6' x ¼"	1.5 BBBT*	He	30	210	2	260	8	EM-238 EM-242 EM-239 Couch #2 EM-240 EM-241	Phenanthrene, anthracene, fluoranthene, pyrene, chrysene, benz[a]anthracene, triphenylene
6' x ¼"	2.5 BBBT*	He	30	300	-	300	-	EM-338 EM-241 EM-239 EM-242 EM-240 Couch #2	Benzo[a]pyrene, benzo[e]pyrene
2' x ¼"	1.5 BPhBT*	He	30	280	-	280	-	EM-438 EM-241 EM-239 EM-242 EM-24 Couch #2	Benzo[ghi] perylene, anthanthrene, coronene
6' x ¼"	5 SP-2100 + 1 BMBT*	He	30	200	0	260	8	Couch 5 (9A) Couch 5 (9B)	Phenanthrene, anthracene, fluoranthene, pyrene
6' x ¼"	1.5 BPhBT*	He	30	270	0	290	8	Couch 5 (9A) Couch 5 (9B)	Benz[a] anthracene, chrysene, triphenylene, benzo[a] pyrene, benzo[e] pyrene
2' x ¼"	1.5 BPhBT*	He	30	290	-	290	-	Couch 5 (9A) Couch 5 (9B)	Benzo[ghi] perylene, anthracene, coronene

*On 100/120 Supelcoport.

Source: EPA-600/2-80-069

Table 7.5: Concentration of Polynuclear Aromatic Hydrocarbons in Liquid Fuels

	Concentration, μg/gal*					
Fuel	Phenanthrene	Anthracene	Fluoranthene	Pyrene	Triphenylene	Benz[a]anthracene
1	6,800	530	45	68	14	ND (<2)**
2	8,300	760	180	290	27	ND (<2)
3	2,600,000	16,000	95,000	38,000	53,000	34,000
4	1,300,000	38,000	49,000	30,000	17,000	13,000
5	30,000	760	610	200	280	210
6	3,100,000	33,000	61,000	53,000	16,000	13,000
7	800,000	6,400	20,000	7,600	6,100	2,300
8	2,000,000	11,000	76,000	120,000	42,000	28,000
9A	19,000	1,200	6,800	21,000	6,100	2,600
9B	42,000	1,200	7,800	10,000	4,500	2,000

	Concentration, μg/gal*					
Fuel	Chrysene	Benzo[e]pyrene	Benzo[a]pyrene	Benzo[g,h,i] perylene	Anthanthrene	Coronene
1	11	6	ND (<2)	ND (<2)	ND (<2)	ND (<2)
2	12	43	ND (<2)	ND (<2)	ND (<2)	ND (<2)
3	110,000	570	160	240	ND (<20)	ND (<20)
4	16,000	290	260	110	ND (<20)	ND (<20)
5	310	38	280	ND (<10)	ND (<20)	ND (<20)
6	24,000	200	170	68	ND (<20)	ND (<20)
7	5,300	64	76	ND (<10)	ND (<20)	ND (<20)
8	53,000	680	330	110	ND (<20)	ND (<20)
9A	4,500	1,400	800	30	ND (<20)	ND (<20)
9B	4,500	1,200	720	38	ND (<20)	ND (<20)

*Average from duplicate determination.

**ND (<X): Not detected; X represents the detection limit for the compound in the sample analyzed.

Source: EPA-600/2-80-069

PNA IN U.S. WATERS

The information in this section is based on:

> *Method Development and Monitoring of Polynuclear Aromatic Hydrocarbons in Selected U.S. Waters,* EPA Report 600/1-77-052, November 1977, prepared by J. Saxena, D.K. Basu and J. Kozuchowski of Syracuse Research Corporation for U.S. Environmental Protection Agency.

A method for concentration of trace quantities of the six representatives of polynuclear aromatic hydrocarbon (PAH) family has been developed and successfully applied to PAH monitoring in finished and raw waters.

PAH are collected by passing water through polyurethane foam plugs. Water is heated to 62°±2°C prior to passage and flow rate is maintained at approximately 250 ml/min to obtain quantitative recoveries.

Preconcentration Using Foam Plugs

The ability of various foam plugs to retain BaP from aqueous solution under static conditions was determined by allowing preweighed foam plugs to equilibrate for 4 hr in 150 ml volumes of ^{14}C-BaP solution of different concentrations. The activity of the initial and final solution was determined by counting a 5 ml aliquot of the water in Insta-gel (Packard Instrument Co.). From the difference in the ^{14}C-activity in initial and final water sample, the amount of ^{14}C-BaP sorbed by foam plugs was calculated.

The adsorption of BaP on the container was taken into consideration by allowing equilibration of the solution with the glass container prior to introducing the foam plugs and taking the residual activity in water as the initial BaP concentration.

In flow system experiments, a foam plug was wetted with distilled water, squeezed to expel air and placed in a Chromaflex column. Each plug was washed with 20 ml acetone, 50 ml benzene, again 20 ml acetone, and finally with 250 ml distilled water. In a typical experiment, four liters of water was spiked with ^{14}C-BaP to a concentration of 0.1 ppb and allowed to stand for 30 min to allow equilibration with glass surface. The equilibrated solution was drawn through the column at a constant flow rate of 250±10 ml/min by means of two oscillating type pumps connected in series and controlled by a Variac. The flow rate was continuously monitored with the help of a Gilmont flowmeter.

Prior to reaching the foam column, the water was brought to a desired temperature by passing it through a custom made glass coil (25 cm x 6 mm) which was immersed in a Haake thermostated circulator.

The coil was placed inside the reservoir housing through the opening of the cover plate. For continuously monitoring the temperature of water during passage through the column, a water trap equipped with a thermometer was introduced in the line between the exit end of the column and the pump.

The concentration of radioactive BaP in spiked water and effluent after passage through the foam plug was determined by extracting a known aliquot (100 and

500 ml, respectively) with 20 ml benzene and counting the ^{14}C-activity in the solvent layer. To account for change in the concentration of BaP in spiked water with time, two samples, one at the beginning and the other at the end of the run, were taken and the values were averaged.

The BaP retained on the foam was eluted with 20 ml acetone followed by 75 ml benzene. In experiments where two foams were used in the same column, the volume of eluent acetone and benzene was increased to 30 and 125 ml, respectively. Soxhlet extraction of the plugs did not elute more BaP than that recovered by batch column extraction and, therefore, its use was not considered necessary. The calculation of the efficiency of retention of BaP on the foam plugs was based on the concentration of BaP detected in spiked water.

The amount of BaP retained on the glass bottle, heating coil, and connecting tubings was determined by washing them individually with enough acetone necessary to avoid emulsion formation and enough benzene to leach out all the activity.

In every case a 5 ml aliquot of the extracted solution was used for radioactive counting with 10 ml of scintillation fluid. The scintillation fluid was composed of 5 g of 2,5-diphenyloxazole (PPO) and 0.15 g of 2,2'-phenylenebis(5-phenyl)oxazole (POPOP) per liter of toluene. Water samples were counted, when necessary, in Insta-gel.

In several experiments the foam plugs were coated with selective chromatographic phases according to the procedure of Uthe et al (43), and tested for their sorption ability towards BaP.

The stability of BaP on foam plugs was studied by passing 2 liters of BaP spiked tap water (0.1 ppb) through the foam plugs and storing the plugs with sorbed BaP for various lengths of time. In order to prevent photodegradation of BaP, the plugs were stored in Chromaflex columns covered with aluminum foil.

Following the storage periods, BaP from the foam plugs was eluted as before and counted for ^{14}C-activity. The eluent from the storage experiments was concentrated and subjected to thin layer chromatography for determining if any degradation of BaP had occurred. Cellulose plates were spotted with the concentrate and developed with benzene as solvent. The plates were scanned for radioactive spots on a Nuclear Chicago Actigraph III.

Most experiments were repeated 2 to 3 times and results are expressed as mean of these values. The deviations from the mean were usually in the range of 3 to 6%.

Sorption Characteristics Under Static and Flow Condition: Preliminary assessment of retention characteristics of various foam plugs was based on the ability of the plugs to retain BaP from aqueous solution under static conditions. Musty and Nickless (44) used methylene blue adsorption from aqueous solution for determining the relative effectiveness of foam plugs for removing pesticides. A more direct method, such as used in the present case, has been used by Lawrence and Tosine (45).

Table 7.6: Characteristics of Various Types of Foams Used

Type	Source	Trade Name	Chemical Nature	Color	Plug Dimension* (mm)	Foam Density (kg/m^3)
A	Scientific Products, Inc.**	DiSPo plugs	Polyester	White	50 x 38	~24
B	Scientific Products, Inc.***	DiSPo plugs	Polyester	White	50 x 38	~22
C	VWR Scientific	Identiplugs	Polyester	White	45 x 45	~25
D	Thomas E. Forrest Co. Inc.	UU34	Polyester	Green	45 x 45†	24

*Diameter x length. **Batch purchased in 1974. ***Batch purchased in 1976.
†Plugs were cut from 45 mm thick flexible polyurethane foam sheet.

Source: EPA-600/1-77-052

The retention of BaP by various foam plugs increased linearly with increase in BaP concentration as shown in Figure 7.3. The foam plugs differed only to a small extent in their sorption characteristics under static conditions. It was found that type D foam sorbed the greatest amount of BaP, followed by foams B, C and A. The polymer linkage in the foam-ester or ether, and foam density did not appear to be related to the sorption properties.

Figure 7.3: Sorption of BaP by Various Foam Plugs Under Static Conditions

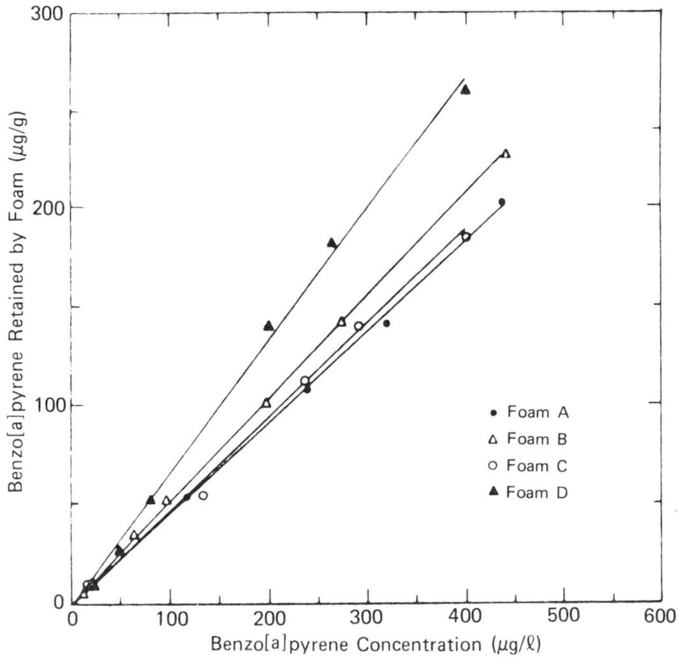

Source: EPA-600/1-77-052

The results presented in Table 7.7 show the recoveries of BaP from 4 ℓ of spiked-distilled and tap water obtained with various foam plugs in a continuous flow system. The results are in agreement with the recoveries obtained under static conditions. Foam type D was most effective in retaining BaP and, again, the retention abilities of various foam plugs were within a narrow range.

The recovery from tap water with various foam plugs at ambient temperature ranged between 58 and 66%. Under similar conditions, the recovery from spiked-distilled water was considerably higher (89 to 96%). The decreased retention from tap water was probably not due to competition for sites by contaminants since the retention of BaP from distilled water was not affected by prior exposure to tap water. The lower efficiency appeared to be linked to the presence of suspended particles, since the tap water dosed with BaP following Millipore filtration (0.45 μ) gave retention values equivalent to distilled water.

Table 7.7: Recovery of BaP from Spiked Water by Continuous Flow System*

	...... Percent Recovery from......	
Foam Plug	Tap Water	Distilled Water
A	62	89
B	58	91
C	65	91
D	66	96

*Water temperature, 23°C; volume, 4 ℓ; flow rate, 150±10 ml/min; column diameter, 25 mm; BaP conc., 0.1 ppb.

Source: EPA-600/1-77-052

Figure 7.4: Effect of Water Temperature on Recovery of BaP on Foam Plugs*

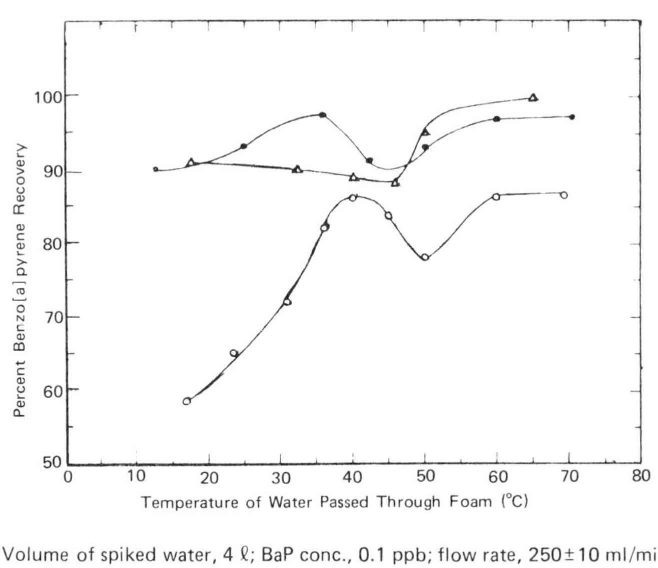

*Volume of spiked water, 4 ℓ; BaP conc., 0.1 ppb; flow rate, 250±10 ml/min; o, unfiltered tap water; ●, filtered tap water; △, distilled water.

Source: EPA-600/1-77-052

Effect of Water Temperature on BaP Recovery: The most dramatic effect on the recovery of benzo[a]pyrene from spiked tap water was observed when the temperature of the water was varied (Figure 7.4). The relationship between BaP recovery from tap water and temperature was found to be biphasic. The percent retention steadily increased with increase in temperature up to 40°C, but decreased with further increase in temperature. When the temperature was increased beyond 50°C, the increase in BaP retention was resumed until a plateau was reached starting at 60°C. The recovery of BaP at temperature >60°C was approximately 87%.

PAH Separation and Analysis

The collection is followed by elution of foam plugs with organic solvent, purification by partitioning with solvents and column chromatography on Florisil. PAH analysis was performed using two techniques: gas liquid chromatography-FID detection and thin layer chromatography-fluorescence detection.

The relative merits of these and other methods of PAH analysis have been discussed by many researchers (46)(47). The TLC procedure used was of Borneff (48) and recommended by the World Health Organization (49). The method was employed after slight modifications. It permits qualitative as well as quantitative determination of PAH directly on the TLC plate and thus losses associated with removal of PAH for fluorescence measurement are eliminated.

PNA ANALYSIS OF INLAND WATER AND SEDIMENT

The information in this section is based on:

> *Multicomponent Polycyclic Aromatic Hydrocarbon Analysis of Inland Water and Sediment*, NTIS CONF-7805130-1, 1978, prepared by W.H. Griest of Oak Ridge National Laboratory for the U.S. Department of Energy.

The paper by W.H. Griest reports the development of multicomponent PAH analytical methods for inland water and sediment. The method consists of separate steps of PAH extraction, isolation and concentration, and identification and quantification.

PAH Extraction

A gross organic extract containing PAH must first be obtained from the sample matrix. Ideally, the method should provide quantitative recoveries with a minimum of time and manpower. Traditionally, water samples are shaken (50) with an immiscible organic solvent and sediments are Soxhlet-extracted (51) with an organic solvent. Radio-labeled PAH tracers were utilized to determine optimum conditions for extraction of PAH from sediment and water.

One-liter samples of water were spiked in separate experiments with known activities of carbon-14 labeled naphthalene (^{14}C-Nap) and carbon-14 labeled benzo[a]pyrene (^{14}C-BaP) and were extracted with 100 ml portions of water-equilibrated cyclohexane in 2 ℓ separatory funnels equipped with Teflon stopcocks.

Liquid scintillation counting of an aliquot of the pooled cyclohexane extract indicated quantitative recovery of both tracers in three extractions. Four extractions of water samples with 0.1 volume of cyclohexane are carried out routinely to strive for complete recovery of all aqueous PAH.

Sediments pose a greater extraction problem, but extended Soxhlet-extraction with acetone is effective in recovering PAH. One hundred gram aliquots of wet sediment were thoroughly mixed in separate experiments with ^{14}C-BaP or ^{14}C-NaP packed with 43 x 123 cm cellulose thimbles and were Soxhlet-extracted with 250 ml of acetone. Periodically, aliquots of the acetone extract were withdrawn for tracer recovery measurements.

Whereas aliquots of the water extracts could be directly analyzed by liquid scintillation spectrometry, the scintillation quenching from gross amounts of organic compounds in the raw sediment extracts necessitated a simple purification of an aliquot on a short-layered Florisil/alumina column (3.5 x 1.1 cm o.d. each) with benzene prior to liquid scintillation tracer recovery measurements. Soxhlet extraction for 68 hours was sufficient to quantitatively recover both tracers. For convenience, such extractions are routinely extended to 72 hours.

Separate aliquots of sediment are dried at 110°C to constant weight to allow calculation of a dry/wet ratio for expression of PAH analytical results in terms of the dry weight of the sediment sample. Aliquots of sediment for PAH measurements are not dried; thus losses of volatile PAH are reduced by this separate dry weight measurement.

PAH Isolation and Concentration

The PAH in the raw water and sediment extracts are next isolated by a two-step adsorption column chromatography procedure. More complex isolation procedures (52) incorporating liquid-liquid partitioning prior to adsorption column chromatography are not necessary to prepare a PAH isolate suitable for analysis by gas chromatography (GC). Cyclohexane extracts of water are concentrated to 10 ml and are applied to 10 g of Florisil in a 25 ml burette and are eluted with 150 ml of 6:1 hexane:benzene (v/v convention will be used throughout). Acetone extracts of sediment are reduced to 10 ml, diluted to 60 ml with water, and extracted four times with 20 ml of cyclohexane. These cyclohexane extracts are then treated identically to the water extracts described above.

The entire eluate from the Florisil column is then reconcentrated to 10 ml and passed through 20 g of neutral, deactivated (4% moisture) alumina, using a hexane:benzene step gradient of 6:1 (100 ml) and 2:1 (300 ml). The PAH isolate is obtained in two fractions defined by the tracers: a diaromatic fraction containing mainly alkyl naphthalenes, acenaphthenes and acenaphthalenes and a polyaromatic fraction consisting of 3 to 7 ring PAH and their alkyl derivatives. Each fraction is separately reduced to a known volume by evaporative concentration with dry-flowing nitrogen under reduced pressure and temperature.

PAH Identification and Quantification

Qualitative identification of PAH in the purified PAH fractions is achieved by comparison of GC retention times and mass spectra with those of authentic PAH standards. Routine GC analyses are performed on a 6.6 m x 3 mm o.d. glass column packed with 3% Dexsil 400 on 100/120 mesh Supelcoport. Temperature programming from 100° to 320°C at 1°C/min allows elution of PAH ranging from 2 rings (e.g., naphthalene) through 7 rings (e.g., coronene) in approximately 4 hours.

Difficult-to-resolve PAH isomers which are separated with fairly good success include phenanthrene/anthracene, benz[a]anthracene/chrysene, and benzo[e]-pyrene/benzo[a]pyrene.

PAH concentrations are calculated by comparing GC peak areas with those of an external standard. Recovery corrections for losses of PAH in isolation and handling are estimated by liquid scintillation counting 0.1 volume aliquots of the fractions in 10 ml of scintillator solution prepared by dissolving 15 g of 2,5-diphenyloxazole and 190 mg of 1,4-bis[5-phenyloxazole] benzene in 1 gal of reagent grade toluene.

FLUORESCENT SPOT TEST FOR PNA DETECTION

The information in this section is based on:

> *Evaluation of Sensitized Fluorescence for Polynuclear Aromatic Hydrocarbon Detection,* EPA Report 600/7-79-207, August 1979, prepared by T.R. Smith of TRW Defense and Space Systems Group for the U.S. Environmental Protection Agency.

The original fluorescence spot test was developed by Arthur D. Little, Inc. (53) under contract to the EPA and is based on the phenomenon of sensitized fluorescence. The objective of the effort of T.R. Smith was to determine if the test could be used as a screening aid for samples submitted for PAH analysis by gas chromatography/mass spectrometry (GC/MS). The procedure can potentially provide a low cost screening technique to predetermine the presence of PAH.

Spot Test Procedure

No preparation of the sample is required. The sample should be in a solution to allow easy application to the test substrate. Solids and gases can be analyzed with some modification of the application technique (53).

A Chromatovue ultraviolet cabinet (Model C-70) was used to expose the samples. The 254 nm lamp source was used. A 7.0 cm circular ashless Whatman No. 42 filter was used as the support for the samples. The sensitizer used was naphthalene at a concentration of 60 $\mu g/\mu l$ in methylene chloride.

The spot test is carried out as follows:

(1) Draw three circles in pencil on a piece of Whatman No. 42 filter paper as close to one another as feasible.

(2) Using a 10 μl syringe, spot the filter paper with the sensitizer (naphthalene). 1 μl of sensitizer should be spotted on the leftmost circle and 1 μl of sensitizer should be spotted on the center circle. The spots should not be allowed to overlap. The substrate should be supported so as not to touch any surface.

(3) Allow the solvent to evaporate from the filter paper.

(4) 1 μl of sample is then applied to the rightmost and center circles. Again, allow the solvent to evaporate and do not allow the substrate to touch any surface until the solvent has evaporated.

(5) The filter paper is placed in the Chromatovue cabinet.

(6) Compare the spots visually with the unaided eye.

(7) If the center spot (sample plus sensitizer) fluoresces brighter than the sensitizer spot (leftmost), the presence of PAH is suspected. Also, any significant difference in color between the sensitizer-only spot versus sensitizer-sample spot is also considered a positive indicator for PAH.

(8) Samples which give positive results by the spot test should be submitted for GC/MS analysis to identify the source of the fluorescence.

Detection Limit and Sample Analysis

In order to determine the level of detection attainable with the equipment available, a set of standards were prepared and tested. Phenanthrene and benzo[a]-pyrene were chosen as the test standards. Phenanthrene and benzo[a]pyrene were chosen because both are considered representative of the PAH likely to occur in combustion effluent samples.

Their differences in structure may also cause differences in their detectability by the PAH spot test. Additionally, the range of aromatic fused ring compounds detectable by GC/MS is covered by these two compounds and the presence of benzo[a]pyrene, because of its carcinogenicity, must be detectable by the spot test at low concentrations.

The levels of standards prepared and the results of the analyses are presented in Table 7.8. The limit of detection of the spot test is extrapolated to be approximately 100 pg for these two materials. That is, a 1 μl sample containing 100 pg of PAH is easily detected. This observation differs by an order of magnitude from the detection limit of 10 pg previously reported (53).

Table 7.8: Results of Spot Test Analysis of Standards

Phenanthrene		Benzo[a]pyrene	
Amount	**Test Results**	**Amount**	**Test Results**
80 ng	Positive	490 ng	Positive
8 ng	Positive	49 ng	Positive
160 pg	Positive	147 pg	Positive
80 pg	Negative	98 pg	Negative
		49 pg	Negative

Source: EPA-600/7-79-207

The difference in the reported detection limit is probably due to the point at which the investigator decides to declare a significantly distinguishable difference in the fluorescent intensity of the sensitizer alone versus the sensitizer-sample spot. Since this decision is subjective, it was decided to choose a level that was easily discernable by personnel not familiar with the test.

A lower detection limit is achievable, but it would require the operator to make judgements based on subtleties of intensity and hue and would require extensive operator training and experience.

Another factor which must be taken into consideration when detection limits are compared is the ultraviolet lamp intensity. Any difference in lamp intensity will result in a difference of fluorescence intensity and, consequently, the detectable difference between the sensitizer-only and the sensitizer-sample spots would be concommitantly effected.

It is an accepted fact that lamp intensity varies with age and it is not unusual to encounter significant differences in lamp intensities in an apparatus such as the Chromatovue C-70. Differences in lamp intensity as a source of variation of the detection limit was not investigated as a part of this task.

Results

A series of industrial samples which gave negative results for both the PAH spot test and GC/MS analysis were spiked with a known amount of benzo[a]pyrene (100 pg/μl) and retested by the spot test. As expected, the samples after spiking yielded a positive indication by the PAH spot test.

Over twenty samples from industrial sites have been tested by both GC/MS and the fluorescence spot test. A partial lisitng of these samples and their results are given in Table 7.9.

Table 7.9: Results of Spot Test Analysis of Samples from Industrial Sites

Sample	Spot Test	GC/MS
. . . . **Coke Oven Extracts**		
A	Positive	Positive
B	Positive	Positive
. . **Combustion Effluent Samples (Commercial Oil Burner)**. .		
A	Negative	Negative
B	Positive	Positive
C	Positive	Positive
. . **Combustion Effluent Samples (Coal Fired Power Plant)**. .		
A	Negative	Negative
B	Negative	Negative
C	Positive	Negative
D	Negative	Negative
E	Negative	Negative
F	Negative	Negative
G	Negative	Negative
H	Positive	Negative

Source: EPA-600/7-79-207

Approximately 10% of the samples which gave a positive result of the spot test showed no PAH by GC/MS. As is illustrated, the spot test has a level of detection which is lower than that of GC/MS when the GC/MS is used in the traditional scanning mode. Taking this fact into consideration, it is not unreasonable to expect a small percentage of the samples to prove positive by the spot test and negative by GC/MS. But what is more important, is the fact that in all cases where the spot test gave negative results, the GC/MS also found no PAH. That is, no false negatives were encountered.

PORTABLE FLUOROMETRIC MONITOR TO DETECT PNA

The information in this section is based on:

> *Portable Fluorometric Monitor to Detect Polynuclear Aromatic Hydrocarbon Contamination of Work Area Surfaces,* NTIS CONF-790855-2, 1979, prepared by D.D Schuresko of Oak Ridge National Laboratory for the U.S. Department of Energy.

The need for instrumentation to monitor PNA buildup on workplace surfaces as, for example, results from aerosol condensation, has also been recognized (54)(55). Such instrumentation must be capable of: (a) functioning during actual plant operation in varied working environments; (b) detecting material spilled on different work surfaces, including machinery, plumbing, construction materials and on personnel and clothing; and (c) being easily and reliably operated by all plant personnel.

The fluorescence spotter described by D.D. Schuresko (U.S. Patent applied for) has these capabilities. This instrument, which consists of a hand-held optics unit connected via an umbilical cable to an electronics module, enables remote monitoring of work area surfaces at distances $\leqslant$1 m. The electronics module, which normally operates with ac power can also be operated on rechargeable batteries; thus the entire instrument plus battery pack is portable.

It is anticipated that portable fluorescence spotters will be used extensively to detect surface contamination in changing areas, lunchrooms, shops, control rooms and other "clean" areas in coal conversion facilities, and to monitor skin contamination of coal plant workers. The currently used method of detecting contamination is to turn off ambient lighting and scan the suspected area with a black light.

In contrast, the newly developed spotter (a) can be operated outdoors in direct sunlight or indoors in the presence of strong background illumination; (b) provides a quantitative measure of the amount of fluorescent material; (c) will discriminate between the fluorescence of organic materials and some inorganic compounds based on their fluorescence lifetimes; and (d) does not present a vision hazard to personnel.

The spotter induces and detects the fluorescence of PNA that characteristically absorb light in the 350 to 440 nm region of the spectrum and fluoresce with high efficiency in the blue-green region of the spectrum. Multiring heteroatom aromatic compounds, including acridines, are also detected, although they generally absorb light and fluoresce at longer wavelengths than do their pure hydrocarbon counterparts.

The spotter has been laboratory tested with pure compounds and with several coal and oil shale products and wastes. When operated in a laboratory area illuminated by fluorescent lighting, the spotter can typically detect 0.1 μg of perylene (as a dilute solution in cyclohexane) located 50 cm distant. Considering that 10 μg doses of several PNA will induce changes in the metabolism of cultured mammalian cells (56)(57), it is evident that the high sensitivity of the spotter is appropriate to its intended task.

In this regard, the spotter might prove to be particularly useful in quantitatively measuring PNA skin contamination caused by contact with coal-derived liquids, thus providing accurate data for correlation with the incidence of skin lesions.

REFERENCES

(1) Schaad, R., *Chromatography Reviews,* 13, 61 (1970).
(2) Cook, J.W., Hewett, C.L. and Heiger, I., *J. Chem. Soc.,* 395 (1933).
(3) Cook, J.W. and Hewett, C.L., *J. Chem. Soc.,* 398 (1933).
(4) La Budde, J.A. and Heidelberger, C.J., *JACS,* 80, 1225 (1958).
(5) Berenblum, I. and Schoenthal, R., *Cancer Research,* 6, 699 (1946).
(6) Cook, J.W., Ludwiczak, R.S. and Schoenthal, R., *J. Chem. Soc.,* 1112 (1950).
(7) Hood, L.V. and Winefordner, J.D., *Anal. Chim. Acta,* 42, 199 (1968).
(8) Sawicki, E., Corey, R.C., Dooley, A.E., Gisclard, J.B., Monkman, J.L., Neligan, J.B. and Ripperton, L.P., *Health Lab. Sci.,* 7, 56, 60, 68 (1970).
(9) Matsushita, H., Esumi, Y. and Yamada, K., *Bunseki Kagaku,* 19, 951 (1970).
(10) Sawicki, E., *Talanta,* 16, 1231 (1969).
(11) Wehry, E.L., Mamantov, G., Kemmerer, R.R., Stroupe, R.C., Tokousbalides, P.T., Hinton, E.R., Hembree, D.M., Dickinson, R.B, Jr., Garrison, A.A., Bilotta, P.V. and Gore, R.R., *Carcinogenesis, Vol. 3: Polynuclear Aromatic Hydrocarbons,* pp. 193-202, P.W. Jones and R.I. Freudenthal (editors), Raven Press, New York, NY (1978).
(12) VoDinh, T., *Anal. Chem.,* 50, 396 (1978).
(13) Sawicki, E., Stanley, T.W. and Elbert, W.C., *J. Chromatog.,* 20, 348 (1965).
(14) Sawicki, E., Corey, R.C., Dooley, A.E., Gisclard, J.B., Monkman, J.L., Neligan, R.E. and Ripperton, L.A., *Health Lab. Sci.,* 7, Supplement 3, 31 (1970).
(15) Searl, T.D., Cassidy, F.J., King, W.H. and Brown, R.A., *Anal. Chem.,* 42, 954 (1970).
(16) *Ultraviolet Atlas of Organic Compounds,* 5 Vols., Plenum Press, New York, NY (1967-1968).
(17) Ungnade, H.E. (editor), *Organic Electronic Spectral Data,* Interscience Publishers, Inc., New York, NY (1960-present).
(18) Wilmhurst, S.R., *J. Chromatog.,* 17, 50 (1965).
(19) Deamio, L. and Corn, M., *Anal. Chem.,* 38, 131 (1966).
(20) Golay, J.J.E., *Theory and Practice of Gas Liquid Partition Chromatography with Coated Capillaries in Gas Chromatography,* V.J. Coates, H.J. Noebels and I.S. Ragerson (editors), Academic Press, New York, NY (1958).
(21) Lee, M.L., Vassilaros, D.L., White, C.M. and Novotny, M., *Anal. Chem.,* 51, 768 (1979).
(22) Janini, G.M., Johnston, K. and Zielinski, W.L., Jr., *Anal. Chem.,* 47, 670 (1975).
(23) Oi, W. and Inabe, E., *Yakugaku Zasshi,* 87, 741 (1967).
(24) Marmon, D.M., *J. Ass. Off. Analytical Chem.,* 57, 495 (1974).
(25) Nagy, F. and Sohar, J., *Chem. Zvesti,* 16, 389 (1962).
(26) Martrka, M., Kroupa, J. and Spevak, A., *Coll. CS. Chem. Commun.,* 36, 1379 (1971).
(27) Lao, R.C., Thomas, R.S. and Monkman, J.L., "Chemistry, Metabolism, and Carcinogenesis," *Carcinogenesis: A Comprehensive Survey, Vol. 1: Polynuclear Aromatic Hydrocarbons,* R. Freudenthal and P.W. Jones (editors), Raven Press, NY (1976).
(28) Fox, M.A. and Staley, S.W., *Anal. Chem.,* 48, 992 (1976).
(29) Dong, M., Locke, D.C. and Ferrand, E., *Anal. Chem.,* 48, 368 (1976).
(30) Klimisch, H.J., *Anal. Chem.,* 45, 1960 (1973).
(31) Zielinski, W.L., Jr., Johnston, K. and Muschik, G.M., *Anal. Chem.,* 48, 907 (1976).
(32) Janini, G.M., Muschik, G.M. and Zielinski, W.L, Jr., *Anal. Chem.,* 48, 809 (1976).
(33) Janini, G.M., Muschik, G.M., Schroer, J.A. and Zielinski, W.L., Jr., *Anal. Chem.,* 48, 1879 (1976).
(34) *GC Reporter,* 2(2), 4 (1977).
(35) Supelco, Inc., *Supelco Technical Bulletin,* 773, Bellefonte, Penn. (1977).
(36) Barrall, E.M., II and Johnson, J.F., *Liquid Crystals and Plastic Crystals, Vol. 2,* G.W. Gray and P.A. Winsor (editors), John Wiley & Sons, Inc., (1974).

(37) Janini, G.M., Shaikh, B. and Zielinski, W.L., Jr., *J. Chromatog.*, 132, 136 (1977).
(38) Robertson, D.J., Elwood, J.H. and Groth, R.H., *Chemical Composition of Exhaust Particles from Gas Turbine Engines.* Final report prepared by United Technologies Corporation, East Hartford, Conn., under Contract No. 68-02-2450 to the U.S. Environmental Protection Agency, Research Triangle Park, NC, EPA-600/2-79-041 (February 1979).
(39) Braddock, J.N. and Gabele, P.A., "Emissions Patterns of Diesel-Powered Passenger Cars–Part II," SAE Paper No. 770168, Detroit, MI (February 1977).
(40) Hare, C.T. and Bradow, R.L., "Characterization of Heavy-Duty Diesel Gaseous and Particulate Emissions, and Effects of Fuel Composition," SAE Paper No. 790490, Detroit, MI (February 1979).
(41) Hare, C.T. and Baines, T.M., "Characterization of Particulate and Gaseous Emissions from Two Diesel Automobiles as a Function of Fuel and Driving Cycle," SAE Paper No. 790424, Detroit, MI (February 1979).
(42) Matsushita, H., Esumi, Y. and Suzuki, A., "Analysis of Polynuclear Hydrocarbons in Petroleum," *Bunseki Kagaku,* 21(3):331-7 (1972).
(43) Uthe, J.F., Reinker, J. and Gesser, H.D., "Extraction of Organochlorine Pesticides from Water by Porous Polyurethane Coated with Selective Absorbent," *Environ. Letters,* 3:117 (1972).
(44) Musty, P.R. and Nickless, G., "The Extraction and Recovery of Chlorinated Insecticides and Polychlorinated Biphenyls from Water Using Porous Polyurethane Foams," *J. Chromatogra.,* 100:83 (1974).
(45) Lawrence, J. and Tosine, H.M., "Adsorption of Polychlorinated Biphenyls from Aqueous Solutions and Sewage," *Environ. Sci. Technol.,* 10:381 (1976).
(46) Acheson, M.A., Harrison, R.M., Perry, R. and Welling, R.A., "Factor Affecting the Extraction and Analysis of Polynuclear Aromatic Hydrocarbons in Water," *Water Res.,* 10:207 (1976).
(47) Sawicki, E., Stanley, T.W., Elbert, W.C., Meeker, J. and McPherson, S.P., "Comparison of Methods for the Determination of Benzo[a]pyrene in Particulates from Urban and Other Atmospheres," *Atmos. Environ.,* 1:131 (1967).
(48) Borneff, J., *Polycyclic Aromatic Hydrocarbons in Water,* Summary of the German work on PAH.
(49) World Health Organization, *International Standards for Drinking Water,* 3rd ed., Geneva (1971).
(50) Strosher, M.T. and Hodgson, G.W., "Polycyclic Aromatic Hydrocarbons in Lake Waters and Associated Sediments: Analytical Determination by Gas Chromatography-Mass Spectroscopy," *Water Quality Parameters,* ASTM-STP-573, pp. 259–270 (1975).
(51) Giger, W. and Schaffner, C., "Determination of Polycyclic Aromatic Hydrocarbons in the Environment by Glass Capillary Gas Chromatography," *Anal. Chem.,* 50, pp. 243–249 (1978).
(52) Kubota, H., Griest, W.H. and Guerin, M.R., "Determination of Carcinogens in Tobacco Smoke and Coal-Derived Samples–Trace Polynuclear Aromatic Hydrocarbons," *Trace Substances in Environmental Health,* Vol. IX, pp. 281–289, D.D. Hemphill (editor), University of Missouri, Columbia (1975).
(53) *Sensitized Fluorescence for the Detection of Polycyclic Aromatic Hydrocarbons,* EPA-600/7-78-182, NTIS No. PB 287-181 IAS (September 1978).
(54) Brookhaven National Laboratory, *Report of the Working Group on Assessing Industrial Hygiene Monitoring Needs for the Coal Conversion and Oil Shale Industries,* Otto White, Chairman (August 1978).
(55) Schuresko, D.D. and Jones, G., "A Portable Fluorometric Monitor to Detect PNA Contamination of Work Area Surfaces," *Proceedings of the Symposium on Assessing the Industrial Hygiene Monitoring Needs of the Coal Conversion and Oil Shale Industries,* Brookhaven National Laboratory (November 1978).
(56) Schuresko, D.D., et al, unpublished results.
(57) Nebert, D.W. and Gelboin, H.V., *J. Biol. Chem.* 243(23), 6242 (1968).

ADDENDUM

Laser Technique to Detect PAH

It was recently reported (U.S. Department of Energy, "Laser Technique to Help Detect Cancer-Causing Industrial Pollutants," *Energy Insider,* 3:21, October 13, 1980) that a new laser technique to identify PAH has been developed by scientists at Lawrence Livermore National Laboratory (LLNL). The system can detect concentrations as low as one part per billion in a sample containing hundreds of chemicals. This is one thousand times more sensitive than high performance liquid chromatography used alone. The system also is estimated to cost ten times less than GC/MS systems of comparable sensitivities.

LLNL is using the technique on samples of fly ash from two coal-fired power plants in the Western United States to determine why the ash from one plant showed cancer-causing potential while the ash from the other did not.

Since the ash contains hundreds of chemical compounds, it is essential to separate them to find any differences. Using this laser technique, the PAH present in the fly ash can be separated and detected.

SOURCES UTILIZED

EPA-440/5-80-069

Ambient Water Quality Criteria for Polynuclear Aromatic Hydrocarbons, prepared by the U.S. Environmental Protection Agency, October 1980.

EPA-600/1-77-052

Method Development and Monitoring of Polynuclear Aromatic Hydrocarbons in Selected U.S Waters, prepared by J. Saxena, D.K. Basu and J. Kozuchowski of Syracuse Research Corporation for U.S. Environmental Protection Agency, November 1977.

EPA-600/2-80-069

Quantitative Analysis of Polynuclear Aromatic Hydrocarbons in Liquid Fuels, prepared by Radian Corporation for U.S. Environmental Protection Agency under EPA Contract No. 68-02-2446, April 1980.

EPA-600/7-79-191

Measurement of Polycyclic Organic Matter for Environmental Assessment, prepared by P.L. Levins, C.E. Rechsteiner, Jr. and J.L. Stauffer of Arthur D. Little, Inc. for U.S. Environmental Protection Agency under EPA Contract No. 68-02-2150, 1979.

EPA-600/7-79-207

Evaluation of Sensitized Fluorescence for Polynuclear Aromatic Hydrocarbon Detection, prepared by T.R. Smith of TRW Defense and Space Systems Group for U.S. Environmental Protection Agency under Contract No. 68-02-2689, 1979.

NTIS CONF-770130-2

Energy Sources of Polycyclic Aromatic Hydrocarbons, prepared by M.R. Guerin of Oak Ridge National Laboratory for U.S. Environmental Protection Agency and the U.S. Energy Research and Development Administration under contract with the Union Carbide Corporation, 1977.

NTIS CONF-770361

Proceedings of the Second ORNL Workshop on Exposure to Polynuclear Aromatic Hydrocarbons in Coal Conversion Processes. I Medical Surveillance; II Industrial Experiences, Personnel Protection and Monitoring, compiled by R.B. Gammage of Oak Ridge National Laboratory, December 1977.

NTIS CONF-780227-4

Cell-Mediated Mutagenesis by Chemical Carcinogens, prepared by Eliezer Huberman and Robert Langenbach of Oak Ridge National Laboratory for National Cancer Institute and the Department of Energy, 1978.

NTIS CONF-7805130-1

Multicomponent Polycyclic Aromatic Hydrocarbon Analysis of Inland Water and Sediment, prepared by W.H. Griest of Oak Ridge National Laboratory for U.S. Department of Energy under Contract No. W-7405-eng-26, 1978.

NTIS CONF-790415-3

Synthetic Fossil Fuel Technologies: Health Problems and Intersociety Cooperation, prepared by R.B. Gammage and J.E. Turner of Oak Ridge National Laboratory for U.S. Department of Energy under Contract No. W-7405-eng-26, 1979.

NTIS CONF-790486-3

Critical Pathways of Polycyclic Aromatic Hydrocarbons in Aquatic Environments, prepared by S.E. Herbes, G.R. Southworth, D.L. Shaeffer, W.H. Griest and M.P. Maskarinec of Oak Ridge National Laboratory, 1979.

NTIS CONF-790855-2

Portable Fluorometric Monitor to Detect Polynuclear Aromatic Hydrocarbon Contamination of Work Area Surfaces, prepared by D.D. Schuresko of Oak Ridge National Laboratory for U.S. Department of Energy under Contract No. W-7405-eng-26, 1979.

NTIS CONF-800303-13

Biologically Important Compounds in Synfuels Processes, presented by B.R. Clark, C.-h. Ho, W.H. Griest and M.R. Guerin of Oak Ridge National Laboratory at the ACS Symposium on Environmental Control in Synfuel Processes, Houston, Texas, March 25, 1980.

NTIS COO-4347-1

Formation and Transformation of Polycyclic Organic Matter from Coal Combustion, prepared by David F.S. Natusch of Colorado State University under Contract No. EE-77-S-02-4347, 1978.

NTIS EPRI-EA-787-SY

Polycyclic Organic Materials and the Electric Power Industry, prepared by J.M. Allen, A. Levy, P.W. Jones and R.I. Freudenthal of Battelle Columbus Laboratories for Electric Power Research Institute, December 1978.

ORNL/EIS-95

Environmental, Health and Control Aspects of Coal Conversion: Volume 2, An Information Overview, prepared by H.M. Braunstein, E.D. Copenhaver and H. A. Pfuderer of Oak Ridge National Laboratory for Energy Research and Development Administration under contract No. W-7405-eng-26, April 1977.

Health Assessment Document for Polycyclic Organic Matter, External Draft No. 2, prepared by the Syracuse Research Corporation for U.S. Environmental Protection Agency, June 1980.

WASTEWATER REUSE AND RECYCLING TECHNOLOGY 1980

by Gordon Culp, George Wesner, Robert Williams and Mark V. Hughes, Jr.

Pollution Technology Review No. 72

This volume describes and evaluates the technology for water reuse and recycling; as well as describing the magnitude of the potential for reusing and recycling industrial, agricultural, and municipal wastewaters.

Wastewaters available for reuse and recycling include discharges from municipalities, industries, steam electric plants and agricultural irrigation return water. The potential uses for reclaimed water are agricultural irrigation, landscape irrigation, cooling in steam electric plants, and other industrial uses.

The beneficial uses for water considered for reuse or recycling are those projected to require either the greatest quantities of water or those that are impacted because of projected water shortages. This book identifies the areas requiring large quantities of water, water-short areas, and areas having potential conflicts among the uses of available water.

The major portion of this volume contains an extensive evaluation of current wastewater treatment technology including water quality criteria, primary treatment and many secondary treatment methods such as trickling filters, extended aeration, rotating biological contactors, coagulation-sedimentation, etc. Listed below is a summarized table of contents illustrating the **chapter headings and some important subtitles.**

ISBN 0-8155-0829-8 **838 pages**

HAZARDOUS CHEMICALS DATA BOOK 1980

Edited by G. Weiss

Environmental Health Review No. 4

Instant information for decision-making in emergency situations by personnel involved with chemical accidents is the prime purpose of this compilation of about 1,350 hazardous chemicals. The book, prepared in clear, concise, easy-to-locate format, should become an invaluable source on any library or laboratory shelf; is intended for use by scientists, engineers, managers, transportation personnel, or anyone who may have contact or require data on a particular chemical.

There is a large amount of pertinent data provided for each chemical, and examples of a few of the headings are:

Toxicity and Health Hazards
Hazards Relating to Water Pollution
Spill and Leak Procedures
Special Protective Equipment
Flammability and Explosion Hazards
Chemical Reactivity
Extinguishing Media
Special Precautions

The volume is set up in two sections, alphabetically arranged. The first section is based on the Department of Transportation's Chemical Hazard Response Information System. The second section is based on Material Safety Data Sheets obtained from the Oak Ridge National Laboratory. The use of a convenient all-inclusive index furnishes the responsible individual with the desired information in a minimum of time.

A partial list of 54 hazardous chemicals taken from the index of about 1,350 chemicals, is shown below.

Acetal
Acetaldehyde
Acetamide
Acetic Acid
Acetic Anhydride
Acetone
Acetone Cyanohydrin
Acetonitrile
Acetonylacetone
Acetophenone
Acetylacetone
Acetyl Bromide
Acetyl Chloride
Acetylene
Acetyl Methylamine
Acetyl Methylurea
Acetyl Peroxide Solution
Acetyl Thiourea
Acetyl Urea
Acridine
Acrolein
Acrylamide
Acrylic Acid
Acrylonitrile
Adipic Acid
Adiponitrile
Aldrin
Alkylbenzenesulfonic Acids
Allyl Alcohol
Allylamine
Allyl Bromide
Allyl Chloride
Allyl Chloroformate
Allyltrichlorosilane
Aluminum Chloride
Aluminum Fluoride
Aluminum Nitrate
Aluminum Sulfate
p-Aminoazobenzene
p-Aminoazobenzene Hydrochloride
2-Amino-1,4-Dimethylbenzene
4-Amino-1,3-Dimethylbenzene
5-Amino-1,3-Dimethylbenzene
o-Aminodiphenyl
Aminoethylethanolamine
2-Aminopyridine
Ammonia, Anhydrous
Ammonium Acetate
Ammonium Benzoate
Ammonium Bicarbonate
Ammonium Bifluoride
Ammonium Carbonate
Ammonium Chloride
Ammonium Citrate
plus about 1,300 other chemicals

ISBN 0-8155-0831-X 1188 pages

OZONE AND CHLORINE DIOXIDE TECHNOLOGY FOR DISINFECTION OF DRINKING WATER 1980

Edited by J. Katz

Pollution Technology Review No. 67
Chemical Technology Reveiw No. 164

Disinfection is the most important stage in the treatment of drinking water supplies. Recent findings on potential health risks of by-products formed during chlorination have led to extensive federal regulations on allowable levels of chlorinated by-products and, consequently, the consideration of alternatives for this established disinfectant.

Two of the best known and most practical alternate technologies are considered here—ozone and chlorine dioxide. Both are strong oxidizing agents and have been in wide use in Europe and Canada; both have the advantage that they can be generated on-site.

Part I of the book reviews disinfectant alternatives and their microbiological and environmental aspects, and Part II is an international technology assessment, with special attention given to oxidation products resulting from ozone or chlorine dioxide treatment. Significant advances noted in the engineering of water supply treatment equipment and systems are identified and discussed in detail. A condensed table of contents is given below.

ISBN 0-8155-0809-3 **659 pages**

PRIORITY TOXIC POLLUTANTS
Health Effects and Allowable Limits 1980

Edited by Marshall Sittig

Enivronmental Health Review No. 1

This book is a practical manual with its contents arranged alphabetically in encyclopedic form. It is intended to provide specific information on the 65 priority toxic pollutants (actually reflecting 129 individual compounds), their derivatives or degradation products, or intermedia transfers.

Compelled by a consent decree obtained in a federal court by public interest groups, the Environmental Protection Agency has promulgated criteria on allowable limits and profuse guidelines on how to interpret them and also how to determine further such criteria.

Here, in ready reference form, are the essentials of the criteria and guidelines which so far have been published for the control of priority toxic pollutants. Each typical pollutant is characterized with descriptions of its:

- Occurrence
- Physical Properties
- Chemical Properties
- Uses
- Toxic Effects
- Current Levels of Exposure
- Special Groups at Risk
- Existing Guidelines & Standards
- Summary of Proposed Criteria
- Bases for the Human Health Criteria
- Pertinent References

A list of the 65 original priority pollutants follows here, however all of the 129 compounds are covered in this book.

1. **Acenaphthene**
2. **Acrolein**
3. **Acrylonitrile**
4. **Aldrin/Dieldrin**
5. **Antimony & Compounds**
6. **Arsenic & Compounds**
7. **Asbestos**
8. **Benzene**
9. **Benzidine**
10. **Beryllium & Compounds**
11. **Cadmium & Compounds**
12. **Carbon Tetrachloride**
13. **Chlordane**
14. **Chlorobenzenes**
15. **Chlorinated Ethanes**
16. **Chloroalkyl Ethers**
17. **Chlorinated Naphthalenes**
18. **Chlorinated Phenols**
19. **Chloroform**
20. **2-Chlorophenol**
21. **Chromium & Compounds**
22. **Copper & Compounds**
23. **Cyanides**
24. **DDT & Metabolites**
25. **Dichlorobenzenes**
26. **Dichlorobenzidine**
27. **Dichloroethylene**
28. **2,4-Dichlorophenol**
29. **Dichloropropane/propene**
30. **2,4-Dimethylphenol**
31. **Dinitrotoluenes**
32. **Diphenylhydrazines**
33. **Endosulfan**
34. **Endrin**
35. **Ethylbenzene**
36. **Fluoranthene**
37. **Haloethers**
38. **Halomethanes**
39. **Heptachlor**
40. **Hexachlorobutadiene**
41. **Hexachlorocyclohexane**
42. **Hexachlorocyclopentadiene**
43. **Isophorone**
44. **Lead & Compounds**
45. **Mercury & Compounds**
46. **Naphthalene**
47. **Nickel & Compounds**
48. **Nitrobenzene**
49. **Nitrophenols**
50. **Nitrosamines**
51. **Pentachlorophenol**
52. **Phenol**
53. **Phthalate Esters**
54. **Polychlorinated Biphenyls**
55. **Polynuclear Aromatics**
56. **Selenium & Compounds**
57. **Silver & Compounds**
58. **Tetrachlorodibenzo-p-dioxin**
59. **Tetrachloroethylene**
60. **Thallium & Compounds**
61. **Toluene**
62. **Toxaphene**
63. **Trichloroethylene**
64. **Vinyl Chloride**
65. **Zinc & Compounds**

ISBN 0-8155-0797-6 **370 pages**

HEALTH HAZARDS AND POLLUTION CONTROL IN SYNTHETIC LIQUID FUEL CONVERSION 1980

Edited by Perry Nowacki

Pollution Technology Review No. 68
Environmental Health Review No. 2
Chemical Technology Review No. 165
Energy Technology Review No. 57

This review addresses the environmental, health and pollution control aspects of coal liquefaction, oil shale processing, and tar sands conversion. The emphasis throughout is on well-documented processes and projects in the pilot scale or early commercial phases. After a preliminary discussion of composition and structural features peculiar to coal, oil shale, and tar sands, the major processing options, proprietary implementations, and associated pollution sources are outlined. Toxicity, carcinogenicity, synergetics, interactions, and residual effects of involved elements, trace elements, particulates, or compounds are treated at length.

The book will be of interest to engineers and biomedical and environmental scientists involved in the assessment and control of pollution hazards in synthetic liquid fuel conversion. A condensed table of contents is listed below with **chapter headings and important subtitles.**

ISBN 0-8155-0810-7 **511 pages**

PESTICIDE MANUFACTURING AND TOXIC MATERIALS CONTROL ENCYCLOPEDIA 1980

Edited by Marshall Sittig

Chemical Technology Review No. 168
Environmental Health Review No. 3
Pollution Technology Review No. 69

This book contains a total of 514 pesticide materials arranged in an alphabetical and encyclopedic fashion by the common or generic name of each pesticide. It is a thorough revision of our previous *Pesticides Process Encyclopedia* published in 1977, plus additional material relative to toxic materials control.

The data on manufacturing processes were drawn primarily from the patent literature, while the data on product toxicity, emissions and product use came mostly from published and unpublished reports released by the Environmental Protection Agency.

This book is definitely of interest to pesticide manufacturers, chemical raw material suppliers, formulators, growers, farmers and food processors. It should also prove useful to chemists, lawyers, industrial hygienists and environmentalists.

It contains much useful extrinsic information, e.g. *allowable tolerance limits, animal and human toxicities,* and similar data not easily ascertained elsewhere.

The use of pesticides leads to healthier plants and bigger crops, and exports of pesticides could provide fast growth for U.S. producers in the coming years.

An indication of the comprehensive nature of this one-volume encyclopedia is given here:

ISBN 0-8155-0814-X **810 pages**